# APPLETON & LANGE'S REVIEW OF

# ANATOMY

## FOURTH EDITION

**Royce L. Montgomery, Ph.D.**
Associate Professor of Anatomy
The University of North Carolina at Chapel Hill
Chapel Hill, North Carolina

**Mary C. Singleton, Ph.D.**
Professor Emeritus of Anatomy and Physical Therapy
The University of North Carolina at Chapel Hill
Chapel Hill, North Carolina

**Gerald A. Montgomery, M.D.**
New Mexico Orthopedic Medicine
Albuquerque, New Mexico
Adjunct Assistant Professor of Anatomy
The University of North Carolina at Chapel Hill
Chapel Hill, North Carolina

**APPLETON & LANGE**
Norwalk, Connecticut/San Mateo, California

0-8385-0213-X

Prentice Hall International (UK) Limited, *London*
Prentice Hall of Australia Pty. Limited, *Sydney*
Prentice Hall Canada, Inc., *Toronto*
Prentice Hall Hispanoamericana, S.A., *Mexico*
Prentice Hall of India Private Limited, *New Delhi*
Prentice Hall of Japan, Inc., *Tokyo*
Simon & Schuster Asia Pte. Ltd., *Singapore*
Editora Prentice Hall do Brasil Ltda., *Rio de Janeiro*
Prentice Hall, *Englewood Cliffs, New Jersey*

**Library of Congress Cataloging-in-Publication Data**

Montgomery, Royce L.
    Appleton & Lange's review of anatomy / Royce L. Montgomery, Mary
C. Singleton, Gerald A. Montgomery. — 4th ed.
        p.  cm.
    Rev. ed. of Human anatomy review. 3rd ed. 1982.
    ISBN 0-8385-0213-X
    1. Anatomy, Human—Examinations, questions, etc.   I. Singleton,
Mary C.  II. Montgomery, Gerald A.  III. Montgomery, Royce L. Human
anatomy review.   IV. Title.   V. Title: Review of anatomy.
    [DNLM: 1. Anatomy—examination questions. QS 18 M788h]
    QM32.M65   1989
    611'.0076—dc19
    DNLM/DLC
    for Library of Congress                                    88-39234
                                                               CIP

Production Editor: Charles F. Evans
Acquisitions Editor: R. Craig Percy

PRINTED IN THE UNITED STATES OF AMERICA

# Contents

# Preface

The acquisition of knowledge is a gradual process and subject mastery demands much time and energy. In the course of his education, the student is exposed to many facts and new concepts through reading and attending lectures. He then reinforces his knowledge by discussing difficult points with instructors and peers. Assimilation and integration of factual and conceptual material is later tested by objective means.

Self-teaching, self-testing, and subject reviews are important aspects of the learning process and become increasingly important in direct proportion to the increase in the total body of knowledge.

This book is not intended to be a replacement for time-tested methods of gaining knowledge. It is a comprehensive subject review which can best be utilized after reading textbooks and attending lectures and prior to taking school or board examinations.

Proper use of this review book will assist the reader further by indicating areas of weakness and familiarizing him with common types of questions. The reader is encouraged to refer to the explanations provided in the answer keys and to return to original reference sources when areas of doubt arise.

We have completely rewritten this fourth edition with one objective: to insure that you will receive an excellent review of gross anatomy. All questions are new and designed specifically to provide you with ample discussion material on which to base your review. As a result of this process, we believe this anatomy review book to be of the highest caliber and of great assistance in your review.

The format of the questions has been revised to adhere to the guidelines set forth by the National Board of Medical Examiners.

# Introduction

This book has been designed to help you review anatomy for your course, National Medical Boards Parts I and III, the Foreign Medical Graduate Examination in the Medical Sciences, and the FLEX. Here, in one package, is a comprehensive review resource with over 1400 Board-type multiple-choice questions with referenced, paragraph-length discussions of each answer. In addition, the last 116 questions have been set aside as a Practice Test for self-assessment purposes. The entire book has been designed to help you assess your areas of relative strength and weakness.

## ORGANIZATION OF THIS BOOK

This book is divided into ten chapters. Eight chapters provide a review of the major areas of anatomy. The ninth provides diagrammatic labeling exercises. The last chapter is a Practice Test, which integrates all of these areas into one simulated examination.

This introduction provides information on question types, question-taking strategies, various ways you can use this book, and specific information on the Part I NBME examination.

### Questions

The NBME Part I contains four different types of questions (or "items," in testing parlance). In general, about 50% of these are "one best answer–single item" questions, 30% are "multiple true–false items," 10% are "one best answer–matching sets," and 10% are "comparison–matching set" questions. In some cases, a group of two or three questions may be related to a situational theme. In addition, some questions have illustrative material (graphs, x-rays, tables) that require understanding and interpretation on your part. Moreover, questions may be of three levels of difficulty: (1) rote memory question, (2) memory question that requires more understanding of the problem, and (3) a question that requires understanding *and* judgment. In view of the fact that the NBME is moving toward the judgment question and away from the rote memory question, it is the judgment question that we have tried to emphasize throughout this text. Finally, some of the items are stated in the negative. In such instances, we have printed the negative word in capital letters (e.g., "All of the following are correct EXCEPT;" "Which of the following choices is NOT correct;" and "Which of the following is LEAST correct").

*One best answer–single item question.* This type of question presents a problem or asks a question and is followed by five choices, only **one** of which is entirely correct. The directions preceding this type of question will generally appear as below:

DIRECTIONS: **Each of the numbered items or incomplete statements in this section is followed by answers or by completions of the statement. Select the ONE lettered answer or completion that is BEST in each case.**

An example for this item type follows:

1. An obese 21-year-old woman complains of increased growth of coarse hair on her lip, chin, chest, and abdomen. She also notes menstrual irregularity with periods of amenorrhea. The most likely cause is

   (A) polycystic ovary disease
   (B) an ovarian tumor
   (C) an adrenal tumor
   (D) Cushing's disease
   (E) familial hirsutism

In this type of question, choices other than the correct answer may be partially correct, but there can only be one *best* answer. In the question above (taken from *Appleton's Review for National Boards Part II*), the key word is "most." Although ovarian tumors, adrenal tumors, and Cushing's disease are causes of hirsutism (described in the stem of the question), polycystic ovary disease is a much more common cause. Familial hirsutism is not associated with the menstrual irregularities mentioned. Thus, the *most* likely cause of the manifestations described can only be "(A)-polycystic ovary disease."

## TABLE 1. STRATEGIES FOR ANSWERING ONE BEST ANSWER–SINGLE ITEM QUESTIONS*

1. Remember that only one choice can be the correct answer.
2. Read the question carefully to be sure that you understand what is being asked.
3. Quickly read each choice for familiarity. (This important step is often not done by test takers.)
4. Go back and consider each choice individually.
5. If a choice is partially correct, tentatively consider it to be incorrect. (This step will help you lessen your choices and increase your odds of choosing the correct answer.)
6. Consider the remaining choices and select the one you think is the answer. At this point, you may want to quickly scan the stem to be sure you understand the question and your answer.
7. Fill in the appropriate circle on the answer sheet. (Even if you do not know the answer, you should at least guess. Your score is based on the number of correct answers, so **do not leave any blanks**.)

*Note that steps 2 through 7 should take an average of 50 seconds total. The actual examination is timed for an average of 50 seconds per question.

*One best answer–matching sets.* These questions are essentially matching questions that are usually accompanied by the following general directions:

**DIRECTIONS (Questions 2 through 4): Each group of items in this section consists of lettered headings followed by a set of numbered words or phrases. For each numbered word or phrase, select the ONE lettered heading that is most closely associated with it. Each lettered heading may be selected once, more than once or not at all.**

Any number of questions (usually two to six) may follow the five headings:

**Questions 2 through 4**

For each adverse drug reaction listed below, select the antibiotic with which it is most closely associated.

(A) Tetracycline
(B) Chloramphenicol
(C) Clindamycin
(D) Cefotaxime
(E) Gentamicin

2. Bone marrow suppression

3. Pseudomembranous enterocolitis

4. Acute fatty necrosis of liver

Note that, unlike the single item questions, the choices in the matching sets questions *precede* the actual questions. As with the single item questions, however, only **one** choice can be correct for a given question.

## TABLE 2. STRATEGIES FOR ANSWERING ONE BEST ANSWER–MATCHING SETS QUESTION*

1. Remember that the *lettered choices* are **followed** by the *numbered questions*.
2. As with the single item questions, these questions have only **one** best answer. Therefore apply steps 2 through 7 of the single item strategies.

*Remember, you only have an average of 50 seconds per question.

*Comparison–matching set questions.* Like the one best answer–matching set questions, comparison–matching set questions are essentially just matching questions. They are usually accompanied by the following general directions and code:

**DIRECTIONS (Questions 5 and 6): Each group of items in this section consists of lettered headings followed by a set of numbered words or phrases. For each numbered word or phrase, select**

A if the item is associated with (A) only,
B if the item is associated with (B) only,
C if the item is associated with both (A) and (B),
D if the item is associated with neither (A) nor (B).

Any number of questions (usually two to six) may follow the four headings:

**Questions 5 and 6**

(A) polymyositis
(B) polymyalgia rheumatica
(C) both
(D) neither

5. Pain is a prominent syndrome

6. Associated with internal malignancy in adults

Note that as with the other matching set questions, the choices precede the actual questions. Once again, only **one** choice can be correct for a given question.

**TABLE 3. STRATEGIES FOR ANSWERING COMPARISON–MATCHING SET QUESTIONS***

1. Remember that the *lettered choices* are **followed** by the *numbered questions.*

2. As with the one best answer questions, these questions have only **one** best answer.

3. Quickly note what the lettered choices are.

4. Carefully read the question to be sure you understand what is being asked or what its relationship to the lettered choices is.

5. Focus on choices (A) and (B), and use the following sequence logically to determine the correct answer:
   a. If you can determine that (A) is incorrect, your answer must be either (B) or (D).
   b. If you can determine that (B) is incorrect, your answer must be either (A) or (D).
   c. If you can determine that both (A) and (B) are incorrect, your answer must be (D).
   d. If you can determine that both (A) and (B) are correct, your answer must be (C).

*Remember, you only have an average of 50 seconds per question.

*Multiple true–false items.* These questions are considered the most difficult (or tricky), and you should be certain that you understand and follow the directions that always accompany these questions:

**DIRECTIONS: For each of the items in this section, ONE or MORE of the numbered options is correct. Choose the answer**

    A if only 1, 2, and 3 are correct.
    B if only 1 and 3 are correct.
    C if only 2 and 4 are correct.
    D if only 4 is correct.
    E if all are correct.

This code is always the same (i.e., "D" would never say "if 3 is correct"), and it is repeated throughout this book in a summary box (see below) at the top of any page on which multiple true–false item questions appear.

| SUMMARY OF DIRECTIONS | | | | |
|---|---|---|---|---|
| **A** | **B** | **C** | **D** | **E** |
| 1, 2, 3 only | 1, 3 only | 2, 4 only | 4 only | All are correct |

A sample question follows:

7. The superficial perineal space contains which of the following?

    (1) crura of the clitoris
    (2) the deep transverse perineal muscle
    (3) the ischiocavernosus muscle
    (4) the anal sphincter

You first need to determine which choices are right and wrong, and then which code corresponds to the correct numbers. In the example above, 1 and 3 are both structures contained in this space, and therefore (B) is the correct answer to this question.

**TABLE 4. STRATEGIES FOR ANSWERING MULTIPLE TRUE–FALSE ITEM QUESTIONS***

1. Carefully read and become familiar with the accompanying directions to this tricky question type.

2. Carefully read the stem to be certain that you know what is being asked.

3. Carefully read each of the numbered choices. If you can determine whether any of the choices are true or false, you may find it helpful to place a "+" (true) or a "−" (false) next to the number.

4. Focus on the numbered choices and your true/false notations, and use the following sequence to logically determine the correct answer:
   a. Note that in the answer code choices 1 *and* 3 are *always* both either true or false together. If you are sure that either one is incorrect, your answer must be (C) or (D).
   b. If you are sure that choice 2 *and either* choice 1 *or* 3 are incorrect, your answer must be (D).
   c. If you are sure that choices 2 and 4 are incorrect, your answer must be (B).

*Remember, you only have an average of 50 seconds per question. Note that the following two combinations cannot occur: choices 1 and 4 both incorrect; choices 3 and 4 both incorrect.

## Answers, Explanations, and References

In each of the sections of this book, the question sections are followed by a section containing the answers, explanations, and references for the questions. This section (1) tells you the answer to each question; (2) gives you an explanation and review of why the answer is correct, background information on the subject matter, and why the other answers are incorrect; and/or (3) tells you where you can find more in-depth information on the subject matter in other books and journals. We encourage you to use this section as a basis for further study and understanding.

**If you choose the correct answer** to a question, you can then read the explanation (1) for reinforcement and (2) to add to your knowledge about the subject matter (remember that the explanations usually tell not only why the answer is correct, but often also why the other choices are incorrect). **If you choose the wrong answer** to a question, you can read the explanation for an instructional review of the material in the question. Furthermore, you can look up the complete source in the list of references, which follows this introduction, and refer to the specific pages cited for a more in-depth discussion.

## Practice Test

The 116-question Practice Test at the end of the book covers and reviews all the topics covered in Chapters 1 through 9. The questions are grouped according to question type (one best answer–single item, one best answer–matching sets, comparison–matching sets, then multiple true–false items), with the subject areas integrated. Specific instructions for how to take the Practice Test are given on page 189.

The Practice Test is followed by a subspecialty list, which will enable you to analyze your areas of strength and weakness to help you focus your review. For example, by checking off your incorrect answers, you may find that a pattern develops in that you are incorrect on most or all of the clinical anatomy questions. In this case, you could note the references (in the Answers and Explanations section) for your incorrect answers and read those sources. You might also want to purchase a clinical anatomy text or review book to do a much more thorough review. We think you will find this subspecialty list very helpful and we urge you to use it.

## HOW TO USE THIS BOOK

There are two logical ways to get the most value from this book. We will call them Plan A and Plan B.

In **Plan A,** you go straight to the Practice Test and complete it according to the instructions given on page 189. Using the subspecialty list, analyze your areas of strength and weakness. This will be a good indicator of your initial knowledge of the subject and will help to identify specific areas for preparation and review. You can now use the first 9 chapters of the book to help you improve your relative weak points.

In **Plan B,** you go through Chapters 1 through 9 checking off your answers, and then comparing your choices with the answers and discussions in the book. Once you have completed this process, you can take the Practice Test and see how well prepared you are. If you still have a major weakness, it should be apparent in time for you to take remedial action.

In Plan A, by taking the Practice Test first, you get quick feedback regarding your initial areas of strength and weakness. You may find that you have a good command of the material, indicating that perhaps only a cursory review of the first 9 chapters is necessary. This, of course, would be good to know early in your exam preparation. On the other hand, you may find that you have many areas of weakness. In this case, you could then focus on these areas in your review—not just with this book, but also with textbooks.

It is, however, unlikely that you will not do some studying prior to taking the National Boards (especially since you have this book). Therefore, it may be more realistic to take the Practice Test after you have reviewed the first 9 chapters (as in Plan B). This will probably give you a more realistic type of testing situation since very few of us just sit down to a test without studying. In this case, you will have done some reviewing (from superficial to in-depth), and your Practice Test will reflect this studying time. If, after reviewing the first 9 chapters and taking

the Practice Test, you still have some weaknesses, you can then go back to the first 9 chapters and supplement your review with your texts.

## SPECIFIC INFORMATION ON THE PART I EXAMINATION

The official source of all information with respect to National Board Examination Part I is the National Board of Medical Examiners (NBME), 3930 Chestnut Street, Philadelphia, PA 19104. Established in 1915, the NBME is a voluntary, nonprofit, independent organization whose sole function is the design, implementation, distribution, and processing of a vast bank of question items, certifying examinations, and evaluative services in the professional medical field.

In order to sit for the Part I examination, a person must be either an officially enrolled medical student or a graduate of an accredited United States or Canadian medical school. It is not necessary to complete any particular year of medical school in order to be a candidate for Part I. Neither is it required to take Part I before Part II.

In applying for Part I, you must use forms supplied by NBME. Remember that registration closes *ten weeks* before the scheduled examination date. Some United States and Canadian medical schools require their students to take Part I even if they are noncandidates. Such students can register as noncandidates at the request of their school. A person who takes Part I as a noncandidate can later change to candidate status and, after payment of a fee, receive certification credit.

### Scoring

Because there is no deduction for wrong answers, you should **answer every question.** Your test is scored in the following way:

1. The number of questions answered correctly is totaled. This is called the raw score.

2. The raw score is converted statistically to a "standard" score on a scale of 200 to 800, with the mean set at 500. Each 100 points away from 500 is one standard deviation.

3. Your score is compared statistically with the criteria set by the scores of the second-year medical school candidates for certification in the June administration during the prior four years. This is what is meant by the term, "criterion-referenced test."

4. A score of 500 places you around the 50th percentile. A score of 380 is the minimum passing score for Part I; this probably represents about the 12th to 15th percentile. If you answer 50% or so of the questions correctly, you will almost certainly receive a passing score.

Remember: You do not have to pass all seven basic science components, although you will receive a standard score in each of them. A score of less than 400 (about the 15th percentile) on any particular area is a real cause for concern as it will certainly drag down your overall score. Likewise, a 600 or better (85th percentile) is an area of great relative strength. (You can use the practice test included in *Appleton's Review* to help determine your areas of strength and weakness well in advance of the actual examination.)

## Physical Conditions

The NBME is very concerned that all their exams be administered under uniform conditions in the numerous centers that are used. Except for several No. 2 pencils and an eraser, you are not permitted to bring anything (books, notes, calculators, etc.) into the test room. All examinees receive the same questions at the same session. The questions, however, are printed in different sequences in several different booklets, and the booklets are randomly dis-

tributed. In addition, examinees are removed to different seats at least once during the test. And, of course, each test is monitored by at least one proctor. The object of these maneuvers is to discourage cheating or even the temptation to cheat.

The number of candidates who fail Part I is quite small; however, individual students as well as entire medical school programs benefit when scores on National Boards are high. No one wants to squeak by with a 350 when a little effort might raise that score to 450. That is why you have made a wise decision to use the self-assessment and review materials available in this, the 4th edition of *Appleton & Lange's Review of Anatomy*.

# References

Basmajian, JV: Grant's Method of Anatomy, 10th ed. Baltimore, The Williams and Wilkins Company, 1980.

Hollinshead, WH, Rossee, C: Textbook of Anatomy, 4th ed. New York, Harper and Row, 1985.

Moore, KL: Clinically Oriented Anatomy. Baltimore, The Williams and Wilkins Company, 1980.

Woodburne, RT: Essentials of Human Anatomy, 7th ed. New York, Oxford University Press, 1983.

# Head and Neck
## Questions

**DIRECTIONS (Questions 1 through 82): Each of the numbered items or incomplete statements in this section is followed by answers or by completions of the statement. Select the <u>ONE</u> lettered answer or completion that is <u>BEST</u> in each case.**

1. The body of the mandible has a sharp inferior margin that ends posteriorly in which of the following?

    (A) mental protuberance
    (B) mental tubercle
    (C) angle of the mandible
    (D) mandibular condyle
    (E) mandibular notch

2. Which of the following muscles is considered to be the principal muscular landmark of the neck?

    (A) mylohyoid
    (B) sternocleidomastoid
    (C) sternohyoid
    (D) stylohyoid
    (E) anterior scalene

3. The digastric muscle is a two-bellied muscle that attaches by an intermediate tendon to which of the following?

    (A) mandibular condyle
    (B) thyroid cartilage
    (C) cricoid cartilage
    (D) styloid process
    (E) hyoid bone

4. The omohyoid, the sternocleidomastoid, and the posterior belly of the digastric muscle forms the boundary for which of the following triangles?

    (A) occipital
    (B) submandibular
    (C) submental
    (D) carotid
    (E) omoclavicular

5. As a rule, the isthmus of the thyroid gland crosses which of the following structures?

    (A) hyoid bone
    (B) second to fourth tracheal rings
    (C) cricoid cartilage
    (D) thyroid cartilage
    (E) inferior belly of the omohyoid muscle

6. The prevertebral layer of cervical fascia forms the floor for which of the following triangles?

    (A) submental
    (B) posterior cervical
    (C) submandibular
    (D) carotid
    (E) muscular

7. Which of the following muscles aids in depressing the corner of the mouth downward and widens the aperture, as in expressions of sadness or fright?

    (A) orbicularis oris
    (B) buccinator
    (C) mylohyoid
    (D) mentalis
    (E) platysma

8. The cervical branch of the facial nerve innervates which of the following muscles?

    (A) sternocleidomastoid
    (B) geniohyoid
    (C) sternothyroid
    (D) platysma
    (E) masseter

9. Which of the following veins unites with the retromandibular to form the external jugular vein?

    (A) posterior auricular
    (B) superficial temporal
    (C) transverse facial
    (D) internal jugular
    (E) facial

10. Which of the following veins crosses perpendicularly the superficial surface of the sternocleidomastoid beneath the platysma muscle?

   (A) internal jugular
   (B) anterior jugular
   (C) posterior jugular
   (D) external jugular
   (E) retromandibular

11. Which of the following nerves is a dorsal ramus of the second cervical nerve?

   (A) great auricular
   (B) greater occipital
   (C) lesser occipital
   (D) transverse cervical
   (E) supraclavicular

12. Which of the following nerves is formed by contributions from the ventral rami of cervical nerves three and four?

   (A) supraclavicular
   (B) greater occipital
   (C) great auricular
   (D) transverse cervical
   (E) occipitalis tertius

13. The superficial layer of cervical fascia splits into two sheets to enclose which of the following muscles?

   (A) sternothyroid
   (B) anterior scalene
   (C) trapezius
   (D) mylohyoid
   (E) semispinalis capitis

14. Which of the following ligaments is formed from a thickening of the deep parotid fascia?

   (A) temporomandibular
   (B) stylohyoid
   (C) stylomandibular
   (D) sphenomandibular
   (E) nuchal

15. Which of the following fascial layers gives rise to the axillary sheath?

   (A) superficial layer of cervical fascia
   (B) prevertebral
   (C) carotid sheath
   (D) buccopharyngeal
   (E) pretracheal

16. The sheath of the thyroid gland is formed from which of the following fascial layers?

   (A) carotid sheath
   (B) prevertebral
   (C) superficial layer of the cervical fascia
   (D) pretracheal
   (E) alar

17. Which of the following structures is located within the cervical visceral fasciae?

   (A) cervical sympathetic trunk
   (B) pharynx
   (C) external jugular vein
   (D) common carotid artery
   (E) hypoglossal nerve

18. The largest and most important interfascial interval in the neck is which of the following spaces?

   (A) suprasternal
   (B) retropharyngeal
   (C) submandibular
   (D) lateral pharyngeal
   (E) parotid

19. The sternohyoid muscle is innervated by which of the following nerves?

   (A) hypoglossal
   (B) ansa cervicalis
   (C) transverse cervical
   (D) supraclavicular
   (E) vagus

20. The superior thyroid artery is usually the first branch of which of the following arteries?

   (A) common carotid
   (B) external carotid
   (C) internal carotid
   (D) subclavian artery
   (E) maxillary artery

21. The inferior thyroid artery is a branch of which of the following arteries?

   (A) dorsal scapular
   (B) costocervical
   (C) external carotid
   (D) thyrocervical
   (E) vertebral

22. The middle thyroid vein empties into which of the following veins?

   (A) external jugular
   (B) anterior jugular
   (C) posterior jugular
   (D) internal jugular
   (E) vertebral

23. Which of the following structures is embedded in the anterior sheath of the carotid sheath?

   (A) sympathetic trunk
   (B) thyrocervical trunk
   (C) vertebral artery

(D) prevertebral fascia
(E) superior ramus of the ansa cervicalis

24. The common carotid artery usually bifurcates into the external and internal carotids at the level of which of the following structures?

(A) jugular notch
(B) cricoid cartilage
(C) upper border of the thyroid cartilage
(D) neck of the mandible
(E) sternoclavicular joint

25. Which of the following arteries enters the cranium to become the principal artery of the brain?

(A) external carotid
(B) internal carotid
(C) maxillary
(D) vertebral
(E) common carotid

26. Which of the following arteries passes obliquely upward deep to the posterior belly of the digastric and the stylohyoid muscles running deep to the submandibular gland?

(A) lingual
(B) facial
(C) maxillary
(D) superior thyroid
(E) occipital

27. Which of the following arteries arises from the posterior aspect of the external carotid at the level of the upper border of the posterior belly of the digastric?

(A) facial
(B) occipital
(C) lingual
(D) posterior auricular
(E) ascending pharyngeal

28. The vagus nerve leaves the skull through which of the following foramina?

(A) jugular
(B) internal acoustic
(C) foramen spinosum
(D) foramen ovale
(E) foramen lacerum

29. The cell bodies of the superior ganglion of the vagus nerve are concerned primarily with which of the following components of the nerve?

(A) general visceral efferent
(B) general somatic afferent
(C) general somatic efferent
(D) general visceral afferent
(E) special visceral afferent

30. Which of the following nerves innervate the cricothyroid and the inferior constrictor muscle of the pharynx?

(A) inferior cervical cardiac
(B) external branch of the superior laryngeal
(C) inferior laryngeal
(D) recurrent laryngeal
(E) superior cervical cardiac

31. The superior deep cervical lymph nodes occupy which of the following cervical triangles?

(A) carotid
(B) omoclavicular
(C) submandibular
(D) occipital
(E) submental

32. Which of the following ganglia is commonly located at the level of the second cervical vertebra?

(A) stellate ganglion
(B) inferior cervical ganglion
(C) vertebral ganglion
(D) middle cervical ganglion
(E) superior cervical ganglion

33. Which of the following ganglia is commonly located at the level of the cricoid cartilage?

(A) superior ganglion of the vagus
(B) inferior ganglion of the glossopharyngeal
(C) otic
(D) middle cervical
(E) submandibular

34. Which of the following ganglia is commonly located at the base of the transverse process of the seventh cervical vertebrae?

(A) pterygopalatine
(B) submandibular
(C) cervicothoracic
(D) vertebral
(E) geniculate

35. The posterior belly of the digastric muscle is innervated by which of the following nerves?

(A) trigeminal
(B) facial
(C) vagus
(D) ansa subclavia
(E) hypoglossal

36. Which of the following arteries pass through the brachial plexus either above or below the middle trunk?

(A) costocervical
(B) suprascapular
(C) transverse scapular
(D) dorsal scapular
(E) vertebral

37. The subclavian vein joins the internal jugular vein to form the brachiocephalic vein at which of these structures?

(A) the outer border of the first rib
(B) behind the acromioclavicular joint
(C) behind the coracoclavicular joint
(D) in front of the coracohumeral ligament
(E) behind the sternal end of the clavicle

38. The costocervical trunk usually gives rise to which of the following arteries?

(A) highest intercostal
(B) inferior thyroid
(C) suprascapular
(D) transverse cervical
(E) ascending cervical

39. Which of the following muscles is an essential muscular landmark of the neck?

(A) longus colli
(B) longus capitis
(C) rectus capitis
(D) scalenus anterior
(E) scalenus posterior

40. Which of the following laryngeal cartilages has a triangular base with vocal and muscular processes?

(A) cricoid
(B) corniculate
(C) arytenoid
(D) cuneiform
(E) epiglottis

41. Which of the following structures contributes to the formation of the vocal ligaments?

(A) thyrohyoid membrane
(B) cricotracheal ligament
(C) quadrangular membrane
(D) conus elasticus
(E) hyoepiglottic ligament

42. Which of the following structures constitutes the vestibular ligament of the false vocal fold?

(A) quadrangular membrane
(B) median cricothyroid ligament
(C) thyrohyoid membrane
(D) thyroepiglottic ligament
(E) cricotracheal ligament

43. The space between the apposed vocal folds and arytenoid cartilages is known as the

(A) glottis
(B) rima glottidis
(C) vestibule
(D) rima vestibuli
(E) piriform recess

44. Which of the following muscles of the larynx is abductor of the vocal ligament?

(A) posterior cricoarytenoid
(B) lateral cricoarytenoid
(C) transverse arytenoid
(D) thyroarytenoid
(E) cricothyroid

45. Which of the following muscles of the larynx increases tension on the vocal folds?

(A) cricothyroid
(B) lateral cricoarytenoid
(C) posterior cricoarytenoid
(D) thyroarytenoid
(E) transverse arytenoid

46. The vocalis muscles are composed of those internal fibers of which of the following muscles?

(A) lateral cricoarytenoid
(B) cricothyroid
(C) thyroarytenoid
(D) posterior cricoarytenoid
(E) oblique arytenoid

47. The principal sensory nerve of the larynx is the

(A) recurrent laryngeal
(B) inferior laryngeal
(C) superior laryngeal
(D) glossopharyngeal
(E) cervical sympathetic trunk

48. Which of the muscles of the larynx is innervated by the external branch of the superior laryngeal nerve?

(A) lateral cricoarytenoid
(B) posterior cricoarytenoid
(C) thyroarytenoid
(D) transverse arytenoid
(E) cricothyroid

49. The superior laryngeal artery is a branch of which of the following arteries?

(A) lingual
(B) superior thyroid

(C) costocervical trunk
(D) thyrocervical trunk
(E) transverse cervical

50. The trachea begins at the level of which of the following structures?

(A) hyoid bone
(B) thyroid cartilage
(C) fourth cervical vertebra
(D) cricoid cartilage
(E) second cervical vertebra

51. The carina is part of which of the following structures?

(A) hyoid bone
(B) epiglottis
(C) trachea
(D) larynx
(E) pharynx

52. The pharynx terminates at the level of which of the following structures?

(A) hyoid bone
(B) second cervical vertebra
(C) thyroid cartilage
(D) cricoid cartilage
(E) jugular notch

53. The pharyngobasilar fascia contributes to which of the following layers of the pharyngeal wall?

(A) mucous membrane
(B) submucosa
(C) longitudinal muscle layer
(D) circular muscle layer
(E) buccopharyngeal fascia

54. The middle pharyngeal constrictor arises from which of the following structures?

(A) pterygomandibular raphe
(B) cricoid cartilage
(C) thyroid cartilage
(D) torus tubarius
(E) hyoid bone

55. Which of the following muscles enters the pharyngeal wall in the gap between the origins of the middle and superior pharyngeal constrictor muscles?

(A) stylopharyngeus
(B) palatopharyngeus
(C) salpingopharyngeus
(D) thyrohyoid
(E) sternohyoid

56. Which of the following muscles of the pharynx is innervated by the glossopharyngeal nerve?

(A) superior pharyngeal constrictor
(B) salpingopharyngeus
(C) stylopharyngeus
(D) palatopharyngeus
(E) middle pharyngeal constrictor

57. Which of the following ganglia is a peripheral ganglion in the course of the parasympathetic innervation of the parotid gland?

(A) ciliary
(B) pterygopalatine
(C) submandibular
(D) otic
(E) geniculate

58. Which of the following nerves supplies parasympathetic fibers through the otic ganglion to the parotid gland?

(A) vagus
(B) glossopharyngeal
(C) facial
(D) hypoglossal
(E) accessory

59. Which of the following nerves innervates the genioglossus muscle?

(A) hypoglossal
(B) ansa cervicalis
(C) glossopharyngeal
(D) vagus
(E) trigeminal

60. The esophagus begins at the level of which of the following structures?

(A) hyoid bone
(B) thyroid cartilage
(C) fourth cervical vertebra
(D) cricoid cartilage
(E) jugular notch

61. Which of the following nerves is a cutaneous branch of the maxillary division of the trigeminal nerve?

(A) lacrimal
(B) infratrochlear
(C) auriculotemporal
(D) buccal
(E) superior labial

62. The levator anguli oris muscle is innervated by which of the following nerves?

(A) auriculotemporal
(B) facial
(C) ansa cervicalis
(D) inferior alveolar
(E) inferior palpebral nerve

63. The pterygomandibular raphe is a ligamentous band that stretches between the pterygoid hamulus and which of the following structures?

    (A) spine of the sphenoid
    (B) hyoid bone
    (C) mental protuberance
    (D) posterior end of the mylohyoid line
    (E) posterior nasal spine

64. The angular artery is the terminal part of which of the following arteries?

    (A) superficial temporal
    (B) posterior auricular
    (C) maxillary
    (D) facial
    (E) occipital

65. The infraorbital artery is one of the terminal branches of which of the following arteries?

    (A) facial
    (B) transverse facial
    (C) superficial temporal
    (D) lingual
    (E) maxillary

66. The mental artery is a terminal branch of which of the following arteries?

    (A) superficial temporal
    (B) transverse facial
    (C) inferior alveolar
    (D) facial
    (E) lingual

67. The retromandibular vein is formed by the confluence of the superficial temporal and which of the following veins?

    (A) transverse facial
    (B) maxillary vein
    (C) facial
    (D) lingual
    (E) submental

68. The parotid duct penetrates which of the following muscles?

    (A) masseter
    (B) medial pterygoid
    (C) buccinator
    (D) superior pharyngeal constrictor
    (E) levator anguli oris

69. The deep portion of the parotid fascia forms which of the following?

    (A) stylomandibular ligament
    (B) spenomandibular ligament
    (C) pterygomandibular raphe
    (D) carotid sheath
    (E) buccopharyngeal fascia

70. The facial nerve enters the temporal bone by way of which of the following openings?

    (A) carotid canal
    (B) foramen lacerum
    (C) stylomastoid foramen
    (D) internal acoustic meatus
    (E) jugular foramen

71. Which of the following nerves is usually the first extracranial branch of the facial nerve?

    (A) cervical branch
    (B) marginal mandibular branch
    (C) buccal branch
    (D) zygomatic branch
    (E) posterior auricular

72. The fourth layer of the scalp is represented by which of the following?

    (A) an aponeurotic
    (B) a muscular
    (C) a dense connective
    (D) a periosteal
    (E) a loose connective

73. The blood vessels of the scalp are located primarily in which of the following layers?

    (A) skin
    (B) dense subcutaneous layer
    (C) aponeurotic layer
    (D) loose connective tissue layer
    (E) periosteal layer

74. The supraorbital artery is a branch of which of the following arteries?

    (A) superficial temporal
    (B) transverse facial
    (C) maxillary
    (D) facial
    (E) ophthalmic

75. The muscles of the scalp are innervated by which of the following nerves?

    (A) supraorbital nerve
    (B) auriculotemporal nerve
    (C) temporal and auricular branches of the facial
    (D) greater occipital
    (E) lesser occipital

76. The masseteric fascia is formed from which of the following?

    (A) superficial layer of cervical fascia
    (B) carotid sheath

(C) prevertebral fascia
(D) buccopharyngeal
(E) pretracheal fascia

77. The sphenomandibular ligament is a thickening of which of the following?

(A) carotid sheath
(B) pterygoid fascia
(C) prevertebral fascia
(D) pretracheal fascia
(E) buccopharyngeal

78. The innervation of the masseter muscle is provided by which of the following nerves?

(A) buccal branch of the facial
(B) buccal branch of the trigeminal
(C) maxillary division of the trigeminal
(D) inferior alveolar nerve
(E) mandibular division of the trigeminal

79. The sphenomeniscus inserts into which of the following structures?

(A) mandibular condyle
(B) articular tubercle
(C) postglenoid tubercle
(D) articular disk
(E) lingula

80. The jaws are opened by forward traction on the neck of the mandible by which of the following muscles?

(A) masseter
(B) temporalis
(C) lower portion of the lateral pterygoid
(D) upper fibers of the medial pterygoid
(E) sphenomeniscus

81. Which of the following muscles positions or stabilizes the condyle and disk against the articular eminence during closing movements of the mandible?

(A) temporalis
(B) masseter
(C) medial pterygoid
(D) anterior belly of the digastric
(E) sphenomeniscus

82. The medial pterygoid muscle assists which of the following muscles in protrusion of the mandible?

(A) mylohyoid
(B) lateral pterygoid
(C) geniohyoid
(D) temporalis
(E) sphenomeniscus

DIRECTIONS (Questions 83 through 102): Each group of items in this section consists of lettered headings followed by a set of numbered words or phrases. For each numbered word or phrase, select the ONE lettered heading that is most closely associated with it. Each lettered heading may be selected once, more than once, or not at all.

Questions 83 through 86

For each insertion below, select the muscle with which it is associated.

(A) temporalis
(B) sphenomeniscus
(C) masseter
(D) medial pterygoid
(E) inferior belly of the lateral pterygoid

83. articular disk

84. pterygoid fovea

85. lateral surfaces of the coronoid process, ramus, and angle of the mandible

86. anterior border and medial surface of the coronoid process

Questions 87 through 90

For each muscle listed below, select the nerve that provides branches for its innervation.

(A) masseteric
(B) medial pterygoid
(C) meningeal
(D) inferior alveolar
(E) auriculo temporal

87. tensor veli palatine

88. mylohyoid

89. tensor tympani

90. anterior belly of the digastric

Questions 91 through 94

For each artery listed below, select the structure associated with the artery.

(A) mandibular foramen
(B) mandibular notch
(C) foramen spinosum
(D) infraorbital canal
(E) sphenopalatine foramen

91.  middle meningeal

92.  inferior alveolar

93.  masseteric

94.  anterior superior alveolar

**Questions 95 through 98**

For each function associated with the tongue, select the nerve that carries those fibers.

    (A) trigeminal
    (B) facial
    (C) hypoglossal
    (D) glossopharyngeal
    (E) vagus

95.  taste to the anterior two thirds of the tongue

96.  motor fibers to the styloglossus

97.  general sensation to the posterior one third of the tongue

98.  general sensation to the anterior two thirds of the tongue

**Questions 99 through 102**

For each muscle listed below, select the nerve that provides the innervation for that muscle.

    (A) vagus
    (B) glossopharyngeal
    (C) trigeminal
    (D) facial
    (E) hypoglossal

  99.  levator veli palatini

100.  tensor veli palatini

101.  palatoglossus

102.  palatopharyngeus

**DIRECTIONS (Questions 103 through 122): Each group of items in this section consists of lettered headings followed by a set of numbered words or phrases. For each numbered word or phrase, select**

    **A if the item is associated with (A) <u>only</u>,**
    **B if the item is associated with (B) <u>only</u>,**
    **C if the item is associated with <u>both</u> (A) <u>and</u> (B),**
    **D if the item is associated with <u>neither</u> (A) <u>nor</u> (B).**

**Questions 103 through 106**

    (A) the palatoglossus
    (B) the palatopharyngeus
    (C) both
    (D) neither

103.  forms the palatine arches

104.  inserts into the side of the tongue

105.  innervated by the vagus nerve

106.  takes origin on the torus tubaris

**Questions 107 through 110**

    (A) the frontal sinus
    (B) the maxillary sinus
    (C) both
    (D) neither

107.  the largest of the paranasal sinuses

108.  dependent drainage of the sinus requires the laying of the head on one side

109.  its roof is ridged by the infraorbital canal

110.  the sinus opens into the sphenoethmoidal recess

**Questions 111 through 114**

    (A) the greater petrosal nerve
    (B) the deep petrosal nerve
    (C) both
    (D) neither

111.  consists of postganglionic sympathetic fibers

112.  consists of preganglionic parasympathetic fibers

113.  composed of general visceral efferent fibers

114.  composed of general somatic efferent fibers

**Questions 115 through 118**

    (A) the pterygopalatine ganglion
    (B) the otic ganglion
    (C) both
    (D) neither

115.  fibers of the greater petrosal nerve synapse

116.  fibers of the deep petrosal nerve synapse

117.  fibers of the lesser petrosal nerve synapse

118.  a preganglionic sympathetic synapse

**Questions 119 through 122**

    (A) foramen rotundum
    (B) sphenopalatine foramen
    (C) both
    (D) neither

119. maxillary nerve enters the pterygopalatine by way of the

120. maxillary nerve enters the orbit by way of the

121. maxillary nerve sends fibers to the nasal cavity by way of the

122. maxillary nerve sends fibers to the soft palate by way of the

**DIRECTIONS (Questions 123 through 176): For each of the items in this section, ONE or MORE of the numbered options is correct. Choose the answer**

    A if only 1, 2, and 3 are correct,
    B if only 1 and 3 are correct,
    C if only 2 and 4 are correct,
    D if only 4 is correct,
    E if all are correct.

123. The superior and inferior ophthalmic veins may communicate with which of the following?

    (1) cavernous sinus
    (2) pterygoid plexus
    (3) supraorbital vein
    (4) superficial temporal vein

124. Which of the following statements apply to the eyeball?

    (1) cornea is the principal refractive structure of the eye
    (2) eyeball has three layers
    (3) eyeball is essentially elongated
    (4) cornea is transparent

125. Which of the following statements apply to the fibrous outer layer of the eye?

    (1) the sclera constitutes the posterior five sixths of the outer layer of the eye
    (2) it is continuous in front with the cornea
    (3) the transparent cornea constitutes the anterior one sixth of the outer layer
    (4) it is perforated by vorticose veins

126. Which of the following statements apply to the choroid layer of the eye?

    (1) it is brown in color
    (2) it consists of a dense capillary plexus
    (3) the ciliary body is part of the choroid layer
    (4) the iris is part of the choroid layer

127. Which of the following statements correctly apply to the lens?

    (1) lens is a transparent biconvex body
    (2) shape of the lens is modified by the ciliary muscle
    (3) lens is held in place by the suspensory ligament
    (4) lens is the principal refractive structure of the eye

128. Which of the following statements correctly apply to the external acoustic meatus?

    (1) mastoid air cells are posterior to the bony meatus
    (2) external acoustic meatus lies behind the condyle of the mandible
    (3) outer portion of the canal is cartilaginous
    (4) external acoustic meatus is a canal that is approximately 6 inches long

129. The middle ear includes which of the following structures?

    (1) tragus
    (2) anthelix
    (3) cochlea
    (4) auditory ossicles

130. Which of the following statements correctly apply to the tympanic membrane?

    (1) it is set oblique into the external acoustic meatus
    (2) it is composed of three layers
    (3) it forms the lateral wall of the middle ear
    (4) the umbo is at the most indrawn part of the membrane

131. Which of the following statements correctly apply to the stapedius muscle?

    (1) it is innervated by the facial nerve
    (2) it inserts onto the malleus
    (3) it takes origin from the pyramidal eminence
    (4) it is attached to the incus

132. Which of the following arteries provide branches to the middle ear?

    (1) maxillary
    (2) posterior auricular
    (3) ascending pharyngeal
    (4) middle meningeal

133. The nerves of the middle ear include branches from which of the following?

    (1) auriculotemporal
    (2) glossopharyngeal
    (3) vagus
    (4) carotid sympathetic

| SUMMARY OF DIRECTIONS | | | | |
|:---:|:---:|:---:|:---:|:---:|
| A | B | C | D | E |
| 1, 2, 3 only | 1, 3 only | 2, 4 only | 4 only | All are correct |

134. The internal ear consists of which of the following structures?

    (1) cochlea
    (2) semicircular canals
    (3) utricle
    (4) saccule

135. The bony labyrinth consists of which of the following?

    (1) stapes
    (2) incus
    (3) malleus
    (4) vestibule

136. Which of the following statements correctly apply to the cochlear duct?

    (1) it is the membranous part of the bony cochlea
    (2) it is triangular in cross-section
    (3) it is bounded below by a fibrous extension of the osseous spiral lamina, the basilar membrane
    (4) it is bounded above by the more delicate vestibular membrane

137. Which of the following arteries send branches to the labyrinth?

    (1) maxillary
    (2) basilar
    (3) facial
    (4) posterior auricular

138. Which of the following structures traverse the internal acoustic meatus?

    (1) facial nerve
    (2) vestibulocochlear nerve
    (3) labyrinthine artery
    (4) labyrinthine vein

139. The nervus intermedius contains which of the following?

    (1) large motor fibers for the muscles of facial expression
    (2) taste fibers from the anterior two thirds of the tongue
    (3) general sensory fibers from the posterior one third of the tongue
    (4) parasympathetic and visceral afferent fibers for the submandibular gland

140. Which of the following statements correctly apply to the geniculate ganglion?

    (1) it is the sensory ganglion of the trigeminal nerve
    (2) it contains motor neurons
    (3) it is located in the pterygoid canal
    (4) it is located at the junction of the internal acoustic meatus and the facial canal

141. Which of the following statements correctly apply to the chorda tympani nerve?

    (1) it arises from the facial nerve
    (2) it passes through the tympanic cavity
    (3) it passes forward over the medial surface of the tympanic membrane
    (4) it passes through the petrotympanic fissure

142. The peripheral processes of the cochlear nerve is located in which of the following structures?

    (1) saccule
    (2) utricle
    (3) ampulla of the posterior semicircular duct
    (4) organ of corti

143. Which of the following statements correctly apply to the skull?

    (1) cranium consists of eight bones
    (2) bones of the face total 14
    (3) skull contains both irregular and flat bones
    (4) bones of the face include the frontal

144. The lambda marks the junction between which of the following sutures?

    (1) sagittal
    (2) coronal
    (3) lambdoidal
    (4) metopic

145. The floor of the temporal fossa is formed by parts of which of the following bones?

    (1) frontal
    (2) parietal
    (3) temporal
    (4) sphenoid

146. Which of the following statements correctly identify boundaries of the infratemporal fossa?

    (1) superiorly, by the infratemporal crest of the sphenoid bone
    (2) inferiorly, by the alveolar border of the maxilla
    (3) laterally, by the ramus of the mandible
    (4) medially, by the medial pterygoid plate

147. Which of the following foramina open into the infratemporal fossa?

    (1) foramen rotundum
    (2) jugular foramen
    (3) stylomastoid
    (4) foramen ovale

148. Which of the following muscles are associated with the medial pterygoid plate?

    (1) stapedius
    (2) tensor tympani
    (3) palatopharyngeus
    (4) tensor veli palatini

149. Which of the following structures pass through the foramen ovale?

    (1) mandibular division of the trigeminal nerve
    (2) middle meningeal artery
    (3) accessory meningeal artery
    (4) deep auricular artery

150. Which of the following structures pass through the foramen spinosum?

    (1) sphenopalatine artery
    (2) posterior superior alveolar nerve
    (3) greater petrosal nerve
    (4) middle meningeal artery

151. The spine of the sphenoid bone gives attachment to which of the following structures?

    (1) sphenomandibular ligament
    (2) salpingopharyngeus muscle
    (3) tensor veli palatini muscle
    (4) masseter muscle

152. The stylomastoid foramen transmits which of the following structures?

    (1) glossopharyngeal nerve
    (2) occipital artery
    (3) vagus nerve
    (4) facial nerve

153. The jugular foramen transmits which of the following structures?

    (1) internal jugular vein
    (2) glossopharyngeal nerve
    (3) vagus nerve
    (4) accessory nerve

154. Which of the following nerves are associated with the tympanic and mastoid canaliculi?

    (1) glossopharyngeal
    (2) facial
    (3) vagus
    (4) accessory

155. Which of the following structures pass through the foramen magnum?

    (1) vertebral arteries
    (2) anterior and posterior spinal arteries
    (3) spinal cord and its meningeal coverings
    (4) vagus nerve

156. Which of the following bones form the floor of the anterior cranial fossa?

    (1) frontal
    (2) ethmoid
    (3) sphenoid
    (4) maxilla

157. Which of the following statements correctly apply to the cavernous sinuses?

    (1) it has nerves in its outer wall
    (2) it has a nerve and a major artery coursing through it
    (3) it lies on either side of the body of the sphenoid
    (4) it is formed between the meningeal and periosteal layers of the dura

158. The confluens of sinuses represents the junction of which of the following?

    (1) superior sagittal sinus
    (2) sphenoparietal sinus
    (3) straight sinus
    (4) basilar plexus

159. The meningeal arteries arise from which of the following arteries?

    (1) maxillary
    (2) anterior ethmoidal
    (3) ascending pharyngeal
    (4) occipital

160. Which of the meningeal vessels supply approximately four fifths of the dura mater?

    (1) accessory
    (2) posterior
    (3) anterior
    (4) middle

161. The sensory nerves of the dura mater are derived from which of the following nerves?

    (1) trigeminal
    (2) glossopharyngeal
    (3) vagus
    (4) facial

**SUMMARY OF DIRECTIONS**

| A | B | C | D | E |
|---|---|---|---|---|
| 1, 2, 3 only | 1, 3 only | 2, 4 only | 4 only | All are correct |

(1) the ciliary processes and ring are drawn toward the corneoscleral junction
(2) reduced tension on the fibers of the suspensory ligament
(3) the lens increases its curvatures
(4) greater refractive power for vision of close object

162. The oculomotor nerve is located between which of the following arteries?

(1) middle cerebral
(2) posterior cerebral
(3) anterior cerebral
(4) superior cerebellar

163. Which of the following statements correctly apply to the pia of the brain?

(1) it follows the contours of the brain
(2) it invests the brain and does not dip into its sulci, except for the longitudinal cerebral fissure
(3) it is described as a vascular membrane
(4) it is a thick, dense, fibrous layer

164. Which of the following structures pass through the foramen lacerum?

(1) greater petrosal nerve
(2) condyloid emissary vein
(3) deep petrosal
(4) accessory nerve

165. The superior orbital fissure provides communication with the orbit in the interval between which of the following structures?

(1) petrous portion of the temporal bone
(2) greater wing of the sphenoid
(3) orbital plate of the frontal
(4) lesser wing of the sphenoid

166. The clinoid processes are associated with which of the following structures?

(1) lesser wings of the sphenoid
(2) sella turcica
(3) dorsum sellae
(4) arcuate eminence

167. The vorticose veins drain into which of the following veins?

(1) posterior ciliary
(2) facial
(3) ophthalmic
(4) diploic

168. Which of the following statements correctly apply to the contracting ciliary muscle?

169. Which of the following structures are located in the iris?

(1) ciliary muscle
(2) sphincter pupillae muscle
(3) ora serrata
(4) dilator pupillae muscle

170. Which portion of the retina is characterized by nervous elements?

(1) pars ciliaris retinae
(2) pars iridica retinae
(3) ora serrata
(4) pars optica retinae

171. Which of the following statements correctly apply to the vitreous body?

(1) it is transparent
(2) it is a semigelatinous material
(3) it is adherent to the ora serrata
(4) it lies behind the cornea and anterior to the iris

172. The nasociliary nerve gives rise to which of the following nerves?

(1) posterior ethmoidal
(2) anterior ethmoidal
(3) infratrochlear
(4) supratrochlear

173. The fasciae of the orbit include which of the following?

(1) bulbar
(2) periorbita
(3) muscular
(4) dura

174. Which of the following statements correctly apply to the lacrimal sac?

(1) it lies behind the medial palpebral ligament
(2) it is lodged in the ethmoid bone
(3) it lies in front of the lacrimal portion of the orbicularis
(4) it is located in the lacrimal lake

**175.** Which of the following statements correctly apply to the anterior superior alveolar nerve?

(1) it arises from the supraorbital nerve
(2) it gives branches to the maxillary sinus
(3) it arises within the infraorbital groove
(4) it supplies the upper cuspid and incisor teeth

**176.** Which of the following statements correctly apply to the maxillary division of the trigeminal?

(1) it is entirely sensory
(2) it supplies the skin of the cheek
(3) it supplies the lower eyelid
(4) it supplies the upper lip

# Answers and Explanations

1. **(C)** The prominence of the chin is the mental protuberance. The body of the mandible has a sharp inferior margin that ends posteriorly in the angle of the mandible. The ramus continues upward toward the ear and ends in the mandibular condyle. *(Woodburne, p 145)*

2. **(B)** The principal muscular landmark of the neck is the sternocleidomastoid. The infrahyoid and suprahyoid muscles are located within the anterior cervical triangle in which the sternocleidomastoid forms one boundary. *(Woodburne, p 145)*

3. **(E)** The digastric muscle is a two-bellied muscle that attaches by an intermediate tendon to the hyoid bone. The digastric muscle completes the boundaries of the submandibular triangle with the base formed by the mandible. *(Woodburne, p 145)*

4. **(D)** The omohyoid muscle passes downward and laterally from the hyoid bone to disappear behind the sternocleidomastoid muscle and thus subdivides this area of the anterior triangle into two triangles. The upper triangle, bordered by the omohyoid, the sternocleidomastoid, and the posterior belly of the digastric muscle, is known as the carotid triangle. *(Woodburne, p 146)*

5. **(B)** Usually the isthmus of the thyroid gland crosses the second, third, and fourth tracheal rings. *(Woodburne, p 146)*

6. **(B)** The posterior triangle has as its floor the prevertebral layer of cervical fascia covering the splenius, levator scapulae, scalenus medius, and scalenus posterior muscles. *(Woodburne, p 146)*

7. **(E)** The platysma muscle draws the corner of the mouth downward and widens the aperture, as in expressions of sadness or fright. *(Woodburne, p 147)*

8. **(D)** The platysma muscle is innervated by the cervical branch of the fascial nerve. *(Woodburne, p 147)*

9. **(A)** The external jugular vein is formed a little below and behind the angle of the mandible by the union of the retromandibular vein and the posterior auricular vein. *(Woodburne, p 147)*

10. **(D)** The external jugular crosses perpendicularly the superficial surface of the sternocleidomastoid muscle directly under the platysma muscle. *(Woodburne, p 147)*

11. **(B)** The greater occipital nerve is a dorsal ramus of the second cervical nerve. The lesser occipital, great auricular, transverse cervical and supraclavicular nerves are cutaneous branches of the ventral rami of the cervical plexus. *(Woodburne, p 148)*

12. **(A)** The supraclavicular nerves are branches of a large trunk formed by contributions from the ventral rami of cervical nerves 3 and 4. *(Woodburne, p 151)*

13. **(A)** The superficial layer of cervical fascia splits into two sheets to enclose the sternocleidomastoid and trapezius muscles. *(Woodburne, p 151)*

14. **(C)** The stylomandibular ligament is a thickening of the deep parotid fascia extending from the tip of the styloid process to the angle of the mandible. *(Woodburne, p 152)*

15. **(B)** As the lower cervical spinal nerves emerge between the anterior and middle scalene muscles and are extended laterally over the first rib as the brachial plexus, they carry with them a prolongation of the prevertebral fascia known as the axillary sheath. *(Woodburne, p 152)*

16. **(D)** The sheath of the thyroid gland is a well-differentiated portion of the pretracheal fascia and encloses the gland on all sides. *(Woodburne, p 153)*

17. **(B)** In the central part of the neck lie the cervical viscera—pharynx, esophagus, larynx, trachea, thyroid, and parathyroid glands—enclosed within a

cylindrical cervical visceral fasciae. *(Woodburne, p 153)*

**18.** **(B)** The largest and most important interfascial interval in the neck is the retropharyngeal space. This is an areolar interval between the buccopharyngeal fascia anteriorly and the prevertebral fascia posteriorly. *(Woodburne, p 155)*

**19.** **(B)** The sternohyoid muscle is innervated by one or more branches of the superior ramus of the ansa cervicalis, which enter its lateral margin. *(Woodburne, p 155)*

**20.** **(B)** The superior thyroid artery is usually the first branch of the external carotid artery. *(Woodburne, p 157)*

**21.** **(D)** The inferior thyroid artery is the largest branch of the thyrocervical trunk of the subclavian artery. *(Woodburne, p 158)*

**22.** **(D)** The middle thyroid vein crosses the common carotid artery and empties into the lower end of the internal jugular vein. *(Woodburne, p 158)*

**23.** **(E)** The superior ramus of the ansa cervicalis nerve, conducting motor impulses to the infrahyoid muscles from the upper cervical nerves, lies on the anterior sheath of the carotid sheath. *(Woodburne, p 159)*

**24.** **(C)** The external carotid artery extends from the upper border of the thyroid cartilage to the neck of the mandible, where it divides into the superficial temporal and maxillary arteries. *(Woodburne, p 160)*

**25.** **(B)** The internal carotid has no branches in the neck but continues within the carotid sheath to the base of the skull, where it enters the cranium to become the principal artery of the brain. *(Woodburne, p 160)*

**26.** **(B)** The facial artery arises immediately above the lingual. It passes obliquely upward, deep to the posterior belly of the digastric and the stylohyoid muscles, over which it arches to lie in a groove on the deep surface of the submandibular gland. *(Woodburne, p 160)*

**27.** **(D)** The posterior auricular artery arises from the posterior aspect of the external carotid at the level of the upper border of the posterior belly of the digastric muscle. It ascends through the parotid fossa to the notch between the external acoustic meatus and the mastoid process. *(Woodburne, p 161)*

**28.** **(A)** The vagus nerve leaves the skull through the jugular foramen, contained in the same dural sheath with the accessory nerve. *(Woodburne, p 163)*

**29.** **(B)** The cell bodies of the superior ganglion of the vagus are concerned primarily with the general somatic afferent (cutaneous) component of the nerve. *(Woodburne, p 163)*

**30.** **(B)** The external branch of the superior laryngeal nerve descends on the inferior constrictor of the pharynx to the lower border of the thyroid cartilage, where it terminates in the cricothyroid muscle. It also supplies the inferior constrictor muscle of the pharynx. *(Woodburne, p 164)*

**31.** **(A)** The deep cervical lymph nodes, as is generally true elsewhere in the body, are arranged along the blood vessels, the arteries or veins. The superior deep cervical nodes thus occupy the carotid triangle of the neck. *(Woodburne, p 165*

**32.** **(E)** The superior cervical ganglion is the largest of the three cervical. It is usually broad, flattened, and tapered at the ends and lies in front on the transverse process of the second cervical vertebra. *(Woodburne, p 168)*

**33.** **(D)** The middle cervical ganglion commonly lies at the level of the cricoid cartilage in the bend of the inferior thyroid artery. *(Woodburne, p 169)*

**34.** **(C)** The cervicothoracic ganglion represents a combination of the inferior cervical and the first or the first several thoracic ganglia and lies anterior to the base of the transverse process of the seventh cervical vertebra and is usually posteromedial to the origin of the vertebral artery. *(Woodburne, p 170)*

**35.** **(B)** The anterior belly receives its innervation at its lateral border from the nerve to the mylohyoid, a branch of the mandibular division of the trigeminal nerve. The posterior belly is supplied by a branch of the facial nerve. *(Woodburne, p 171)*

**36.** **(D)** The dorsal scapular artery has an intimate relation to the brachial plexus, passing posteriorly through it, most frequently either above or below the middle trunk. *(Woodburne, p 181)*

**37.** **(E)** The subclavian vein is the continuation of the axillary vein at the outer border of the first rib. The subclavian vein joins the internal jugular vein to form the brachiocephalic vein behind the sternal end of the clavicle. *(Woodburne, p 181)*

**38.** **(A)** The costocervical trunk arises from the posterior aspect of the subclavian artery behind the anterior scalene muscle. It passes backward over the cervical pleura and the apex of the lung to the neck of the first rib, where it divides into the highest intercostal and the deep cervical. *(Woodburne, p 181)*

**39. (D)** The anterior scalene is one of the essential muscular landmarks of the neck. The phrenic nerve is formed at its lateral margin, descends on the muscle under the prevertebral fascia, and enters the thorax by passing its medial border. The roots of the brachial plexus and the subclavian artery emerge between the anterior and middle scalene muscles, and the subclavian vein passes anterior to the insertion of the anterior scalene. *(Woodburne, p 182)*

**40. (C)** The base of the arytenoid cartilage is triangular of the three borders of the pyramid; one forms the sharp, anteriorly directed vocal process to which the vocal ligament is attached. The laterally directed angle ends in the muscular process on which insert the posterior and lateral cricoarytenoid muscles. *(Woodburne, p 185)*

**41. (D)** The median part of the conus elasticus is the median cricothyroid ligament. The lateral parts of the conus end above in parallel thickenings, known as vocal ligaments, uniting the thyroid and the arytenoid cartilages. *(Woodburne, p 186)*

**42. (A)** The quadrangular membrane is a submucosal sheet of connective tissue ending in the aryepiglottic fold posterosuperiorly and inferiorly in a free margin that constitutes the vestibular ligament of the false vocal (vestibular) fold. *(Woodburne, p 186)*

**43. (B)** The vocal folds and the space between them are designated as the glottis and are the part of the larynx most directly concerned in the production of sounds. The rima glottidis, the space between the apposed vocal folds and arytenoid cartilages, is the narrowest part of the laryngeal cavity. *(Woodburne, p 187)*

**44. (A)** The posterior cricoarytenoid muscles are the abductors of the vocal folds; drawing the muscular processes backwards turns the vocal processes in a lateral direction. *(Woodburne, p 188)*

**45. (A)** The cricothyroid muscle draws the thyroid cartilage downward and forward toward the cricoid, which increases the distance between the thyroid and the arytenoid cartilages (mounted on the cricoid). Thus the anterior and posterior attachments of the vocal ligaments are carried further apart, and the tension of the vocal folds is made greater. *(Woodburne, p 188)*

**46. (C)** The vocalis muscles are composed of those internal fibers of the thyroarytenoid muscles most closely related to the vocal ligament. Such fibers are considered to be chiefly responsible for the control of pitch through their ability to regulate the vibrating part of the vocal ligaments. *(Woodburne, p 189)*

**47. (C)** The superior laryngeal nerve is the principal sensory nerve of the larynx and sends branches to the surfaces of the epiglottis and to the aryepiglottic fold. *(Woodburne, p 190)*

**48. (E)** The long, slender external branch of the superior laryngeal nerve descends along the oblique line of the thyroid cartilage to the cricothyroid muscle, which it supplies. All of the other intrinsic muscles of the larynx are innervated by the inferior laryngeal nerve. *(Woodburne, p 190)*

**49. (B)** The arteries of the larynx are the superior laryngeal, the inferior laryngeal, and the cricothyroid. The superior laryngeal artery is a branch of the superior thyroid artery, which pierces the thyrohyoid membrane in company with the internal branch of the superior laryngeal nerve. *(Woodburne, p 190)*

**50. (D)** The trachea begins at the lower border of the cricoid cartilage at the level of the sixth cervical vertebra and terminates at the sternal angle (upper border of the fifth thoracic vertebra). *(Woodburne, p 191)*

**51. (C)** The last tracheal cartilage is thick and broad in the middle, and its lower border is prolonged downward and backward in a hooked process. This is the carina, which forms a keel-like projection between the origins of the right and left bronchi. *(Woodburne, p 191)*

**52. (D)** The pharynx is approximately 12 cm long: it begins at the base of the skull and terminates below at the level of the lower border of the cricoid cartilage. *(Woodburne, p 191)*

**53. (B)** The submucosa is a strong fibrous sheet, known as the pharyngobasilar fascia, which is uncovered by the muscular layer of the pharynx in its uppermost part and is especially strong there. *(Woodburne, p 192)*

**54. (E)** The middle constrictor muscle arises from the upper border of the greater horn of the hyoid bone, from the lesser horn, and from the stylohyoid ligament. *(Woodburne, p 193)*

**55. (A)** The stylopharyngeus muscle enters the pharyngeal wall in the gap between the origins of the middle and superior pharyngeal constrictors. *(Woodburne, p 194)*

**56. (C)** The pharyngeal plexus provides the innervation of the muscles of the pharynx, with the exception of the stylopharyngeus. The pharyngeal plexus is formed by the pharyngeal and vagal nerves and of the superior cervical sympathetic ganglion. *(Woodburne, p 195)*

57. **(D)** The otic ganglion is a peripheral ganglion in the course of the parasympathetic innervation of the parotid gland; the postganglionic neurons arising in the ganglion distribute by way of the auriculotemporal branch of the trigeminal nerve to the gland. *(Woodburne, p 196)*

58. **(B)** The tympanic nerve, a branch of the glossopharyngeal, supplies parasympathetic fibers through the otic ganglion, to the parotid gland and sensory fibers, to the mucous membrane of the middle ear. *(Woodburne, p 196)*

59. **(A)** The terminal branches of the hypoglossal nerve distribute to the styloglossus, hyoglossus, genioglossus, and the intrinsic muscles of the tongue. *(Woodburne, p 197)*

60. **(D)** The esophagus begins at the level of the lower border of the cricoid cartilage and ends below the diaphragm opposite the eleventh thoracic vertebra. *(Woodburne, p 197)*

61. **(E)** The superior labial branches, three or four in number, descend to the skin and mucous membrane of the upper lip and gingiva and to the labial glands. The lacrimal and infratrochlear nerves are cutaneous branches of the ophthalmic division of the trigeminal. Both the buccal and the auriculotemporal are cutaneous branches of the mandibular division of the trigeminal nerve. *(Woodburne, p 199)*

62. **(B)** The levator anguli oris muscle elevates the angle of the mouth, at the same time drawing it in a medial direction, and is innervated by a buccal branch of the facial nerve. *(Woodburne, p 202)*

63. **(D)** The pterygomandibular raphe is a ligamentous band that, stretched between the pterygoid hamulus superiorly and the posterior end of the mylohyoid line of the mandible inferiorly, separates the fibers of the buccinator from those of the superior pharyngeal constrictor muscle. *(Woodburne, p 203)*

64. **(D)** The angular artery is the terminal part of the facial artery. Ascending to the medial angle of the orbit, it supplies orbicularis oculi and the lacrimal sac. *(Woodburne, p 204)*

65. **(E)** The infraorbital artery is one of the terminal branches of the maxillary artery and reaches the face by traversing the infraorbital groove and canal in company with the infraorbital vein and nerve. *(Woodburne, p 205)*

66. **(C)** The mental artery is a terminal branch of the inferior alveolar artery that, as a branch of the first part of the maxillary artery, passes through a canal in the mandible to supply the lower teeth. *(Woodburne, p 205)*

67. **(B)** The retromandibular vein is formed in the upper portion of the parotid gland, deep to the neck of the mandible, by the confluence of the superficial temporal vein and the maxillary vein. *(Woodburne, p 206)*

68. **(C)** Crossing the masseter muscle and the buccal fat pad, the duct turns deeply at the anterior border of the buccal pad of fat and penetrates the buccinator muscle, opening in the interior of the mouth opposite the second upper molar tooth. *(Woodburne, p 207)*

69. **(A)** The parotid fascia is an extension into the face of the superficial layer of the cervical fascia. The deep portion of the parotid fascia also forms a thickening, the stylomandibular ligament, which passes downward and lateralward from the styloid process to the angle of the mandible. *(Woodburne, p 207)*

70. **(D)** The facial nerve enters the temporal bone by way of the internal acoustic meatus and leaves it through the stylomastoid foramen. *(Woodburne, p 208)*

71. **(E)** The posterior auricular is the first extracranial branch of the facial nerve. It ascends between the external acoustic meatus and the mastoid process. *(Woodburne p 209)*

72. **(E)** The third layer of the scalp is musculoaponeurotic. The second layer is a dense connective layer. The fourth layer is both loose and scanty, which permits a wide spread of fluid accumulations in this layer. *(Woodburne, p 210)*

73. **(B)** The blood vessels of the scalp are firmly held by the strong fibers of the subcutaneous connective tissue and are prevented thereby from retracting or contracting their lumina when severed. *(Woodburne, p 210)*

74. **(E)** The supraorbital artery is a branch of the ophthalmic artery, which is a branch of the internal carotid. *(Woodburne, p 212)*

75. **(C)** The occipital portion of the muscle is innervated by the posterior auricular branch of the facial nerve. The temporal branch of the facial nerve supplies the frontal portion of the muscle. The posterior auricular branch of the facial nerve supplies the posterior muscle; the anterior and superior auricular muscles are innervated by the temporal branch of the facial. *(Woodburne, p 214)*

76. **(A)** The superficial layer of cervical fascia, after attachment to the inferior border of the mandible,

continues into the face as both the parotid and masseteric fascia. *(Woodburne, p 215)*

77. **(B)** The sphenomandibular ligament is a thickening of the pterygoid fascia, which extends between the spine of the sphenoid bone and the lingula of the mandible. *(Woodburne, p 215)*

78. **(E)** The nerve to the masseter muscle is a branch of the mandibular division of the trigeminal nerve, which reaches it by passing through the mandibular notch and penetrating its deep surface. *(Woodburne, p 216)*

79. **(D)** The upper head of the lateral pterygoid is designated as the sphenomeniscus and inserts into the articular disk of the temporomandibular joint and into the upper part of the neck of the condyle. *(Woodburne, p 217)*

80. **(C)** The jaws are opened by forward traction on the neck of the mandible by the lower portion of the lateral pterygoid muscles, assisted by the digastric, mylohyoid, and geniohyoid muscles. *(Woodburne, p 218)*

81. **(E)** Electromyographic studies suggest that the sphenomeniscus portion of the lateral pterygoid muscle positions or stabilizes the condyle and disk against the articular eminence during closing movements of the mandible. *(Woodburne, p 218)*

82. **(B)** The medial pterygoid muscle assists the lateral pterygoid in protrusion of the mandible; the posterior fibers of the temporalis retract it, assisted by the suprahyoid group of muscles. *(Woodburne, p 218)*

83. **(B)**

84. **(E)**

85. **(E)**

86. **(A)**

83–86. The sphenomeniscus inserts into the articular disk of the temporomandibular joint. The inferior belly of the lateral pterygoid inserts into the pterygoid fovea. The masseter inserts into the lateral surfaces of the coronoid process, ramus, and angle of the mandible. The temporalis inserts into the anterior border and medial surface of the coronoid process of the mandible. *(Woodburne, pp 216–217)*

87. **(B)**

88. **(E)**

89. **(B)**

90. **(E)**

87–90. The medial pterygoid nerve provides the nerve to the tensor veli palatini muscle, which enters this muscle near its origin, and the nerve to the tensor tympani muscle. The mylohyoid nerve arises from the inferior alveolar nerve just before the latter enters the mandibular foramen and descends in the mylohyoid groove to provide the motor innervation of the mylohyoid muscle and the anterior belly of the digastric muscle. *(Woodburne, pp 222–224)*

91. **(C)**

92. **(A)**

93. **(B)**

94. **(D)**

91–94. The middle meningeal artery enters the middle cranial fossa by way of the foramen spinosum. The inferior alveolar artery enters the mandibular foramen with the inferior alveolar nerve. The masseteric artery passes through the mandibular notch with the masseteric nerve and supplies the masseter muscle. The anterior superior alveolar branches arise in the infraorbital canal and descend in the alveolar canals to the upper incisive and cuspid teeth. *(Woodburne, pp 224–225)*

95. **(B)**

96. **(C)**

97. **(D)**

98. **(A)**

95–98. Taste to the anterior two thirds of the tongue is provided by the facial nerve. The hypoglossal nerve provides the motor innervation to all the extrinsic and intrinsic muscles of the tongue. The glossopharyngeal nerve provides general sensation to the posterior one third of the tongue. The trigeminal nerve provides general sensation to the anterior two thirds of the tongue. *(Woodburne, p 232)*

99. **(A)**

100. **(C)**

101. **(A)**

102. **(A)**

99–102. The tensor veli palatini muscle is supplied by the mandibular division of the trigeminal nerve. All others of the palatal muscles are supplied by the contribution of the vagus nerve to the pharyngeal plexus. *(Woodburne, p 235)*

**103.** **(C)**

**104.** **(A)**

**105.** **(C)**

**106.** **(D)**

**103–106.** The palatine tonsil is a collection of lymphatic tissue underlying the mucous membrane between the palatine arches. These are the palatoglossal and palatopharyngeal folds of mucous membrane and their corresponding muscles. The palatoglossus muscle inserts into the side of the tongue, some of its fibers spreading over the dorsum and others blending with fibers of the transversus linguae muscle. All palatal muscles, with the exception of the tensor veli palatini, are supplied by the contribution of the vagus nerve to the pharyngeal plexus. Both the palatoglossus and palatopharyngeus attach to the palatine aponeurosis. *(Woodburne, pp 194, 234–235)*

**107.** **(B)**

**108.** **(B)**

**109.** **(B)**

**110.** **(D)**

**107–110.** The maxillary sinuses are the largest of the paranasal sinuses. Each maxillary sinus extends up to the orbital surface of the maxilla, where its roof is ridged by the infraorbital canal. The drainage of the maxillary sinus is very poor in the erect posture, and dependent drainage of the sinuses requires the laying of the head on one side. The maxillary sinuses communicate with the nasal cavity by an opening that passes from its upper medial wall to the lower part of the hiatus semilunaris. The frontal sinus communicates with the nasal cavity by way of the frontonasal duct, which empties into the anterior part of the middle meatus. *(Woodburne, pp 239–240)*

**111.** **(B)**

**112.** **(A)**

**113.** **(C)**

**114.** **(B)**

**111–114.** The greater petrosal nerve is composed of general visceral efferent fibers for the parasympathetic supply of the lacrimal, nasal, and palatine glands and of sensory fibers from the soft palate. The greater petrosal nerve unites with the deep petrosal nerve to form the nerve of the pterygoid canal. The deep petrosal nerve is a branch of the internal carotid plexus, the continuation of the cervical sympathetic trunk that consists of postganglionic sympathetic fibers. *(Woodburne, p 243)*

**115.** **(A)**

**116.** **(D)**

**117.** **(B)**

**118.** **(D)**

**115–118.** At the pterygopalatine ganglion the fibers of the greater petrosal nerve synapse; contrarily, those of the deep petrosal nerve pass directly into the maxillary nerve branches. Carried indistinguishably in the branches of the maxillary nerve, these parasympathetic and sympathetic postganglionic fibers pass to the lacriminal gland and to the glands of the nasal cavity, palate, and upper pharynx. The lesser petrosal nerve is a continuation of the tympanic nerve beyond the plexus. The lesser petrosal nerve ends in the otic ganglion, which is a peripheral ganglion in the course of the parasympathetic innervation of the parotid gland. *(Woodburne, pp 196, 242–243)*

**119.** **(A)**

**120.** **(D)**

**121.** **(B)**

**122.** **(D)**

**119–122.** Emerging through the foramen rotundum, the maxillary nerve enters the deep, triangular pterygopalatine fossa. From the pterygopalatine fossa, the maxillary nerve enters the orbit by way of the inferior orbital fissure as the infraorbital nerve. The posterosuperior lateral nasal branches pass through the sphenopalatine foramen into the posterior part of the nasal cavity. The greater and lesser palatine nerves reach the soft palate by way of the greater and lesser palatine foramina. *(Woodburne, pp 240–242)*

**123.** **(A)** The superior ophthalmic vein communicates with the supraorbital vein and usually joins with the inferior ophthalmic vein at the medial end of the superior orbital fissure. The inferior ophthalmic vein proceeds to the back of the orbit and communicates through the inferior orbital fissure with the pterygoid plexus. Occasionally it empties separately into the cavernous sinus. *(Woodburne, p 253)*

124. **(C)** The eyeball is essentially a sphere. Its form is modified by a small anteriorly bulging segment, the transparent cornea. In its essential structure the eyeball can be resolved into three layers. A lens lies behind the cornea and is the principal refractive structure of the eye. *(Woodburne, p 253)*

125. **(E)** The sclera is a firm fibrous cup that constitutes the posterior five sixths of the outer layer of the eye. It is continuous in front with the cornea at the corneoscleral junction. The sclera is perforated by vorticose veins. The transparent cornea constitutes the anterior one sixth of the fibrous tunic. *(Woodburne, pp 253–254)*

126. **(E)** The choroid layer is brown in color due to the pigment cells of its outermost layer. It consists of a dense capillary plexus. The ciliary body is an elevated zone of the anterior portion of the choroid layer. The iris is a thin contractile membrane having a central aperture, the pupil. *(Woodburne, p 254)*

127. **(E)** The lens is a transparent biconvex body, flatter on its anterior surface and more curved posteriorly. The shape of the lens is modified by the ciliary muscle, as the eye focuses on objects at differing distances. The lens is held in place by a series of straight fibrils, constituting the suspensory ligament of the lens. The lens lies behind the cornea and is the principal refractive structure of the eye. *(Woodburne, pp 253,256)*

128. **(A)** The mastoid air cells are posterior to the bony meatus. The external acoustic meatus lies behind the condyle of the mandible. The cartilaginous outer part of the canal constitutes about one third of its length. The external acoustic meatus is a canal, approximately 2.5 cm long. *(Woodburne, p 257)*

129. **(D)** The middle ear is a narrow cavity in the temporal bone; here the energy of sound waves is converted into mechanical energy through a chain of ossicles. *(Woodburne, p 257)*

130. **(E)** The tympanic membrane is set obliquely into the external acoustic meatus and is composed of three layers: the modified skin of the external meatus on its outer surface, the mucous membrane of the cavity internally, and an intermediate fibrous stratum. The center point, the umbo, is at the most indrawn part of the membrane. *(Woodburne, p 259)*

131. **(B)** The pyramidal eminence is hollow, and its walls give rise to the fibers of the stapedial muscle, the central tendon that emerges at an aperture on the summit of the eminence and inserts in the posterior surface of the neck of the stapes. A small branch of the facial nerve is conducted into the muscle. *(Woodburne, p 260)*

132. **(E)** The arteries of the middle ear are numerous. The two larger ones are the anterior tympanic branch of the maxillary artery to the tympanic membrane and the stylomastoid branch of the posterior auricular artery to the tympanic antrum and mastoid cells. Smaller arteries include the inferior tympanic branch of the ascending pharyngeal artery and the superficial petrosal branch of the middle meningeal artery. *(Woodburne, p 261)*

133. **(C)** The nerves of the middle ear are branches of the tympanic plexus, formed by the tympanic branch of the glossopharyngeal nerve and by superior and inferior caroticotympanic nerves from the carotid sympathetic plexus. *(Woodburne, p 261)*

134. **(E)** The internal ear provides the essential organs of hearing and equilibrium. It consists of the cochlea, for the auditory sense, and a series of intercommunicating channels (the semicircular ducts, the utricle, and the saccule), for the sense of balance and position. *(Woodburne, p 261)*

135. **(D)** The bony labyrinth consists of three parts: the vestibule, the semicircular canals, and the cochlea. *(Woodburne, p 262)*

136. **(E)** The cochlear duct is the membranous part of the bony cochlea. The cochlear duct is triangular in cross-section, being bounded below by a fibrous extension of the osseous spiral lamina, the basilar membrane, and above by the more delicate vestibular membrane. *(Woodburne, p 262)*

137. **(C)** The principal artery is the labyrinthine, a branch of the basilar artery that traverses the internal acoustic meatus and divides into branches that accompany the cochlear and vestibular nerves. The stylomastoid artery, a branch of the posterior auricular, also provides some blood supply. *(Woodburne, p 263)*

138. **(E)** The facial nerve and the vestibulocochlear nerve traverse the internal acoustic meatus. Both the labyrinth artery and vein also traverse the internal acoustic meatus. *(Woodburne, p 263)*

139. **(C)** The facial nerve is composed of two unequal roots: the largest motor root supplies the muscles of facial expression; the smaller root is known as nervus intermedius. It contains taste fibers from the anterior two thirds of the tongue, fibers of general

sensation from the external acoustic meatus, and parasympathetic and visceral fibers for the submandibular, sublingual, lacrimal, nasal, and palatine glands. *(Woodburne, p 263)*

140. **(D)** The geniculate ganglion is the sensory ganglion of the facial nerve. It is located at the abrupt bend taken by the nerve as it turns from the internal acoustic meatus into the posteriorly directed facial canal. *(Woodburne, p 263)*

141. **(E)** The chorda tympani arises from the facial nerve proximal to the stylomastoid foramen and enters the tympanic cavity. It passes forward over the medial surface of the tympanic membrane and arches across the handle of the malleus. It emerges from the skull through the petrotympanic fissure. *(Woodburne, p 263)*

142. **(D)** The peripheral processes of the cochlear nerve are located in the organ of corti, their cell bodies forming the spiral ganglion of the cochlea, which is situated in the osseous spiral lamina. *(Woodburne, p 264)*

143. **(A)** The skull is a series of flat bones united by interlocking or overlapping sutures in the formation of the cranium and by a group of irregular bones that help to establish the bony framework of the face and the base of the cranium. The cranium, enclosing a cavity accommodating the brain, consists of eight bones; the bones of the face total fourteen. *(Woodburne, p 264)*

144. **(B)** Lambda marks the junction of the interparietal sagittal suture and the highly serrated parieto-occipital lambdoidal suture. *(Woodburne, p 265)*

145. **(E)** The floor of the temporal fossa is formed by parts of the frontal, parietal, and temporal bones and by the great wing of the sphenoid bone. *(Woodburne, p 265)*

146. **(A)** The infratemporal fossa is bounded superiorly by the infratemporal crest of the sphenoid bone and inferiorly by the alveolar border of the maxilla. The ramus of the mandible is its lateral boundary: the lateral pterygoid plate is its medial limit. *(Woodburne, p 267)*

147. **(D)** The foramen ovale and foramen spinosum open into the roof of the infratemporal fossa. *(Woodburne, p 267)*

148. **(D)** At the base of the medial pterygoid plate, it is hollowed out in the oval scaphoid fossa for the origin of the tensor veli palatini muscle; at the inferior extremity of the plate, the tendon of the tensor turns around its recurved hamulus. *(Woodburne, p 270)*

149. **(B)** At the base of the lateral pterygoid plate lies the foramen ovale for the mandibular division of the trigeminal nerve and the accessory meningeal artery. *(Woodburne, pp 270–271)*

150. **(D)** The foramen spinosum transmits the middle meningeal vessels and the meningeal branch of the mandibular nerve. *(Woodburne, p 271)*

151. **(B)** Behind the foramen spinosum is the spine of the sphenoid bone, which gives attachment to the sphenomandibular ligament and the tensor veli palatini muscle. *(Woodburne, p 271)*

152. **(D)** At the base of the styloid process is the stylomastoid foramen, which transmits the facial nerve and the stylomastoid artery. *(Woodburne, p 271)*

153. **(E)** The jugular foramen transmits the jugular vein and the glossopharyngeal, vagus, and accessory nerves. It also transmits the inferior petrosal sinus. *(Woodburne, p 271)*

154. **(B)** On the ridge of bone between the orifice of the carotid canal and the jugular foramen is the tympanic canaliculus for the passage of the tympanic branch of the glossopharyngeal nerve: in the lateral wall of the jugular foramen is the mastoid canaliculus, through which the auricular branch of the vagus enters the temporal bone. *(Woodburne, p 271)*

155. **(A)** Through the foramen magnum pass the spinal cord and its meningeal coverings, the ascending rootlets of the accessory nerve, the vertebral arteries, and the anterior and posterior spinal arteries. *(Woodburne, p 271)*

156. **(A)** The floor of the anterior cranial fossa is formed by the orbital plates of the frontal bone, the cribriform plate of the ethmoid, and the lesser wings and forepart of the body of the sphenoid bone. *(Woodburne, pp 272–273)*

157. **(E)** The cavernous sinuses lie on either side of the body of the sphenoid. They are formed between the meningeal and periosteal layers of the dura. This sinus has nerves in its outer wall and a nerve and a major artery coursing through it. *(Woodburne, p 278)*

158. **(B)** The confluens of sinuses represents the junction of the superior sagittal, straight, and occipital sinuses with the right and left transverse sinuses. *(Woodburne, p 278)*

159. **(E)** The middle meningeal artery is a branch of the maxillary artery. The anterior meningeal artery is a branch of the anterior ethmoidal artery. The posterior meningeal ethmoidal artery is a branch of the ascending pharyngeal, occipital and vertebral arteries. *(Woodburne, p 279)*

160. **(D)** The middle meningeal artery supplies four fifths of the dura mater. *(Woodburne, p 279)*

161. **(B)** The sensory nerves of the dura mater are derived from all three divisions of the trigeminal nerve and from the vagus nerve. *(Woodburne, p 280)*

162. **(C)** The posterior cerebral artery is separated from the superior cerebellar artery by the oculomotor nerve. *(Woodburne, p 281)*

163. **(B)** The pia mater enmeshes the blood vessels on their surface and is thus described as a vascular membrane. It follows the contours of the brain and spinal cord. The arachnoid membrane loosely invests the brain and does not dip into its sulci except for the longitudinal cerebral fissure. *(Woodburne, p 275)*

164. **(B)** Both the greater and deep petrosal nerves pass through the foramen lacerum. *(Woodburne, p 275)*

165. **(C)** The superior orbital fissure provides communication with the orbit in the interval between the lesser and the greater wings of the body of the sphenoid bone. This fissure transmits many of the nerves entering the orbit together with the ophthalmic vein. *(Woodburne, p 274)*

166. **(A)** The anterior clinoid processes of each side project backward from the lesser wings of the sphenoid. The sella turcica is bounded in front by a pair of small projections, the middle clinoid processes, and behind by a flat plate of bone, the dorsum sellae, on the upper angles of which are the posterior clinoid processes. *(Woodburne, p 274)*

167. **(B)** The veins of the vascular tunic are almost all tributary to the vorticose veins, which drain into the posterior ciliary and the ophthalmic veins. *(Woodburne, p 255)*

168. **(E)** The ciliary muscle is the muscle of accommodation. By its contraction, the ciliary process and ring are drawn toward the corneoscleral junction, and the wedge-shaped ring is bulged toward its center. This reduces the tension on the fibers of the suspen-

sory ligament, thereby allowing the natural elasticity of the lens to increase its curvatures, resulting in greater refractive power for vision of close objects. *(Woodburne, p 254)*

169. **(C)** Within the loose stroma of the iris are two involuntary muscles. The sphincter pupillae muscle is formed of circular fibers, and the dilatory pupillae muscle consists of radiating fibers. *(Woodburne, p 255)*

170. **(D)** Three regions are differentiated in the retina. The pars optica retinae is that portion characterized by nervous elements. At the ora serrata the nervous elements of the retina cease. *(Woodburne, p 255)*

171. **(A)** The vitreous chamber of the eye lies behind the lens and ciliary processes. It is filled by the vitreous body, a body of transparent semigelatinous material, which is firmly adherent to the ora serrata. *(Woodburne, p 255)*

172. **(A)** The nasociliary nerve, along the border of the medial rectus, divides into its principal branches: the posterior ethmoidal, anterior ethmoidal, and infratrochlear nerves. *(Woodburne, p 252)*

173. **(A)** The fasciae of the orbit are the periorbita, the bulbar sheath, and the muscular fasciae. *(Woodburne, p 247)*

174. **(C)** The lacrimal sac is the upper dilated portion of the nasolacrimal duct. The sac lies behind the medial palpebral ligament and in front of the lacrimal portion of the orbicularis oculi. *(Woodburne, p 245)*

175. **(C)** The anterior superior alveolar nerve is given off from the infraorbital nerve just before the latter emerges from the infraorbital foramen. It supplies the upper cuspid and incisor teeth and adjacent gingiva. Branches are also given to the maxillary sinus. *(Woodburne, p 242)*

176. **(E)** The maxillary division of the trigeminal nerve is entirely sensory and supplies the skin of the cheek, the side of the nose, the lower eyelid, and the upper lip. *(Woodburne, p 240)*

# Abdominal Region
## Questions

1.  The highest point of the iliac crest is at the transverse level of which of the following structures?

    (A) inguinal ligament
    (B) xiphosternal joint
    (C) fourth lumbar vertebra
    (D) anterior iliac spine
    (E) ischial tuberosity

2.  Which of the following nerves traverses the superficial inguinal ring?

    (A) iliohypogastric
    (B) subcostal
    (C) ilioinguinal
    (D) pudendal
    (E) obturator

3.  The tunica dartos scroti contains which of the following?

    (A) fat
    (B) striated muscle
    (C) the iliohypogastric nerve
    (D) the inferior epigastric artery
    (E) smooth muscle

4.  The musculophrenic artery is a branch of which of the following arteries?

    (A) internal thoracic
    (B) inferior epigastric
    (C) superficial epigastric
    (D) superficial circumflex
    (E) deep external pudendal

5.  The inferior epigastric artery arises from which of the following arteries?

    (A) internal thoracic
    (B) external iliac
    (C) femoral
    (D) obturator
    (E) musculophrenic

6.  The superficial epigastric artery arises from which of the following arteries?

    (A) femoral
    (B) internal iliac
    (C) external iliac
    (D) inferior epigastric
    (E) superior epigastric

7.  The superficial circumflex iliac artery arises from which of the following arteries?

    (A) external iliac
    (B) internal iliac
    (C) superior epigastric
    (D) inferior epigastric
    (E) femoral artery

8.  The superficial external pudendal artery emerges through which of the following?

    (A) deep inguinal ring
    (B) saphenous opening
    (C) lumbar triangle
    (D) arcuate line
    (E) rectus sheath

9.  The upper five slips of origin for the external abdominal oblique muscle interdigitate with which of the following muscles?

    (A) pectoralis major
    (B) rectus abdominis
    (C) latissimus dorsi
    (D) serratus anterior
    (E) subscapularis

10. Which of the following structures contributes to the formation of the lumbar triangle?

    (A) rectus sheath
    (B) crest of the ilium
    (C) ischial tuberosity
    (D) inguinal ligament
    (E) serratus posterior

11. Which of the following structures is NOT a specialization of the external oblique aponeurosis?

    (A) inguinal ligament
    (B) lacunar ligament
    (C) intercrural fibers
    (D) internal spermatic fascia
    (E) medial and lateral crura

12. The thickened, rolled-under portion of the external oblique aponeurosis, which is stretched between the anterior superior spine of the ilium and the pubic tubercle, is known as the

    (A) inguinal ligament
    (B) lacunar ligament
    (C) intercrural fibers
    (D) rectus sheath
    (E) linea alba

13. The lacunar ligament represents the more medial, rolled-under fibers of which of the following?

    (A) medial crura
    (B) reflected inguinal ligament
    (C) fundiform ligament
    (D) pectineal ligament
    (E) inguinal ligament

14. In the lower one fourth of the abdomen, the internal abdominal oblique aponeurosis does which of the following?

    (A) splits to send one sheet anterior and one posterior to the rectus abdominis muscle
    (B) fails to split and passes to the median line entirely posterior to the rectus abdominis muscle
    (C) fails to split and passes to the median line entirely anterior to the rectus abdominis muscle
    (D) disappears
    (E) gives rise to the internal spermatic fascia

15. Among the coverings of the cord and testis, which of the following represents the internal abdominal oblique muscle layer?

    (A) tunica tartos
    (B) external spermatic fascia
    (C) cremaster muscle and fascia
    (D) internal spermatic fascia
    (E) subcutaneous layer

16. The cremaster muscle is innervated by which of the following nerves?

    (A) ilioinguinal
    (B) iliohypogastric
    (C) femoral
    (D) subcostal
    (E) genitofemoral

17. The aponeurosis of the transversus abdominis muscle contributes to which of the following?

    (A) inguinal ligament
    (B) lacunar ligament
    (C) falx inguinalis
    (D) external spermatic fascia
    (E) superficial inguinal ring

18. Which of the following statements correctly applies to the tendinous intersections?

    (A) they are firmly adherent to both layers of the rectus sheath
    (B) the lowest is at the level of the symphysis pubis
    (C) the highest is near the xiphoid process
    (D) the lowest is at the level of the arcuate line
    (E) there are usually ten tendinous intersections

19. Which of the following statements correctly applies to the pyramidalis muscle?

    (A) it is a large, well-developed muscle
    (B) it is always present
    (C) it is contained within the rectus sheath
    (D) it is posterior to the rectus abdominis muscle
    (E) it is innervated by the tenth thoracic nerve

20. The transversalis fascia contributes to which of the following structures?

    (A) deep inguinal ring
    (B) cremaster muscle and fascia
    (C) inguinal ligament
    (D) pectineal ligament
    (E) external spermatic fascia

21. The innervation of the muscles of the abdominal wall is provided by which of the following cord segments?

    (A) thoracic 3 through sacral 3
    (B) thoracic 5 through lumbar 2
    (C) thoracic 1 through sacral 5
    (D) thoracic 7 through lumbar 4
    (E) lumbar 1 through sacral 5

22. The tenth intercostal nerve enters the rectus sheath at the level of which of the following structures?

    (A) xiphoid process
    (B) umbilicus

(C) pyramidalis muscle
(D) pubic tubercle
(E) symphysis pubis

23. The four lumbar arteries arise from which of the following vessels?

(A) internal thoracic
(B) internal iliac
(C) femoral
(D) aorta
(E) external iliac

24. The iliolumbar artery is a branch of which of the following arteries?

(A) internal iliac
(B) external iliac
(C) internal thoracic
(D) femoral
(E) inferior epigastric

25. The remnant of the gubernaculum testis is known as which of the following?

(A) processus vaginalis
(B) tunica vaginalis testis
(C) inguinal canal
(D) scrotal ligament
(E) genitoinguinal ligament

26. Which of the following structures is NOT located in the spermatic cord?

(A) ductus deferens
(B) deferential artery
(C) testicular artery
(D) pampiniform plexus of veins
(E) urethra

27. Which of the following structures does NOT traverse the inguinal canal?

(A) inferior epigastric artery
(B) ilioinguinal nerve
(C) genital branch of the genitofemoral nerve
(D) cremasteric artery
(E) internal spermatic fascia

28. The tunica vaginalis testis is a remnant of which of the following?

(A) urachus
(B) processus vaginalis
(C) scrotal ligament
(D) gubernaculum testis
(E) genitoinguinal ligament

29. Which of the following is a derivative of the dorsal common mesentery?

(A) transverse mesocolon
(B) lienorenal ligament
(C) falciform ligament
(D) hepatogastric ligament
(E) gastrocolic ligament

30. Which of the following is a derivative of the dorsal mesogastrium?

(A) hepatogastric ligament
(B) hepatocolic ligament
(C) phrenic colic ligament
(D) sigmoid mesocolon
(E) coronary ligament

31. Which of the following is a derivative of the ventral mesogastrium?

(A) mesoappendix
(B) gastrolienal ligament
(C) lienorenal ligament
(D) hepatoduodenal ligament
(E) sigmoid mesocolon

32. The primordia of the pancreas, liver, and gall bladder appear as outgrowths of the gut tube during which of the following weeks of development?

(A) second
(B) fourth
(C) eighth
(D) 16th
(E) 21st

33. Which of the following statements correctly applies to the rotation of the gastrointestinal tube?

(A) rotation is clockwise when viewed from the ventral side
(B) rotation occurs during the second week of development
(C) rotation takes place around an axis represented by the superior mesenteric artery
(D) rotation is usually complete by the fourth week of development
(E) rotation only involves the small intestine

34. The esophagogastric junction is located at the level of which of the following vertebrae?

(A) seventh cervical
(B) 11th thoracic
(C) second lumbar
(D) fifth lumbar
(E) sixth thoracic

**35.** The aortic hiatus of the diaphragm is located at the level of which vertebra?

(A) fourth cervical
(B) sixth cervical
(C) fifth thoracic
(D) 12th thoracic
(E) fourth lumbar

**36.** Which of the following arteries is a branch of the celiac trunk?

(A) gastroduodenal
(B) proper hepatic
(C) common hepatic
(D) right gastroepiploic
(E) cystic

**37.** The cystic artery arises from which of the following arteries?

(A) splenic
(B) left gastroepiploic
(C) right gastric
(D) right hepatic
(E) gastroduodenal

**38.** The left gastroepiploic artery arises from which of the following arteries?

(A) left hepatic
(B) gastroduodenal
(C) left gastric
(D) common hepatic
(E) splenic

**39.** Which of the following statements correctly applies to the innervation of the stomach?

(A) the sacral plexus provides the parasympathetic fibers
(B) the anterior vagal trunk carries preganglionic visceral efferents only
(C) the stomach has no sympathetic innervation
(D) the vagal trunks contain preganglionic visceral efferent and general visceral afferent fibers
(E) the postganglionic parasympathetic cell bodies are located in the celiac plexus

**40.** Which of the following statements correctly applies to the spleen?

(A) the sixth, seventh, and eighth ribs are in relation to the spleen
(B) the spleen is retroperitoneal
(C) the spleen develops in the ventral mesogastrium
(D) the spleen normally descends below the costal margin
(E) the spleen rests on the left flexure of the colon

**41.** Which of the following statements correctly applies to the first part of the duodenum?

(A) it is surrounded by the hepatoduodenal ligament
(B) it is related to the caudate lobe of the liver
(C) the common bile duct passes ventral
(D) it is located at the level of the third lumbar vertebra
(E) it has circular folds in its interior

**42.** The second portion of the duodenum is crossed by which of the following structures?

(A) right renal artery
(B) transverse colon
(C) right ureter
(D) portal vein
(E) superior mesenteric vein

**43.** Which of the following arteries crosses the anterior aspect of the third part of the duodenum?

(A) proper hepatic
(B) left colic
(C) superior mesenteric
(D) inferior mesenteric
(E) splenic

**44.** Which of the following statements correctly applies to the greater duodenal papilla?

(A) the location for the terminal opening of the accessory pancreatic duct
(B) it is located in the interior of the third part of the duodenum
(C) it is continued below by the longitudinal fold of the duodenum
(D) it is superior to the lesser duodenal papilla
(E) it opens into the duodenojejunal flexure

**45.** The lower left portion of the head of the pancreas is inserted behind which of the following arteries?

(A) left gastroepiploic
(B) common hepatic
(C) left colic
(D) inferior mesenteric
(E) superior mesenteric

**46.** The superior mesenteric and splenic veins unite to form the portal vein behind which of the following structures?

(A) first part of the duodenum
(B) transverse colon
(C) spleen
(D) neck of the pancreas
(E) duodenojejunal junction

**47.** The tail of the pancreas enters which of the following structures?

(A) epiploic foramen
(B) lienorenal ligament
(C) suspensory ligament of the duodenum
(D) paracolic fossa
(E) left coronary ligament

48. Which of the following structures develops in two parts from a dorsal and ventral primordium?

(A) liver
(B) spleen
(C) gall bladder
(D) pancreas
(E) cecum

49. Which of the following structures is formed from the distal part of the duct of the dorsal primordium and the proximal part of the duct of the ventral primordium?

(A) cystic
(B) hepatic
(C) common hepatic
(D) accessory pancreatic
(E) main pancreatic

50. The pancreatica magna artery is a branch of which of the following arteries?

(A) common hepatic
(B) inferior mesenteric
(C) superior mesenteric
(D) left gastroepiploic
(E) splenic

51. Which of the following structures represents the obliterated remains of the umbilical vein?

(A) ligamentum teres hepatis
(B) ligamentum venosum
(C) ductus arterious
(D) falciform ligament
(E) porta hepatis

52. The hepatoduodenal ligament transmits which of the following structures?

(A) hepatic vein
(B) main pancreatic duct
(C) portal vein
(D) ligamentum venosum
(E) superior mesenteric vein

53. The cystic artery usually arises from which of the following arteries?

(A) splenic
(B) gastroduodenal
(C) right gastroepiploic
(D) right hepatic
(E) celiac trunk

54. The hepatic veins drain into which of the following veins?

(A) portal
(B) coronary
(C) inferior vena cava
(D) superior mesenteric
(E) splenic

55. The portal vein ascends to the liver in the free margin of which of the following structures?

(A) mesocolon
(B) greater omentum
(C) mesentery
(D) lesser omentum
(E) falciform ligament

56. The superior mesenteric artery supplies all of the small intestine except which of the following?

(A) proximal part of the duodenum
(B) duodenojejunal junction
(C) jejunoileal junction
(D) distal end of the ileum
(E) descending portion of the duodenum

57. Which of the following statements correctly apply to the middle colic artery?

(A) it takes origin from the celiac trunk
(B) it supplies the cecum
(C) it anastomoses with the inferior pancreaticoduodenal artery
(D) it is a branch of the superior mesenteric artery
(E) it primarily supplies the left colic flexure

58. Both the vagal parasympathetic innervation and the thoracic splanchnic sympathetic innervation of the gastrointestinal tract terminate at which of the following?

(A) duodenojejunal junction
(B) junction of the middle and left thirds of the transverse colon
(C) jejunoileal junction
(D) ileocecal junction
(E) distal one third of the sigmoid colon

59. Epiploic appendages are located on which of the following structures?

(A) duodenum
(B) stomach
(C) ileum
(D) jejunum
(E) sigmoid

60. Which of the following statements correctly applies to the vermiform appendix?

    (A) it is usually retroperitoneal
    (B) it has a small mesentery
    (C) it receives its blood supply from the inferior mesenteric artery
    (D) it is usually located in a retrocecal position
    (E) it receives its innervation from the pelvic splanchnic

61. The transverse mesocolon is attached posteriorly to which of the following structures?

    (A) hepatoduodenal ligament
    (B) spleen
    (C) second portion of the duodenum
    (D) lesser omentum
    (E) gastrocolic ligament

62. The pampiniform plexus is located in which of the following locations?

    (A) pancreas
    (B) kidney
    (C) spleen
    (D) inguinal canal
    (E) liver

63. The psoas major muscle inserts on to which of the following structures?

    (A) greater trochanter
    (B) anterior superior iliac spine
    (C) crest of the ilium
    (D) lesser trochanter
    (E) ischial spine

64. Which of the following layers of fasciae is associated with the diaphragm?

    (A) alar
    (B) superficial
    (C) transversalis
    (D) buccopharyngeal
    (E) innominate

65. The right suprarenal vein drains into which of the following veins?

    (A) right renal
    (B) inferior mesenteric
    (C) superior mesenteric
    (D) portal
    (E) inferior vena cava

66. Which of the following statements correctly applies to the suprarenal gland?

    (A) cortex is essential to life
    (B) medulla is essential to life

    (C) medulla is concerned with carbohydrate metabolism
    (D) medulla is concerned with the body fluid and electrolyte balance
    (E) medulla receives only a postganglionic innervation

67. Which of the following structures is located in the renal column?

    (A) interlobular arteries
    (B) collecting tubule
    (C) arcuate arteries
    (D) interlobar arteries
    (E) minor calyx

68. Which of the following structures is important in the selective reabsorption of water and dissolved materials back into the circulation?

    (A) glomerular capsule
    (B) renal papilla
    (C) straight and convoluted tubules
    (D) glomerulus
    (E) major calyx

69. The pelvic splanchnic nerves provide parasympathetic fibers to all of the following structures EXCEPT

    (A) bladder
    (B) right colic flexure
    (C) descending colon
    (D) sigmoid colon
    (E) distal one third of the transverse colon

70. Which of the following statements correctly applies to the sigmoid colon?

    (A) it begins at the brim of the pelvis
    (B) it has no mesocolon
    (C) it continues as the rectum at the level of the fifth sacral segment
    (D) it receives its blood supply from the left colic artery
    (E) it has no teniae coli

71. Which of the following statements correctly applies to the small intestine?

    (A) the upper three fifths is considered jejunum
    (B) the lower three fifths contains aggregated lymph nodules
    (C) none of it is retroperitoneal
    (D) the parasympathetic innervation is provided by the pelvic splanchnic nerves
    (E) the blood supply is provided by both the superior and inferior mesenteric arteries

72. Which of the following statements correctly applies to the gall bladder?

(A) the submucosal layer is well developed
(B) the mucous membrane is thrown into circular folds
(C) its epithelium concentrates the contents of the gall bladder
(D) it produces bile
(E) it lies to the left of the falciform ligament

73. Which of the following statements correctly applies to the falciform ligament?

(A) it represents the inferior limit of the common mesentery
(B) it encloses the ligamentum teres of the liver
(C) it extends from the umbilicus to the liver
(D) it contains the common bile duct
(E) it does not extend over the diaphragmatic surface of the liver

74. The anterior surface of the liver lies against all of the following structures EXCEPT

(A) diaphragm
(B) costal margin
(C) xiphoid process
(D) abdominal wall
(E) spleen

75. Which of the following structures is situated between the celiac trunk and the superior mesenteric artery?

(A) duodenum and pancreas
(B) spleen and stomach
(C) transverse colon and ileum
(D) stomach and cecum
(E) pancreas and jejunum

76. Which of the following statements correctly applies to the pancreas?

(A) it extends from the right kidney to the spleen
(B) it is inferior to the stomach
(C) it is crossed by the transverse mesocolon
(D) it overlies the fourth lumbar vertebra
(E) its uncinate process extends behind the inferior mesenteric vessels

77. Which of the following statements correctly apply to the fourth portion of the duodenum?

(A) it is located at the level of the first lumbar vertebra
(B) it is entirely retroperitoneal
(C) the root of the mesentery begins at the duodenojejunal flexure
(D) it is in direct continuity with the pylorus of the stomach
(E) it overlies the hilum of the right kidney

78. All of the following statements correctly apply to the duodenum EXCEPT which?

(A) it is the shortest portion of the small intestine
(B) it is usually the breadth of twelve fingers
(C) it is the fixed portion of the small intestine
(D) it is suspended by a mesentery
(E) it is continuous with the completely peritonealized stomach and duodenum

79. Which of the following statements correctly applies to the innervation of the stomach?

(A) the parasympathetic innervation enhances muscular movement
(B) the sympathetic innervation exerts the greater influence on the secretion of water
(C) the sympathetic innervation exerts the greater influence on the secretion of hydrochloric acid
(D) the parasympathetic innervation has the major influence in the secretion of enzymes
(E) afferents principally accompany the parasympathetic system

80. The inferior boundary of the epiploic foramen is which of the following structures?

(A) inferior vena cava
(B) hepatoduodenal ligament
(C) caudate lobe of the liver
(D) lesser omentum
(E) first part of the duodenum

81. The lesser omentum includes which of the following ligaments?

(A) phrenicocolic
(B) coronary
(C) hepatogastric
(D) gastrocolic
(E) gastrolienal

82. The lesser peritoneal sac is closed off from the greater peritoneal sac except for the communication through which of the following?

(A) aortic hiatus
(B) esophageal hiatus
(C) caval foramen
(D) deep inguinal ring
(E) epiploic foramen

**DIRECTIONS (Questions 83 through 102): Each group of items in this section consists of lettered headings followed by a set of numbered words or phrases. For each numbered word or phrase, select the ONE lettered heading that is most closely associated with it. Each lettered heading may be selected once, more than once, or not at all.**

**Questions 83 through 86**

For each nerve listed below, choose the spinal cord segments associated with that nerve.

   (A) L1
   (B) T5–9
   (C) L2,L3,L4
   (D) T1–5
   (E) S2,S3,S4

83.   the greater splanchnic

84.   the iliohypogastric

85.   the pelvic splanchnic

86.   the obturator

**Questions 87 through 90**

For each of the structures listed below choose the term that identifies that particular structure.

   (A) haustra
   (B) longitudinal folds
   (C) circular folds
   (D) spiral fold
   (E) coronary ligament

87.   gall bladder

88.   liver

89.   transverse colon

90.   jejunum

**Questions 91 through 94**

For each of the structures listed below, choose the source of its origine.

   (A) external abdominal oblique aponeurosis
   (B) internal abdominal oblique aponeurosis
   (C) transverse abdominis muscle
   (D) transversalis fascia
   (E) extraperitoneal fat

91.   cremaster muscle and fascia

92.   inguinal ligament

93.   superficial inguinal ring

94.   deep inguinal ring

**Questions 95 through 98**

For each of the arteries listed below, choose the artery from which it arises.

   (A) superior mesenteric artery
   (B) celiac trunk
   (C) inferior mesenteric
   (D) external iliac artery
   (E) femoral artery

95.   inferior epigastric artery

96.   superior rectal artery

97.   superficial circumflex iliac artery

98.   splenic artery

**Questions 99 through 102**

For each of the veins listed below, choose the vein that it empties into.

   (A) splenic vein
   (B) portal vein
   (C) superior mesenteric vein
   (D) inferior vena cava
   (E) internal iliac vein

99.   hepatic vein

100.   left gastric vein

101.   right colic vein

102.   pancreatic magna

**DIRECTIONS (Questions 103 through 132): Each group of items in this section consists of lettered headings followed by a set of numbered words or phrases. For each numbered word or phrase, select**

A if the item is associated with (A) only,
B if the item is associated with (B) only,
C if the item is associated with both (A) and (B),
D if the item is associated with neither (A) nor (B).

**Questions 103 through 106**

   (A) the second portion of the duodenum
   (B) the jejunum
   (C) both
   (D) neither

103.   contains circular folds

104.   contains the major duodenal papilla

**105.** is innervated by the pelvic splanchnic nerve

**106.** is retroperitoneal

**Questions 107 through 110**

    (A) the genitofemoral nerve
    (B) the iliohypogastric nerve
    (C) both
    (D) neither

**107.** the nerve arises from the tenth thoracic spinal cord segment

**108.** it provides the innervation for the cremaster muscle

**109.** enters the rectus sheath as the level of the arcuate line

**110.** innervates the pyramidalis muscle

**Questions 111 through 114**

    (A) the superior mesenteric artery
    (B) the inferior mesenteric artery
    (C) both
    (D) neither

**111.** provides the blood supply to the kidney

**112.** acts as the axis of rotation for the counterclock rotation of the primitive gut

**113.** gives rise to the inferior rectal artery

**114.** provides the blood supply for the transverse colon

**Questions 115 through 118**

    (A) linea alba
    (B) linea semilunares
    (C) both
    (D) neither

**115.** defines the lateral margins of the rectus muscles

**116.** produced by tendinous bands that interrupt the fibers of the rectus abdominis

**117.** begin at the pubic tubercles and end above at the costal margin

**118.** marks the aponeurotic junction of the abdominal muscles of the two sides

**Questions 119 through 122**

    (A) the lacunar ligament
    (B) the pectineal ligament
    (C) both
    (D) neither

**119.** this ligament is continuous with the inguinal ligament

**120.** a delicate tubulosaccular sheath that represents the continuation of the intercrural fibers along the spermatic cord

**121.** decussating fibers of the opposite side that descend to the pubic crest and tubercle

**122.** the more medial, rolled-under fibers of the inguinal ligament that flatten down into a horizontal shelf

**Questions 123 through 127**

    (A) the myenteric plexus
    (B) the submucosal plexus
    (C) both
    (D) neither

**123.** contains preganglionic sympathetic cell bodies

**124.** contains postganglionic parasympathetic cell bodies

**125.** is located between the circular and longitudinal muscle layers

**126.** peristalsis depends on its integrity

**127.** receives sympathetic postganglionic terminal fibers

**Questions 128 through 132**

    (A) the splenic artery
    (B) the posterior superior pancreaticoduodenal artery
    (C) both
    (D) neither

**128.** gives off a number of pancreatic branches

**129.** enters the lienorenal ligament

**130.** located in the superior border of the pancreas

**131.** anastomoses with a branch of the superior mesenteric artery

**132.** passes from right to left along the greater curvature of the stomach

**DIRECTIONS (Questions 133 through 177): For each of the items in this section, ONE or MORE of the numbered options is correct. Choose the answer**

> A if only <u>1, 2, and 3</u> are correct,
> B if only <u>1 and 3</u> are correct,
> C if only <u>2 and 4</u> are correct,
> D if only <u>4</u> is correct,
> E if <u>all</u> are correct.

133. The anterolateral muscles of the abdominal wall include which of the following muscles?

    (1) external abdominal oblique
    (2) quadratus lumborum
    (3) rectus abdominis
    (4) psoas major

134. Which of the following structures correctly applies to the boundaries of the lumbar triangle?

    (1) posterior margin of the external abdominal oblique
    (2) anterior margin of the latissimus dorsi
    (3) iliac crest
    (4) inferior epigastric artery

135. Specializations of the external abdominal oblique aponeurosis include which of the following?

    (1) inguinal ligament
    (2) medial and lateral crura
    (3) superficial inguinal ring
    (4) intercrural fibers

136. The internal abdominal oblique muscle and fascia give rise to which of the following?

    (1) pectineal ligament
    (2) intercrural fibers
    (3) reflected inguinal ligament
    (4) cremaster muscle and fascia

137. The transversus abdominis muscle contributes to which of the following structures?

    (1) lacunar ligament
    (2) internal spermatic fascia
    (3) lateral crura
    (4) falx inguinalis

138. In the lower one fourth of the rectus sheath the aponeurosis of the internal oblique muscle does which of the following?

    (1) it does not split
    (2) it passes posterior to the rectus abdominis
    (3) it passes anterior to the rectus abdominis
    (4) it splits to pass anterior and posterior to the rectus abdominis

139. Which of the following statements apply to the pyramidalis muscle?

    (1) it is located outside the rectus sheath
    (2) it draws down on the linea alba
    (3) it arises from the inguinal ligament
    (4) it is supplied by a branch of the 12th thoracic nerve

140. Below the arcuate line, the principal layer of the posterior sheath of the rectus abdominis is formed by which of the following?

    (1) transversus abdominis
    (2) internal abdominal oblique aponeurosis
    (3) falx inguinalis
    (4) transversalis fascia

141. Which of the following structures arise from the transversalis fascia?

    (1) falx inguinalis
    (2) deep inguinal ring
    (3) fundiform ligament
    (4) internal spermatic fascia

142. The blood vessels of the abdominal wall are derived from which of the following arteries?

    (1) internal thoracic
    (2) internal iliac
    (3) external iliac
    (4) femoral

143. Which of the following statements correctly apply to the inferior epigastric artery?

    (1) its origin is just medial to the deep inguinal ring
    (2) it arises from the external iliac artery just above the inguinal ligament
    (3) it pierces the transversalis fascia and enters the rectus sheath near the arcuate line
    (4) it gives rise to the cremaster artery

144. Which of the following statements correctly apply to the gubernaculum?

    (1) in the male, its remnant constitutes the scrotal ligament
    (2) in the female, its remnant becomes the ovarian ligament
    (3) in the female, the gubernaculum extends between the uterus and the labium majus to form the round ligament of the uterus
    (4) in the male, it forms the tunica vaginalis testis

145. Which of the following structures are located in the inguinal canal in both males and females?

    (1) round ligament
    (2) ductus deferens
    (3) deferential artery
    (4) ilioinguinal nerve

146. Which of the following statements correctly apply to the primordia of the pancreas, liver, and gall bladder?

    (1) they appear as outgrowths of the gut tube just caudal to the stomach
    (2) they appear during the second week of development
    (3) the liver and gallbladder are ventral outgrowths
    (4) the pancreas is definitively embedded in the ventral mesogastrium

147. Which of the following statements correctly apply to the hepatocolic ligament?

    (1) it is a derivative of the dorsal common mesentery
    (2) it invests the left flexure of the transverse colon
    (3) it is part of the greater omentum
    (4) it is inconstant

148. Which of the following peritoneal folds do not contain an artery?

    (1) ileocecal fold
    (2) mesoappendix
    (3) duodenojejunal fold
    (4) paraduodenal fold

149. Which of the following statements correctly apply to the pyloric sphincter?

    (1) it is formed from the circular muscle layer of the stomach
    (2) it is under hormonal control from the duodenum
    (3) it does not open with every cycle of contraction
    (4) all digestion is completed proximal to the pyloric sphincter

150. Which of the following statements correctly apply to the stomach?

    (1) it is firmly fixed only at its beginning and at its end
    (2) the lesser curvature marks its right border
    (3) its posterior surface forms the anterior wall of the omental bursa
    (4) the stomach is retroperitoneal

151. The stomach receives its blood supply from which of the following arteries

    (1) splenic
    (2) left gastric
    (3) common hepatic
    (4) superior mesenteric

152. The branches of the common hepatic artery include which of the following arteries?

    (1) posterior superior pancreaticoduodenal
    (2) gastroduodenal
    (3) left gastric
    (4) proper hepatic

153. Which of the following statements correctly apply to the spleen?

    (1) it develops in the dorsal mesogastrium
    (2) it is usually palpable through the abdominal wall
    (3) it is entirely surrounded by peritoneum
    (4) it normally descends below the costal margin

154. Which of the following statements correctly apply to the first part of the duodenum?

    (1) it is retroperitoneal
    (2) it is surrounded by a portion of the lesser omentum
    (3) it makes contact with the right kidney
    (4) it is related above to the quadrate lobe of the liver

155. Which of the following statements correctly apply to the second part of the duodenum?

    (1) it overlies the hilum of the right kidney posteriorly
    (2) it rests on the inferior vena cava
    (3) it rests on the psoas muscles
    (4) its medial surface is in contact with the pancreas

156. Which of the following statements correctly apply to the third portion of the duodenum?

    (1) it is completely peritonealized
    (2) it crosses the body of the first lumbar vertebra
    (3) the common bile duct and the pancreatic duct penetrate its posteromedial wall
    (4) the superior mesenteric artery rises above and crosses the anterior aspect of the third part of the duodenum

157. The greater duodenal papilla marks the termination of which of the following ducts?

    (1) common bile
    (2) cystic
    (3) main pancreatic
    (4) accessory pancreatic

SUMMARY OF DIRECTIONS

| A | B | C | D | E |
|---|---|---|---|---|
| 1, 2, 3 only | 1, 3 only | 2, 4 only | 4 only | All are correct |

**158.** The anterior and posterior inferior pancreaticoduodenal arteries may arise from which of the following arteries?

(1) superior mesenteric
(2) splenic
(3) jejunal
(4) gastroduodenal

**159.** The anterior pancreatic arcade is formed by the anastomosing branches of which of the following arteries?

(1) celiac trunk
(2) inferior mesenteric
(3) superior mesenteric
(4) right renal

**160.** The dorsal pancreatic artery may arise from which of the following arteries?

(1) splenic
(2) celiac
(3) superior mesenteric
(4) left renal

**161.** Which of the following structures may be observed on the visceral surface of the liver?

(1) porta hepatis
(2) gallbladder
(3) fissure for the ligamentum teres
(4) fissure for the ligamentum venosum

**162.** The gallbladder does NOT have which of the following?

(1) a serous layer
(2) longitudinal folds
(3) a mucosal layer
(4) a submucosal layer

**163.** Which of the following statements correctly apply to the common bile duct?

(1) it descends in the free border of the lesser omentum
(2) it usually lies to the right of the hepatic artery
(3) it usually lies anterior to the portal vein
(4) it descends behind the first portion of the duodenum

**164.** Which of the following arteries is NOT a branch of the superior mesenteric artery?

(1) inferior pancreaticoduodenal
(2) superior rectal
(3) middle colic
(4) left colic

**165.** The ileocolic artery usually gives rise to which of the following arteries?

(1) appendicular
(2) anterior cecal
(3) posterior cecal
(4) middle colic

**166.** The marginal artery is formed by anastomatic channels from which of the following arteries?

(1) ileocolic
(2) right colic
(3) middle colic
(4) left colic

**167.** Which of the following statements correctly apply to the superior mesenteric vein?

(1) it lies anterior and to the right of the superior mesenteric artery
(2) it crosses the third part of the duodenum
(3) it crosses the uncinate process of the pancreas
(4) it joins the splenic vein behind the neck of the pancreas to form the portal vein

**168.** Which of the following statements correctly apply to the ileocecal valve?

(1) it contains a frenulum
(2) it is a well-defined sphincter
(3) it is formed by the circular muscle layers of the ileum
(4) it is formed by the longitudinal muscle layers of the cecum

**169.** Which of the following statements correctly apply to the descending colon?

(1) it is the largest and most movable part of the large intestine
(2) it passes inferiorly from the left colic flexure to end in the sigmoid colon
(3) it usually has a well-developed mesocolon
(4) it passes along the lateral border of the left kidney over the quadratus lumborum to the iliac fossa

**170.** The inferior mesenteric artery gives rise to which of the following arteries?

(1) left colic
(2) sigmoidal
(3) superior rectal
(4) middle rectal

171. The inferior mesenteric vein begins in which of the following plexuses of veins?

    (1) pampiniform
    (2) pterygoid
    (3) basilar
    (4) rectal

172. The intestinal lymph trunk is formed by the union of efferents from which of the following nodes?

    (1) celiac
    (2) hepatic
    (3) superior mesenteric
    (4) superficial inguinal

173. Which of the following statements correctly apply to the portal vein?

    (1) it is formed behind the neck of the pancreas by the union of the superior mesenteric vein and the splenic vein
    (2) it ascends behind the first part of the duodenum
    (3) it ascends in the free margin of the hepatoduodenal ligament
    (4) it lies behind the common bile duct and the proper hepatic artery

174. Which of the following structures are located at the level of the 12th thoracic or first lumbar vertebra?

    (1) caval foramen
    (2) celiac tunic
    (3) esophageal hiatus
    (4) aortic hiatus

175. The lumbar plexus of nerves includes which of the following nerves or plexuses?

    (1) greater splanchnic nerves
    (2) iliohypogastric nerve
    (3) superior hypogastric plexus
    (4) obturator

176. Which of the following nerves provide the sympathetic innervation for the pelvic viscera?

    (1) pelvic splanchnic nerves
    (2) anterior and posterior vagal trunk
    (3) thoracic splanchnic nerves
    (4) lumbar splanchnic nerves

177. Which of the following nerves carry parasympathetic fibers to the pelvic viscera?

    (1) iliohypogastric
    (2) lumbar splanchnic
    (3) anterior and posterior vagal trunks
    (4) pelvic splanchnic nerves

# Answers and Explanations

1. **(C)** The prominent iliac crest forms the upper limit of the region of the hip: the curve of the crest ends ventrally in the anterior superior spine of the ilium, dorsally in the posterior superior iliac spine. The highest point of the iliac crest is at the transverse level of the body of the fourth lumbar vertebra. *(Woodburne, p 367)*

2. **(C)** The ilioinguinal nerve traverses the inguinal canal to the superficial inguinal ring, where it emerges on the lateral aspect of the spermatic cord. *(Woodburne, p 369)*

3. **(E)** The tunica dartos scroti is directly continuous with the subcutaneous tissue of the abdominal wall. It is without fat and contains smooth muscle intermingled with its areolar tissue. This muscle is the cause of the wrinkling of the skin of the scrotum. *(Woodburne, p 368)*

4. **(A)** The musculophrenic branch of the internal thoracic artery supplies twigs to the skin of the abdomen along the costal arch. *(Woodburne, p 369)*

5. **(B)** The inferior epigastric artery enters the rectus sheath posterior to the rectus sheath from below after arising from the external iliac artery. *(Woodburne, p 369)*

6. **(A)** The superficial epigastric artery arises from the anterior aspect of the femoral artery about 1 cm below the inguinal ligament. Piercing the femoral sheath and cribriform fascia, the artery turns superiorly over the inguinal ligament and runs toward the umbilicus. *(Woodburne, p 369)*

7. **(E)** The superficial circumflex iliac artery arises from the femoral artery about 1 cm below the inguinal ligament. It pierces the fascia lata lateral to the saphenous opening and runs laterally across the upper thigh below and parallel to the inguinal ligament. *(Woodburne, p 370)*

8. **(B)** The superficial external pudendal artery emerges through the saphenous opening and then passes medially and upward across the spermatic cord in the male (or the round ligament of the uterus in the female) to be distributed to the skin of the suprapubic region of the abdomen and to the penis and scrotum (or the labium majus in the female). *(Woodburne, p 370)*

9. **(D)** The upper five slips of origin for the external abdominal oblique muscle interdigitate with those of the serratus anterior muscle; the lower three digitations with the costal attachments of the latissimus dorsi muscle. *(Woodburne, p 371)*

10. **(B)** The posterior margin of the external abdominal oblique is free and forms, with the converging border of the latissimus dorsi muscle and the iliac crest below, the lumbar triangle. *(Woodburne, p 371)*

11. **(D)** Additional specializations of the external oblique aponeurosis are the inguinal ligament, the lacunar ligament, the reflected inguinal ligament, the medial and lateral crura of the superficial inguinal ring, the intercrural fibers, and the external spermatic fascia. *(Woodburne, p 371)*

12. **(A)** The external oblique aponeurosis has, at its inferior extremity, a thickened, rolled-under border that is stretched between the anterior superior spine of the ilium and the pubic tubercle. This is the inguinal ligament. *(Woodburne, pp 371–372)*

13. **(E)** The lacunar ligament represents the more medial, rolled-under fibers of the inguinal ligament which, flattening down into a horizontal shelf, attach to the pecten of the pubis for about 2 cm. *(Woodburne, p 373)*

14. **(C)** Above the umbilicus the posterior layer of internal oblique aponeuroses fuses with the aponeurosis of the transverus abdominis muscle: the anterior layer with that of the external abdominal oblique muscle. Thus an equal split of the abdominal aponeuroses forms the sheath of the rectus abdominis muscle. In the lower one fourth of the abdomen the internal abdominal oblique aponeurosis

fails to split and passes to the median line entirely anterior to the rectus abdomnis. *(Woodburne, p 373)*

15. **(C)** The cremaster muscle and fascia are the representatives of the internal abdominal oblique muscle layer among the coverings of the cord and testis. They form a layer that immediately underlies the external spermatic fascia. *(Woodburne, p 376)*

16. **(E)** The cremaster muscle is innervated by the genital branch of the genitofemoral nerve. This nerve, a derivative of the lumbar plexus, joins the spermatic cord from within the abdomen at the deep inguinal ring, traverses the inguinal canal, and lies under the cremaster muscle. *(Woodburne, p 376)*

17. **(C)** Like the internal oblique, the lowest fibers of the transversus abdominis arch downward to the pubis. They insert into the superior border of the pubis and into the medial 2 cm of the pecten. Together with the lowermost fibers of the internal oblique muscle, which end in the superior border of the pubis, they constitute the falx inguinalis. *(Woodburne, p 376)*

18. **(C)** In the anterior surface of the rectus abdominis, there are three tendinous intersections. They are firmly adherent to the anterior layer of the rectus sheath. The lowest is at the level of the umbilicus; the highest is near the xiphoid process; and the third is halfway between these levels. *(Woodburne, p 377)*

19. **(C)** The pyramidalis muscle is an insignificant muscle, frequently absent and contained in the rectus sheath, where it lies anterior to the inferior portion of the rectus abdominis muscle. It is supplied by a branch of the 12th thoracic nerve. *(Woodburne, p 377)*

20. **(A)** The principal outpouching of the transversalis fascia is the internal spermatic fascia, which invests the ductus deferens and the testicular vessels as they leave the abdominal cavity. The mouth of this outpouching constitutes the deep inguinal ring, located about 1.5 cm above the middle of the inguinal ligament. *(Woodburne, pp 378–379)*

21. **(D)** The innervation of the muscles of the abdominal wall is by ventral rami of spinal nerves thoracic 7 through lumbar 4. This is the same segmental sequence that provides the cutaneous nerves in the region. *(Woodburne, p 378)*

22. **(B)** The tenth intercostal nerve enters the sheath just below the tendinous intersection at the level of the umbilicus. The higher nerves (T7 to 9) are above, and the lower nerves distribute, in sequence, below this level. *(Woodburne, p 379)*

23. **(D)** The lumbar arteries are in series with the posterior intercostal and subcostal vessels. Four in number, they arise from the back of the abdominal aorta at the levels of the bodies of the upper four lumbar vertebrae. *(Woodburne, p 380)*

24. **(A)** The iliolumbar artery is a branch of the internal iliac artery, usually of its posterior trunk. *(Woodburne, p 381)*

25. **(D)** The gubernaculum testis becomes much reduced in the adult, its remnant constituting the scrotal ligament that extends from the inferior pole of the testis and the tail of the epididymis to the skin of the bottom of the scrotum. *(Woodburne, p 382)*

26. **(E)** The spermatic cord contains the ductus deferens, the deferential artery and vein, the testicular artery, the pampiniform plexus of veins, the lymphatics, and the autonomic nerves of the testis. *(Woodburne, p 384)*

27. **(A)** Traversing the inguinal canal are the ilioinguinal nerve, the genital branch of the genitofemoral nerve, and the cremasteric artery. The components of the spermatic cord are invested by the internal spermatic fascia all the way through the canal. *(Woodburne, pp 384–385)*

28. **(B)** The tunica vaginalis testis is the invaginated serous sac that partially covers the testis and that represents the lower closed-off portion of the processus vaginalis of the peritoneum. The visceral layer is closely applied to the testis, the epididymis, and the lower part of the spermatic cord. *(Woodburne, p 385)*

29. **(A)** The derivatives of the dorsal common mesentery include the mesentery, transverse mesocolon, sigmoid mesocolon, and mesoappendix. *(Woodburne, p 388)*

30. **(C)** The derivatives of the dorsal mesogastrium include the greater omentum, the lienorenal ligament, and the phrenicocolic ligament. *(Woodburne, p 388)*

31. **(D)** The derivatives of the ventral mesogastrium include the hepatogastric, hepatoduodenal, hepatocolic, falciform, coronary, right triangular, and left triangular ligaments. *(Woodburne, p 388)*

32. **(B)** The primordia of the pancreas, liver, and gallbladder appear as outgrowths of the gut tube just caudal to the stomach during the fourth week of development. *(Woodburne, p 388)*

33. **(C)** Viewed from the ventral side, the rotation is counterclockwise and brings up cephalically and to

the right the caudal part of the loop, representing the cecum and the large intestine, and turns the originally cephalic limb of the loop (small intestine) downward and to the left. The rotation of the loop takes place around an axis represented by the superior mesenteric artery. *(Woodburne, pp 388–389)*

34.  **(B)** The esophagogastric junction is on the horizontal plane of the tip of the xiphoid process and to the left of the 11th thoracic vertebral body. *(Woodburne, p 400)*

35.  **(D)** The aorta begins its abdominal distribution by passing through the aortic hiatus of the diaphragm in front of the lower border of the 12th thoracic vertebra. *(Woodburne, pp 402–403)*

36.  **(C)** The celiac trunk arises from the abdominal aorta just below the aortic hiatus and at the level of the upper portion of the first lumbar vertebra. It gives rise to the left gastric, common hepatic, and splenic arteries. *(Woodburne, p 403)*

37.  **(D)** The right hepatic artery passes to the right, usually behind the common hepatic duct, to gain the right end of the liver hilum. Here it breaks up into several branches, which enter the right lobe of the liver with radicles of the portal vein and hepatic duct. As it passes between the hepatic duct and the cystic duct, it gives rise to the small cystic artery, which follows the cystic duct to the gallbladder. *(Woodburne, p 404)*

38.  **(E)** The left gastroepiploic artery arises from the splenic artery or an inferior terminal branch and passes toward the greater curvature of the stomach through the gastrolienal ligament. *(Woodburne, p 405)*

39.  **(D)** The anterior and posterior vagal trunks pass through the esophageal hiatus of the diaphragm anterior and posterior to the terminal esophagus, and both lie toward its right side. They contain preganglionic visceral efferent and general visceral afferent fibers. *(Woodburne, p 407)*

40.  **(E)** The ninth, tenth, and 11th ribs are in relation to the spleen. It develops in the dorsal mesogastrium. It is entirely surrounded by peritoneum and normally does not descend below the costal margin but rests on the left flexure of the colon and the phrenicocolic ligament. *(Woodburne, p 408)*

41.  **(A)** The first part of the duodenum is surrounded by the hepatoduodenal ligament. It is related to the quadrate lobe of the liver and the common bile duct passes behind the first part of the duodenum. The first part of the duodenum passes posteriorly along the right side of the body of the first lumbar vertebra. *(Woodburne, p 410)*

42.  **(B)** The middle one third of the second portion of the duodenum is crossed ventrally by the transverse colon. *(Woodburne, p 410)*

43.  **(C)** The superior mesenteric artery arises above and crosses the anterior aspect of the third part of the duodenum, whereas the inferior mesenteric artery arises from the aorta directly below the duodenum. *(Woodburne, p 410)*

44.  **(C)** The greater duodenal papilla, on which are the terminal openings of the common bile duct and the pancreatic duct, is found at the junction of the middle and lower thirds of the second part of the duodenum. The papilla is continued below by the tapered longitudinal fold of the duodenum. The lesser duodenal papilla marks the termination of the accessory pancreatic duct. *(Woodburne, p 411)*

45.  **(E)** The lower left portion of the head of the pancreas is inserted behind the superior mesenteric vessels, forming the uncinate process. *(Woodburne, p 411)*

46.  **(D)** Behind the neck of the pancreas the superior mesenteric and splenic veins unite to form the portal vein. The anterior surface of the neck is covered by peritoneum and lies in the floor of the omental bursa. *(Woodburne, p 412)*

47.  **(B)** The tail of the pancreas is usually blunted and turned upward. It enters the lienorenal ligament and frequently makes contact with the spleen; inferiorly it is in relation with the left flexure of the colon. *(Woodburne, p 413)*

48.  **(D)** The pancreas develops in two parts from a dorsal and a ventral primordium. *(Woodburne, p 413)*

49.  **(D)** The ducts of the two primordia fuse in the upper part of the head of the gland, and the principal path of the pancreatic secretion becomes established through the region of fusion. Thus the pancreatic duct of the adult is formed from the distal part of the duct of the dorsal primordium and the proximal part of the duct of the ventral primordium. *(Woodburne, p 414)*

50.  **(E)** The pancreatica magna artery is the largest of the series of superior pancreatic branches of the splenic artery. It enters the pancreas in the region of the junction of the middle and left thirds of the gland. *(Woodburne, p 416)*

51.  **(A)** Extending from the inferior border to the porta, there is a deep fissure for the ligamentum teres; the ligamentum teres, the obliterated remains of the umbilical vein, passes through this to end in the left branch of the portal vein. *(Woodburne, pp 417–418)*

52.  (C) The hepatoduodenal ligament transmits the hepatic artery, the portal vein, and the common bile duct. *(Woodburne, p 419)*

53.  (D) The cystic artery usually arises from a normal right hepatic artery as that vessel crosses the cystohepatic angle. As its most frequent variations, the cystic artery may arise as a branch of either a right hepatic or a common hepatic artery from the superior mesenteric artery. *(Woodburne, p 425)*

54.  (C) The hepatic veins drain into the inferior vena cava. They are entirely intrahepatic. *(Woodburne, p 425)*

55.  (D) The portal vein, formed behind the neck of the pancreas by the junction of the superior mesenteric and splenic veins, ascends to the liver in the free margin of the lesser omentum posterior to the common bile duct and the proper hepatic artery. *(Woodburne, pp 425–426)*

56.  (A) The superior mesenteric artery supplies all of the small intestine except the proximal part of the duodenum; it also supplies the cecum, the ascending colon, and most of the transverse colon, the embryonic midgut. *(Woodburne, p 429)*

57.  (D) The middle colic artery takes origin from the front of the superior mesenteric artery immediately below the neck of the pancreas. It divides into right and left branches. The right branch anastomoses with the right colic artery and the left branch with the left colic branch of the inferior mesenteric artery. *(Woodburne, p 431)*

58.  (B) Both the vagal parasympathetic innervation and the thoracic splanchnic sympathetic innervation of the gastrointestinal tract terminate with the end of the distribution of the superior mesenteric artery at about the junction of the middle and left thirds of the transverse colon. *(Woodburne, p 432)*

59.  (E) Three surface features serve to distinguish the small intestine from the large. Longitudinal musculature of the large intestine forms three bands called teniae coli. These bands are shorter than the colon and therefore force the wall to bulge between the teniae, forming sacculations, or haustra coli. The third characteristic of the large intestine is the occurrence along its length of epiploic appendages, which are fat-filled tabs of peritoneum that hang down from the serous coat of the large intestine. *(Woodburne, p 434)*

60.  (B) The vermiform appendix has a complete peritoneal investment and a small mesentery, the mesoappendix. It receives its blood supply from the ileocolic artery and is innervated by the vagus and thoracic splanchnic nerves. It is usually located in the pelvis. *(Woodburne, pp 435,430)*

61.  (C) The transverse colon is attached posteriorly by the transverse mesocolon on a line that crosses the second part of the duodenum and passes across the pancreas except for its distal extremity. *(Woodburne, p 436)*

62.  (D) The testicular veins arise from the testis and the epididymis. They form the pampiniform plexus which, consisting of eight to ten anastomosing veins, ascends along the ductus deferens. The plexus, as a constituent of the spermatic cord, traverses the inguinal canal and ends near the deep inguinal ring by forming two accompanying veins of the testicular artery. *(Woodburne, p 464)*

63.  (D) The psoas major muscle ends in the iliopsoas tendon, which inserts on the lesser trochanter of the femur. This tendon also receives most of the fibers of the iliac muscle. *(Woodburne, p 465)*

64.  (C) The diaphragm consists of a central tendon and muscle covered on both surfaces by a membranous layer of fascia. For the thoracoabdominal diaphragm, the inferior fascia is provided by the upper portion of the parietal abdominopelvic fascia, the transversalis fascia. Its superior fascia is the parietal thoracic fascia (endothoracic fascia), which lines the interior of the thoracic cavity. *(Woodburne, p 449)*

65.  (E) The right suprarenal vein empties directly into the posterior surface of the inferior vena cava. On the left side (after union with the inferior phrenic vein) it is a tributary of the renal vein. *(Woodburne, p 449)*

66.  (A) The cortex is essential to life. It secretes hormones that influence the fluid and electrolyte balance in the body. It is also concerned with carbohydrate metabolism. The medulla receives only a preganglionic innervation and produces epinephrine and norepinephrine. *(Woodburne, p 448)*

67.  (D) Traversing each renal column is a principal branch of a renal artery, the interlobar artery; its name is a reflection of the lobar character of a single medullary pyramid and the cortical tissue on all sides of it. *(Woodburne, p 443)*

68.  (C) The double capillary network in relation to the tubular nephron forms the essential mechanism of the kidney, for the glomerulus and the glomerular capsule provide for filtration of the blood plasma, and the second capillary plexus around the tubules provides for selective reabsorption of water and the return of dissolved materials to the circulation. *(Woodburne, p 443)*

69. **(B)** The preganglionic parasympathetic fibers of the pelvic splanchnic nerves distribute to the viscera of the pelvis and the perineum as components of this plexus. However, their supply of the left colic flexure, the descending colon, and the sigmoid colon is by independent routes. *(Woodburne, p 439)*

70. **(A)** The sigmoid colon is characterized by its S-shaped loop and by the presence of a mesocolon. It begins at the brim of the pelvis on the left side and ends at the median line opposite the third segment of the sacrum. *(Woodburne, p 437)*

71. **(B)** The upper two fifths (8 feet) is considered jejunum; the lower three fifths (12 feet) is ileum. Aggregated lymph nodules occur in patches along the antimesenteric border of the ileum. *(Woodburne, p 429)*

72. **(C)** The gallbladder serves as a reservoir for bile, and its epithelium concentrates it by extracting water. The gallbladder has only serous, fibromuscular, and mucosal layers. The mucous membrane has a honeycomb appearance due to the elevation of folds in low criss-crossing ridges. *(Woodburne, pp 422–423)*

73. **(C)** The falciform ligament represents the inferior limit of the ventral mesogastrium and encloses the ligamentum teres of the liver in its free border. This double layer extends from the umbilicus upward to the liver. Here its layers pass over both the diaphragmatic and visceral surfaces of the liver. *(Woodburne, p 418)*

74. **(E)** The anterior surface of the liver lies against the diaphragm, the costal margin, the xiphoid process, and the abdominal wall. *(Woodburne, p 417)*

75. **(A)** The duodenum and the pancreas are situated between the celiac trunk and the superior mesenteric artery and receive major blood vessels from both. *(Woodburne, p 414)*

76. **(C)** The pancreas lies transversely across the posterior abdominal wall from the duodenum to the spleen and is behind the stomach. The head overlies the second and third lumbar vertebrae, and the uncinate process inserts behind the superior mesenteric vessels. *(Woodburne, p 411)*

77. **(C)** The end of the duodenum is covered by peritoneum and is movable, but most of the fourth segment is retroperitoneal. The root of the mesentery begins at the duodenojejunal flexure. This portion of the duodenum is stabilized by a fibromuscular band, the suspensory ligament of the duodenum. *(Woodburne, pp 410–411)*

78. **(D)** The duodenum is the fixed and retroperitoneal portion of the small intestine. It receives its name from the fact that its length is equal to the breadth of 12 fingers. At its extremities, the duodenum is continuous with the completely peritonealized stomach and jejunum. *(Woodburne, pp. 409–410)*

79. **(A)** The parasympathetic innervation enhances muscular movements, exerting the greater influence on the secretion of water and hydrochloric acid. The sympathetic innervation is important in vasomotor control and has the major influence in the secretion of enzymes. Afferent impulses principally accompany the sympathetic system. *(Woodburne, p 408)*

80. **(E)** The caudate lobe of the liver forms the superior boundary of the epiploic foramen. The posterior boundary is formed by the inferior vena cava. The hepatoduodenal ligament and its contents form the ventral boundary. The inferior boundary is the first part of the duodenum. *(Woodburne, p 392)*

81. **(C)** The lesser omentum, the double layer remaining in the interval between the stomach and the liver, encloses the common bile duct (hepatic diverticulum) and is a continuous sheet that includes both the hepatogastric and the hepatoduodenal ligaments. *(Woodburne, p 392)*

82. **(E)** The adult bursa omentalis (lesser peritoneal sac) is a peritoneal space behind the stomach that is closed off from the major peritoneal cavity (greater peritoneal sac) except for the communication through the epiploic foramen. *(Woodburne, pp 392–393)*

83. **(B)** The greater thoracic splanchnic nerve (T5 to 9 or 10) terminates in the abdomen in the lateral border of the celiac ganglion. The lesser thoracic splanchnic nerve (T10,11) terminates in the aorticorenal ganglion. The least thoracic splanchnic nerve (last thoracic ganglion), when present, ends in the renal plexus. *(Woodburne, pp 453–454)*

84. **(A)** The iliohypogastric arises from the first lumbar nerve together with a frequent contribution from the 12th thoracic nerve. *(Woodburne, p 467)*

85. **(E)** The pelvic splanchnic nerves have their cells of origin in the second, third, and fourth segments of the sacral spinal cord and arise from the corresponding sacral spinal nerves. *(Woodburne, p 456)*

86. **(C)** The obturator nerve is the principal preaxial nerve of the lumbar plexus. It arises from the anterior branches of lumbar nerves 2, 3, and 4 and descends along the medial border of the psoas muscle. *(Woodburne, p 468)*

87. **(D)** The neck of the gallbladder is directed toward the porta hepatis, makes an S-shaped curve, and is continuous with the cystic duct. Crescentic folds of mucous membrane in its interior are spirally arranged and constitute the spiral fold. *(Woodburne, p 422)*

88. **(E)** The layers of the falciform ligament pass over both the diaphragmatic and visceral surfaces of the liver. The layers of the diaphragmatic surface spread to the right and left anterior layers of the coronary ligament until they reach the sharp reversals of peritoneal reflection to the diaphragm known as the right and left triangular ligaments. *(Woodburne, p 418)*

89. **(A)** The longitudinal bands of muscle of the large intestine are about one sixth shorter than the colon, so that the wall is forced to bulge between the teniae. The colon, therefore, presents three rows of sacculations, or haustra coli, alternating between the teniae. *(Woodburne, p 434)*

90. **(C)** The wall of the intestine, like that of the stomach, is composed of four coats: mucosal, submucosal, muscular, and serous. The mucous membrane exhibits circular folds and intestinal villi. The circular folds are well developed in the jejunum but become very small or absent in the ileum. *(Woodburne, p 429)*

91. **(B)** Among the coverings of the spermatic cord and testis, the cremaster muscle and fascia are the representatives of the internal abdominal oblique muscle layer. *(Woodburne, p 376)*

92. **(A)** The external oblique aponeurosis has, at its inferior extremity, a thickened, rolled-under border that is stretched between the anterior superior spine of the ilium and the pubic tubercle. This is the inguinal ligament. *(Woodburne, pp 371–372)*

93. **(A)** The superficial inguinal ring lies at the end of a triangular cleft in the external oblique aponeurosis and is located just above and lateral to the pubic tubercle. The long triangular cleft represents a weakness in the aponeurosis. *(Woodburne, p 373)*

94. **(D)** The principal outpouching of the transversalis fascia is the internal spermatic fascia, which invests the ductus deferens and the testicular vessels as they leave the abdominal cavity. The mouth of this outpouching constitutes the deep inguinal ring. *(Woodburne, pp 378–379)*

95. **(D)** The inferior epigastric artery enters the rectus sheath from below after arising from the external iliac artery. *(Woodburne, p 369)*

96. **(C)** The superior rectal artery is the continuation of the inferior mesenteric artery. *(Woodburne, p 438)*

97. **(E)** The superficial circumflex iliac artery arises from the femoral artery about 1 cm below the inguinal ligament. *(Woodburne, p 370)*

98. **(B)** The splenic artery is the largest branch of the celiac trunk. It arises from the left side of the trunk distal to the left gastric artery. The splenic artery runs a highly tortuous course, partially embedded in the superior border of the pancreas. *(Woodburne, p 404)*

99. **(D)** The hepatic veins have no valves and usually empty individually into the inferior vena cava. *(Woodburne, p 404)*

100. **(B)** The left gastric vein ends in the portal vein. Its circular course along the lesser curvature and inferiorly on the body wall is expressed in the old name "coronary vein." *(Woodburne, p 405)*

101. **(C)** The right colic vein enters the right side of the superior mesenteric vein. *(Woodburne, p 432)*

102. **(A)** The caudal pancreatic and pancreatica magna veins empty into the splenic vein on the back of the pancreas. *(Woodburne, p 416)*

103. **(C)** The internal surface of the small intestine is greatly folded, thus increasing its absorptive and secreting surface. Circular folds project a centimeter or less into the lumen. The greater duodenal papilla, on which are the terminal openings of the common bile duct and the pancreatic duct, is found at the junction of the middle and lower thirds of the second part of the duodenum. The circular folds are well developed in the jejunum but become very small or absent in the ileum. *(Woodburne, pp 411,429)*

104. **(A)** The greater duodenal papilla, on which are the terminal openings of the common bile duct and the pancreatic duct, is found at the junction of the middle and lowest thirds of the second part of the duodenum. *(Woodburne, p 411)*

105. **(D)** The parasympathetic innervation for the small intestine is derived from the celiac division of the posterior vagal trunk. *(Woodburne, p 432)*

106. **(A)** The duodenum is the fixed and retroperitoneal portion of the small intestine. *(Woodburne, p 409)*

107. **(D)** The iliohypogastric nerve arises from the first lumbar nerve, together with a frequent contribution from the 12th thoracic nerve. The genitofemoral nerve arises by the union of branches

from the anterior portions of the first and second lumbar nerves. *(Woodburne, p 467)*

108. **(A)** The genital branch of the genitofemoral nerve enters the inguinal ring. It supplies the cremaster muscle and gives small branches to the skin and fascia of the scrotum and the adjacent part of the thigh. *(Woodburne, p 467)*

109. **(D)** The inferior epigastric artery pierces the transversalis fascia near the arcuate line to enter the rectus sheath. *(Woodburne, p 381)*

110. **(D)** The pyramidalis is supplied by a branch of the 12th thoracic nerve. *(Woodburne, p 377)*

111. **(D)** The renal arteries arise, one on each side on the aorta, at the level of the upper border of the second lumbar vertebra. *(Woodburne, pp 445–446)*

112. **(A)** The counterclock rotation of the primitive gut takes place around an axis represented by the superior mesenteric artery. *(Woodburne, p 389)*

113. **(D)** The superior rectal artery is the continuation of the inferior mesenteric artery and is anastomoses with the terminals of the middle and inferior rectal arteries from the internal iliac system. *(Woodburne, p 438)*

114. **(C)** The inferior mesenteric artery supplies the left one third of the transverse colon, the descending colon, the sigmoid colon, and the greater part of the rectum. The superior mesenteric artery supplies all of the small intestine except the proximal part of the duodenum; it also supplies the cecum, the ascending colon, and most of the transverse colon. *(Woodburne, pp 429,437)*

115. **(B)** Two lineae semilunares define the lateral margins of the rectus muscles. *(Woodburne, p 367)*

116. **(D)** Three transverse lines cross the abdominal wall. These lineae transversae are produced by tendinous bands that interrupt the continuity of the fibers of the rectus abdominis muscle. *(Woodburne, p 368)*

117. **(B)** The lineae similunares begin at the pubic tubercles and end above at the costal margin in the region of the tips of the ninth costal cartilage. *(Woodburne, pp 367–368)*

118. **(A)** The linea alba of the median plane marks the aponeurotic junction of the abdominal muscles of the two sides and separates the vertical bulges of the two rectus abdominis muscles. *(Woodburne, p 367)*

119. **(A)** The lacunar is continuous with the inguinal ligament from the pubic tubercle along the pecten. *(Woodburne, p 373)*

120. **(D)** The external spermatic fascia is a delicate tubulosaccular sheath that represents the continuation of the intercrural fibers along the spermatic cord. *(Woodburne, p 373)*

121. **(D)** The reflected inguinal ligament is interpreted as a group of decussating fibers of the opposite side that descend to the pubic crest and tubercle. *(Woodburne, p 373)*

122. **(A)** The lacunar ligament represents the more medial, rolled-under fibers of the inguinal ligament which, flattening down into a horizontal shelf, attach to the pecten of the pubis. *(Woodburne, p 373)*

123. **(D)** In the intestines there is a nerve and ganglionic plexus between the circular and longitudinal muscle layers, the myenteric plexus, and another in the submucosal connective tissue, the submucosal plexus. The sympathetic neurons reaching the intestinal tract are already postganglionic and contribute to the myenteric and submucosal plexuses only as terminal nerve fibers; they have no cell bodies there. *(Woodburne, p 399)*

124. **(C)** The cell bodies within these plexuses are those of the postganglionic neurons in the two-neuron parasympathetic innervation of the intestinal tract. *(Woodburne, p 399)*

125. **(A)** The myenteric plexus is between the circular and longitudinal muscle layers. *(Woodburne, p 399)*

126. **(A)** Peristalsis depends on the integrity of the myenteric plexus. *(Woodburne, p 399)*

127. **(C)** The sympathetic neurons reaching the intestinal tract are already postganglionic and contribute to the myenteric and submucosal plexuses only as terminal nerve fibers. *(Woodburne, 399)*

128. **(C)** The splenic artery gives off a number of pancreatic branches in its undulating course along the upper border of the pancreas. The posterior superior pancreaticoduodenal artery gives off both duodenal and pancreatic branches. *(Woodburne, pp 404–405)*

129. **(A)** The splenic artery enters the lienorenal ligament over the superior portion of the left kidney and is conducted through this ligament to the hilum of the spleen. *(Woodburne, p 404)*

130. **(A)** The splenic artery runs a highly tortuous course, partially embedded in the superior border of the pancreas. *(Woodburne, p 404)*

131. **(B)** The posterior superior pancreaticoduodenal artery anastomoses with the posterior inferior pancreaticoduodenal branch of the superior mesenteric artery. *(Woodburne, p 404)*

132. **(D)** The right gastroepiploic artery passes from right to left along the greater curvature of the stomach. *(Woodburne, p 404)*

133. **(B)** The muscles of the abdominal wall comprise two groups, anterolateral and posterior. The posterior muscle is the quadratus lumborum. The anterolateral muscles include the rectus abdominis, the pyramidalis, the external abdominal oblique, the internal abdominal oblique, and the transversus abdominis. *(Woodburne, p 371)*

134. **(A)** The posterior margin of the external abdominal oblique is free and forms, with the converging border of the latissimus dorsi muscle and the iliac crest below, the lumbar triangle. *(Woodburne, p 371)*

135. **(E)** Specialization of the external oblique aponeurosis includes the inguinal ligament, the lacunar ligament, the reflected inguinal ligament, the medial and lateral crura of the superficial inguinal ring, the intercrural fibers, and the external spermatic fascia. *(Woodburne, p 371)*

136. **(D)** The cremaster muscle and fascia are the representatives of the internal abdominal oblique muscle. The pectineal ligament, the reflected inguinal ligament, and the intercrural fibers are representatives of the external abdominal oblique. *(Woodburne, pp 373,376)*

137. **(D)** The lowest fibers of the transversus abdominis arch downward to the pubis, where they insert into the superior border of the pubis and into the pecten. Together with the lowermost fibers of the internal oblique muscle, which end in the superior border of the pubis, they constitute the falx inguinalis. *(Woodburne, p 376)*

138. **(B)** In the lower one fourth of the abdominal wall the aponeurosis of the internal oblique muscle does not split, and all three aponeurotic layers pass anteriorly to the rectus abdominis muscle. *(Woodburne, p 377)*

139. **(C)** The pyramidalis muscle is contained in the rectus sheath where it arises from the front of the pubis and the anterior pubic ligament. The muscle draws down on the linea alba. It is supplied by a branch of the 12th thoracic nerve. *(Woodburne, 377)*

140. **(D)** Below the arcuate line the transversalis fascia is the principal layer of the posterior sheath of the rectus abdominis muscle. *(Woodburne, p 378)*

141. **(C)** The principal outpouching of the transversalis fascia is the internal spermatic fascia. The mouth of this outpouching constitutes the deep inguinal ring, located from 1.0 to 1.5 cm above the middle of the inguinal ligament. *(Woodburne, pp 378–379)*

142. **(E)** The blood vessels of the abdominal wall are derived from five systems: the internal thoracic, the lumbar, the femoral, the external iliac, and the internal iliac. *(Woodburne, p 380)*

143. **(E)** The inferior epigastric artery arises from the external iliac immediately above the inguinal ligament. Its origin is just medial to the deep inguinal ring, and near the arcuate line the artery pierces the transversalis fascia and enters the rectus sheath. The cremaster artery arises from the inferior epigastric artery near its origin. *(Woodburne, p 381)*

144. **(B)** The gubernaculum testis becomes much reduced in the adult, its remnant constituting the scrotal ligament. In the female the gubernaculum between the ovary and the uterus becomes the ovarian ligament, and between the uterus and the labium majus it forms the round ligament of the uterus. *(Woodburne, pp 382–383)*

145. **(D)** The inguinal canal in the male contains the elements of the spermatic chord. These are the ductus deferens, the deferential artery and vein, the testicular artery, and the pampiniform plexus of veins. In the female the smaller inguinal canal contains the round ligament of the uterus and the artery and vein of the round ligament. The ilioinguinal nerve emerges at the superficial inguinal ring in both males and females. *(Woodburne, pp 384–385)*

146. **(B)** The primordia of the pancreas, liver, and gallbladder appear as outgrowths of the gut tube just caudal to the stomach during the fourth week of development. During the fourth week of development, the liver and the gallbladder are ventral outgrowths and develop in the ventral mesogastrium. The pancreas, although having two primordia, is definitively embedded in the dorsal mesogastrium. *(Woodburne, p 388)*

147. **(D)** The hepatocolic ligament is inconstant and results from an adhesion between the right flexure of the colon and the duodenum. *(Woodburne, p 392)*

148. **(B)** The ileocecal fold extends from the surface of the terminal ileum to the cecum and crosses the root of the appendix; it carries no important blood vessels. In the free margin of the mesoappendix runs the appendicular artery. The anterior cecal artery is located in the vascular cecal fold. The inferior mesenteric vein occasionally raises a fold of

peritoneum above the duodenojejunal flexure known as the paraduodenal fold. The duodenojejunal fold is a fold of peritoneum with nonstriated muscle. *(Woodburne, pp 397,411)*

149.  **(A)** The pyloric sphincter is formed by the circular layer of the termination of the stomach. It relaxes ahead of contraction waves from the adjacent portion of the stomach but does not open with every cycle of contraction. Among other factors, it is under hormonal control from the duodenum. Digestion is not completed in the stomach. *(Woodburne, p 398)*

150.  **(A)** The stomach is fixed only at its beginning and at its end and therefore the position of the stomach is variable. The lesser curvature marks the right border, and the greater curvature marks the left and inferior borders of the stomach. The posterior surface of the stomach forms the anterior wall of the omental bursa. The stomach is covered by peritoneum, except for the area associated with blood vessels. *(Woodburne, pp 400–401)*

151.  **(A)** The stomach receives its arterial blood supply from all branches of the celiac trunk. The left gastric, common hepatic, and splenic arteries arise from the celiac trunk and provide the arterial supply to the stomach. *(Woodburne, pp 402,403)*

152.  **(C)** The branches of the common hepatic artery are the gastroduodenal, the proper hepatic arteries, and, rarely, the right gastric artery. *(Woodburne, p 403)*

153.  **(B)** The spleen develops in the dorsal mesogastrium and is entirely surrounded by peritoneum. It does not normally descend below the costal margin and therefore is not normally palpable through the abdominal wall. *(Woodburne, p 408)*

154.  **(C)** The first part of the duodenum is completely peritonealized. The hepatoduodenal portion of the lesser omentum surrounds this segment of the duodenum. The first part of the duodenum is related above to the quadrate lobe of the liver as well as to the gallbladder. *(Woodburne, p 410)*

155.  **(E)** The second part of the duodenum overlies the hilum of the right kidney posteriorly; it also rests on the renal vessels, the inferior vena cava, and the psoas muscles. The medial surface of the duodenum is in contact with the pancreas. *(Woodburne, p 410)*

156.  **(D)** The third portion of the duodenum crosses the lower part of the body of the third lumbar vertebra. This portion of the duodenum is retroperitoneal. The superior mesenteric artery arises above and crosses the anterior aspect of this portion of the duodenum. *(Woodburne, p 410)*

157.  **(B)** The greater duodenal papilla marks the termination of the main pancreatic and the common bile ducts. *(Woodburne, p 413)*

158.  **(B)** The origin of the anterior and posterior inferior pancreaticoduodenal arteries is variable. They most frequently arise through the division of the inferior pancreaticoduodenal branch of the superior mesenteric artery; less frequently they come individually from the superior mesenteric artery. The first jejunal artery is sometimes the origin. *(Woodburne, 415)*

159.  **(B)** The principal arteries of the celiac trunk to the duodenum and the pancreas are branches of the gastroduodenal artery and anterior superior pancreaticoduodenal and posterior superior pancreaticoduodenal arteries. The principal arteries from the superior mesenteric artery are corresponding branches of its inferior pancreaticoduodenal artery. Of these paired sets, the anterior vessels anastomose to form the anterior arcade. *(Woodburne, 414)*

160.  **(A)** The dorsal pancreatic artery arises most commonly from the proximal portion of the splenic artery or as a fourth branch of the celiac trunk. It may arise as a branch of the superior mesenteric artery or even the hepatic artery. *(Woodburne, p 415)*

161.  **(E)** The visceral surface of the liver lodges the gallbladder and is deeply indented posteriorly by the inferior vena cava. The porta hepatis is located at about the middle of the visceral surface. The deep fissure for the ligamentum teres extends from the inferior border to the porta hepatis. The deep fissure for the ligamentum venosum extends from the porta hepatis to the posterior surface. *(Woodburne, pp 417,418)*

162.  **(C)** The gallbladder has serous, fibromuscular, and mucosal layers. The neck of the gallbladder has crescentic folds of mucous membrane in its interior that are spirally arranged and that constitute the spiral. *(Woodburne, p 422)*

163.  **(E)** The common bile duct descends in the free border of the lesser omentum to the duodenum. It lies to the right of the hepatic artery and anterior to the portal vein. The duct descends behind the first portion of the duodenum and then crosses the posterior surface of the head of the pancreas. *(Woodburne, p 423)*

164.  **(C)** The branches of the superior mesenteric artery are the inferior pancreaticoduodenal, the intestinal, the ileocolic, the right colic, and the middle colic. It may give rise to the dorsal pancreatic, right hepatic, and common hepatic arteries. *(Woodburne, p 430)*

165. **(A)** The ileocolic artery is the lowermost of the branches that arise from the right side of the superior mesenteric artery. It gives off a colic branch, an anterior cecal branch, a posterior cecal branch, an appendicular branch, and an ileal branch. *(Woodburne, pp 430–431)*

166. **(E)** The anastomotic channels along the large intestine are frequently so large as to constitute a marginal artery that follows the arch of the colon. The right branch of the middle colic anastomose with the right colic, and the left branch anastomose with the left colic branch of the inferior mesenteric. *(Woodburne, pp 430–431)*

167. **(E)** The superior mesenteric vein accompanies the superior mesenteric artery and lies anterior and to its right in the root of the mesentery. Like the artery, the vein crosses the third part of the duodenum and the uncinate process of the pancreas. It terminates behind the neck of the pancreas by joining the splenic vein to form the portal vein. *(Woodburne, p 431)*

168. **(B)** The ileocecal valve is formed by the infolding of the end of the ileum into the cavity of the cecum. The substance of the lips is a duplication of the circular muscle layer of the ileum, which projects into the cecum. The two lips of the valve are continued to either side of the opening as the frenulum of the valve. The ileocecal valve is a questionable sphincter. *(Woodburne, p 435)*

169. **(C)** The descending colon passes inferiorly from the left colic flexure to end in the sigmoid colon at the brim of the pelvis. It is retroperitoneal. It passes along the lateral border of the left kidney. It descends over the quadratus lumborum muscle to the iliac fossa. *(Woodburne, pp 436–437)*

170. **(A)** The inferior mesenteric artery supplies the left one third of the transverse colon, the descending colon, the sigmoid colon, and the greater part of the rectum. It gives off as branches the left colic, sigmoid, and superior rectal arteries. The middle and inferior rectal branches arise from the internal iliac artery. *(Woodburne, pp 437–438)*

171. **(D)** The rectal plexuses interconnect radicles of the inferior rectal, middle rectal, and some superior rectal veins. The inferior mesenteric vein begins in the rectal plexuses. *(Woodburne, p 438)*

172. **(B)** The superior mesenteric nodes drain partly to the celiac nodes above them. The final common path of this lymphatic collection is the intestinal trunk. This trunk is formed by the union of efferents from both the celiac and superior mesenteric nodes. *(Woodburne, p 439)*

173. **(E)** The portal vein is formed behind the neck of the pancreas by the union of the superior mesenteric vein and the splenic vein. The portal vein ascends behind the first part of the duodenum and, in the free margin of the hepatoduodenal ligament, behind the common bile duct and the proper hepatic duct. *(Woodburne, p 440)*

174. **(C)** The aortic hiatus occurs at the lower border of the 12th thoracic vertebra or at the level of the disk below it. The esophageal hiatus occurs at the level of the tenth thoracic vertebra. The vena cava foramen lies at the level of the eighth thoracic vertebra. The celiac trunk arises from the front of the abdominal aorta just below the aortic hiatus at the level of the upper portion of the first lumbar vertebra. *(Woodburne, pp 403,451)*

175. **(C)** The branches of the lumbar plexus include the iliohypogastric, ilioinguinal, genitofemoral, lateral femoral cutaneous, obturator, accessory obturator, and femoral. *(Woodburne, p 467)*

176. **(D)** The lumbar splanchnic nerves, for the most part, provide the sympathetic innervation for the pelvic viscera. *(Woodburne, p 456)*

177. **(D)** That portion of the large intestine that is supplied by the inferior mesenteric artery, and most of the pelvic viscera are innervated by the pelvic splanchnic nerves. These are preganglionic parasympathetic nerves that have their cells of origin in the second, third, and fourth segments of the sacral spinal cord. *(Woodburne, p 456)*

# CHAPTER 3

# Clinical Head and Neck
# and Abdominal
## Questions

**DIRECTIONS (Questions 1 through 9):** Each of the numbered items or incomplete statements in this section is followed by answers or by completions of the statement. Select the ONE lettered answer or completion that is BEST in each case.

1. Hydrocele is a fluid accumulation within which of the following?

    (A) round ligament
    (B) gubernaculum testis
    (C) scrotal ligament
    (D) tunica vaginalis testis
    (E) vas deferens

2. Herniation of abdominal contents into an unobliterated vaginal process and within the coverings of the spermatic cord results in which of the following herniae?

    (A) umbilical
    (B) indirect inguinal
    (C) direct inguinal
    (D) lumbar
    (E) femoral

3. The inferior epigastric artery is lateral to the herniating mass in which of the following herniae?

    (A) direct inguinal
    (B) indirect inguinal
    (C) femoral
    (D) umbilical
    (E) lumbar

4. Meckel's diverticulum is an occasional feature of which of the following structures?

    (A) duodenum
    (B) cecum
    (C) ileum
    (D) jejunum
    (E) liver

5. Femoral herniae descend into the thigh behind which of the following ligaments?

    (A) iliopectineal
    (B) lacunar
    (C) reflected inguinal
    (D) falx inguinalis
    (E) inguinal

6. A fusion fascia has surgical importance because of its

    (A) lack of blood vessels
    (B) lymphatic drainage
    (C) innervation
    (D) position
    (E) venous drainage

7. The transpyloric line is located at the level of which of the following vertebrae?

    (A) fourth lumbar
    (B) second sacral
    (C) first lumbar
    (D) fifth thoracic
    (E) tenth thoracic

8. An annular pancreas is thought to be due to which of the following?

    (A) malrotation of the primitive gut
    (B) abnormal blood supply
    (C) obstructed venous return
    (D) doubling of the ventral pancreatic primordium
    (E) absence of the duodenum

9. The upper one third of the free intestine occupies which of the following areas of the abdomen?

    (A) lower left quadrant
    (B) lower right quadrant
    (C) upper left quadrant
    (D) upper right quadrant
    (E) middle of the abdominal cavity

**DIRECTIONS (Questions 10 through 21): Each group of items in this section consists of lettered headings followed by a set of numbered words or phrases. For each numbered word or phrase, select the ONE lettered heading that is most closely associated with it. Each lettered heading may be selected once, more than once, or not at all.**

**Questions 10 through 13**

For each of the structures listed below, select the vertebral level at which the structure is located.

(A) second lumbar
(B) tenth thoracic
(C) eighth thoracic
(D) fourth lumbar
(E) 12th thoracic

10.  crest of the ilium

11.  vena caval foramen

12.  esophageal hiatus

13.  renal arteries

**Questions 14 through 17**

The collateral circulation of the portal vein involves the veins listed below. Choose the tributaries into which these veins drain.

(A) testicular vein
(B) superior epigastric
(C) femoral
(D) internal iliac
(E) azygos vein

14.  esophageal tributaries of the left gastric

15.  superior rectal

16.  paraumbilical

17.  retroperitoneal

**Questions 18 through 21**

For each characteristic listed below, choose the segment of the intestinal tract with which it is associated.

(A) stomach
(B) duodenum
(C) jejunum
(D) ascending colon
(E) ileum

18.  Teniae coli

19.  Sacculations

20.  Epiploic appendages

21.  Transverse folds

**DIRECTIONS (Questions 22 through 33): Each group of items in this section consists of lettered headings followed by a set of numbered words or phrases. For each numbered word or phrase, select the ONE lettered heading that is most closely associated with it. Each lettered heading may be selected once, more than once, or not at all.**

**Questions 22 through 25**

(A) common bile duct
(B) portal vein
(C) both
(D) neither

22.  located in the porta hepatis

23.  located in the hepatoduodenal ligament

24.  empties into the liver

25.  empties into the first part of the duodenum

**Questions 26 through 29**

(A) The main pancreatic duct
(B) The accessory pancreatic duct
(C) both
(D) neither

26.  it develops from the proximal part of the duct of the dorsal primordium

27.  it empties into the second portion of the duodenum

28.  it joins the common bile duct

29.  it develops from the distal part of the duct of the dorsal primordium and the proximal part of the duct of the ventral primordium

**Questions 30 through 33**

(A) The neurons in the myenteric plexus
(B) The neurons in the submucosal plexus
(C) both
(D) neither

30.  are postganglionic parasympathetic neurons

31.  increase motility in the intestinal tract

32. slow peristalsis

33. send pain afferents to the central nervous system

**DIRECTIONS (Questions 34 through 44):** For each of the items in this section, <u>ONE</u> or <u>MORE</u> of the numbered options is correct. Choose the answer

 A if only <u>1, 2, and 3</u> are correct,
 B if only <u>1 and 3</u> are correct,
 C if only <u>2 and 4</u> are correct,
 D if only <u>4</u> is correct,
 E if <u>all</u> are correct.

34. Which of the following statements correctly apply to cleft palate?

    (1) it occurs about once in 2,500 births
    (2) it may or may not be associated with cleft lip
    (3) it is the failure of the lateral palatine processes to fuse with one another
    (4) it is the failure of one or both of the lateral palatine processes to fuse with the primary palate

35. Which of the following statements correctly apply to the marginal mandibular branch of the facial nerve?

    (1) it usually courses through the submandibular platysma muscle
    (2) it runs superficial to the platysma muscle
    (3) it runs deep to the masseter
    (4) it supplies the depressor labii inferioris muscle

36. Which of the following veins communicate with the cavernous sinus and is commonly implicated in transmission of infection to the cavernous sinus?

    (1) superficial temporal
    (2) superior ophthalmic
    (3) posterior auricular
    (4) facial

37. A failure of the obliteration of the epithelium of the thyroglossal duct may result in which of the following?

    (1) cryptorchidism
    (2) ectopic submandibular gland
    (3) development of a small thyroid gland
    (4) thyroglossal cyst

38. Which of the following nerves is commonly involved in the excruciating paroxysmal pain associated with tic douloureux?

    (1) marginal mandibular branch of the facial
    (2) maxillary division of the trigeminal
    (3) external branch of the superior laryngeal
    (4) mandibular division of the trigeminal

39. A symmetrical muscular contraction shows up in a patient if asked to smile, show the teeth, frown, or forcibly close the eyes, which is a result of damage to which of the following nerves?

    (1) facial
    (2) trigeminal
    (3) seventh cranial
    (4) fifth cranial

40. Which of the following statements correctly relate to the mandible?

    (1) The commonest fracture occurs at the condylar neck
    (2) The commonest fracture occurs near the mental foramen
    (3) The angle between the body and ramus in the young adult is about 175°
    (4) anterior dislocation of the mandible is relatively common

41. Incisions through the mucous membrane of the floor of the mouth give access to which of the following structures?

    (1) sublingual gland
    (2) submandibular duct
    (3) lingual nerve
    (4) facial nerve

42. Which of the following statements correctly apply to an epidural hemorrhage?

    (1) it is due to rupture of the middle meningeal artery
    (2) it results in increased intracranial pressure
    (3) it expands the potential space between the bone and the periosteal layer of dura
    (4) it results from the rupture of cerebral veins

43. Which of the following statements correctly apply to the subarachnoid hemorrhage?

    (1) it is confined to the subarachnoid space
    (2) it is due to rupture of cerebral vessels
    (3) blood will be found in the cerebrospinal fluid
    (4) it is due to rupture of the middle meningeal artery

44. Which of the following statements correctly apply to the lens?

    (1) the shape of the lens is modified by the ciliary muscle
    (2) it has a tendency to lose its transparency in later years
    (3) it is a transparent biconvex body
    (4) the lens is nearly spherical in the fetus

# Answers and Explanations

1. **(D)** Hydrocele is a fluid accumulation within the tunica vaginalis testis due to secretion of abnormal amounts of serous fluid by the serous membrane. *(Woodburne, p 382)*

2. **(B)** Herniation of abdominal contents into an unobliterated vaginal process and within the coverings of the spermatic cord results in an indirect inguinal hernia. *(Woodburne, p 385)*

3. **(A)** Another type of inguinal hernia, the direct inguinal hernia, occurs less commonly (one third as frequently as indirect herniae in males). The direct inguinal hernia bulges at the superficial inguinal ring and almost never descends into the scrotum. The inferior epigastric artery is lateral to the herniating mass in the direct type. *(Woodburne, p 387)*

4. **(C)** The Meckel's diverticulum is an occasional feature of the ileum. It occurs in about 1% of individuals, somewhere in the last meter of the ileum. It represents the remains of the vitelline duct. *(Woodburne, p 429)*

5. **(E)** Femoral herniae descend into the thigh behind the inguinal ligament, using a tubular compartment of the femoral sheath, the femoral canal. *(Woodburne, p 387)*

6. **(A)** A fusion fascia has surgical importance: because it is a peritoneal remnant along which or across which no blood vessels, nerves, or other accessory structures pass, it can be invaded in the adult body without fear of encountering any important vessels or nerves. *(Woodburne, p 394)*

7. **(C)** The transpyloric line may be marked on the surface of the body at the midpoint between the suprasternal notch and the symphysis pubis and is at the level of the first lumbar vertebra. *(Woodburne, p 400)*

8. **(D)** An annular pancreas is a rare anomaly. It consists of a complete ring of pancreatic tissue around the descending portion of the duodenum. It is considered to be due to a doubling of the ventral primordium of the pancreas or to a splitting of it. *(Woodburne, p 414)*

9. **(C)** The upper one third of the free intestine occupies the left upper quadrant of the abdomen; the middle one third is located in the middle of the abdominal cavity; and the lower one third lies in the pelvis and in the right iliac fossa. *(Woodburne, p 429)*

10. **(D)** The highest point of the iliac crest is at the transverse level of the body of the fourth lumbar vertebra. *(Woodburne, p 367)*

11. **(C)** The vena caval foramen is an opening in the central tendon. It is the highest of the three openings and lies at the level of the eighth thoracic vertebra or the disk below it. *(Woodburne, p 451)*

12. **(B)** The elliptical esophageal hiatus lies in the muscular part of the diaphragm at the level of the tenth thoracic vertebra. It is formed by the divergence and subsequent decussation of the bundles of the right crus. *(Woodburne, p 451)*

13. **(A)** The renal arteries arise, one on each side of the aorta, at the level of the upper border of the second lumbar vertebra. Their origin is about 1 cm below that of the superior mesenteric artery. *(Woodburne, p 446)*

14. **(E)** The esophageal tributaries of the left gastric vein communicate with esophageal veins that empty into the azygos vein of the chest. *(Woodburne, p 440)*

15. **(D)** The rectal plexus of the anal canal and the lower rectum allow communication between tributaries of the superior rectal, middle rectal, and inferior rectal veins. The middle rectal and inferior rectal veins transmit their blood to the inferior vena cava. *(Woodburne, p 440)*

16. **(B)** The paraumbilical veins anastomose with small veins of the anterior abdominal wall which, as radicales of the superior epigastric, inferior epigastric,

thoracoepigastric, and segmental vessels, connect with the superior or the inferior vena cava. *(Woodburne, p 440)*

17. **(A)** The retroperitoneal veins draining the colon (ileocolic, right colic, middle colic, and left colic veins) have anastomotic connections with testicular (or ovarian) veins and especially with small veins of the pararenal fat. *(Woodburne, p 440)*

18. **(D)** Three surface features serve to distinguish isolated loops of the small or the large intestine presenting through an abdominal opening: (1) taeniae coli; (2) sacculations, or haustra coli; (3) epiploic appendages. *(Woodburne, p 434)*

19. **(D)** The colon presents three rows of sacculations, or haustra coli, alternating between the taeniae. *(Woodburne, p 434)*

20. **(D)** The third characteristic of the colon is the occurrence along its length of epiploic appendages. These are fat-filled tabs, or pendants, of peritoneum that project from the serous coat of the large intestine except on the rectum. *(Woodburne, p 434)*

21. **(C)** Little fat exists in the mesentery of the upper jejunum, and "windows" of translucency between the blood vessels of the mesentery are numerous. *(Woodburne, p 429)*

22. **(C)** The porta of the liver is the region of branching and entrance of the hepatic artery and the portal vein and of exit of the hepatic bile ducts. *(Woodburne, p 417)*

23. **(C)** The hepatoduodenal ligament transmits the hepatic artery, the portal vein, and the common bile duct. *(Woodburne, p 419)*

24. **(B)** The hepatic artery conducts arterial blood for the nourishment of the liver tissues, whereas the portal vein carries to the liver blood from the gastrointestinal tract containing certain of the products of digestion. *(Woodburne, p 419)*

25. **(D)** The common bile duct is the excretory duct of the liver and empties into the second portion of the duodenum. It is formed near the porta hepatis by the junction of the common hepatic duct from the liver and the cystic duct from the gallbladder. *(Woodburne, p 419)*

26. **(B)** The ducts of the two primordia fuse in the upper part of the head of the gland, thus the pancreatic duct of the adult is formed from the distal part of the duct of the dorsal primordium and the proximal part of the duct of the ventral primordium. The remaining proximal part of the duct of the dorsal primor-

dium becomes the accessory pancreatic duct in the adult. *(Woodburne, pp 413–414)*

27. **(C)** The greater duodenal papilla marks the termination of the main pancreatic duct and the common bile duct. Typically the accessory pancreatic duct empties at the lesser duodenal papilla, about 2 cm proximal to the greater papilla, in the anteromedial duodenal wall. *(Woodburne, p 413)*

28. **(A)** The greater duodenal papilla marks the termination of the main pancreatic duct and the common bile duct. *(Woodburne, p 413)*

29. **(A)** The pancreatic duct of the adult is formed from the distal part of the duct of the dorsal primordium and the proximal part of the duct of the ventral primordium. *(Woodburne, pp 413–414)*

30. **(C)** The cell bodies within these plexuses are those of the postganglionic neurons in the two-neuron parasympathetic innervation of the intestinal tract. *(Woodburne, p 399)*

31. **(C)** The parasympathetic nerves increase motility in the tract and are important in keeping the chyme moving along. *(Woodburne, p 399)*

32. **(D)** The sympathetic nerves slow peristalsis and thus oppose the emptying mechanism. *(Woodburne, p 399)*

33. **(D)** Pain afferents from the abdominal viscera are carried by the sympathetic nerves. *(Woodburne, p 399)*

34. **(E)** Cleft palate occurs about once in 2,500 births. It may or may not be associated with cleft lip. Fundamentally it is due to failure of the lateral palatine processes to fuse with one another or failure of one or both of the lateral palatine processes to fuse with the primary palate. These fusions normally occur prior to the tenth week of development in man. *(Woodburne, p 235)*

35. **(D)** In about 19% of cases a marginal mandibular branch of the facial nerve curves below the border of the mandible and courses through the submandibular triangle, where it is endangered by incisions along the inferior margin of the mandible posterior to the facial artery. It runs deep to the platysma muscle, runs superficial to the masseter, and supplies the depressor labii inferioris and the mentalis. *(Woodburne, p 209)*

36. **(C)** All tributaries of the superior ophthalmic vein, which drains the eye, empty intracranially into the cavernous sinus. Free passage through these valveless veins and from the lips and face via the facial vein is commonly implicated in transmission of in-

fection to the cavernous sinus, which may result in a serious cavernous sinus thrombosis. *(Woodburne, p 206)*

37.  **(D)** In rare cases there will be a failure of obliteration of the epithelium of the thyroglossal duct and a thyroglossal duct cyst may form. This may occur anywhere between the pyramidal lobe and the tongue. *(Woodburne, p 157)*

38.  **(C)** The cutaneous and mucosal areas of distribution of the mandibular and maxillary divisions are the commonest sites of the excruciating paroxysmal pain of trigeminal neuralgia (tic douloureux). *(Woodburne, p 200)*

39.  **(B)** Facial nerve or facial muscle paralysis is distressing to the sufferer and is usually evident to a discerning observer in loss of normal facial contours. A symmetrical muscular contraction shows up if the patient is asked to smile, show the teeth, frown, or forcibly close the eyes. *(Woodburne, p 210)*

40.  **(C)** The commonest fracture occurs at or just behind the mental foramen. Bilateral anterior dislocation of the mandible is also relatively common. In the adult,

the angle approaches a right angle (110° to 120°). *(Woodburne, p 221)*

41.  **(A)** Incisions through the mucous membrane of the floor of the mouth give access to the sublingual gland, the submandibular duct, and the lingual nerve. *(Woodburne, p 232)*

42.  **(A)** The epidural hemorrhage is due to rupture of the middle meningeal artery. Such blood expands the potential space between the periosteal layer of the dura and the bone. Being under arterial pressure results in increased intracranial pressure. *(Woodburne, p 283)*

43.  **(A)** Subarachnoid hemorrhage is confined to the subarachnoid space and is due to the rupture of cerebral vessels. Blood will be found in the cerebrospinal fluid. *(Woodburne, pp 283–284)*

44.  **(E)** The lens is a transparent biconvex body that is nearly spherical in the fetus. Its shape is modified by the ciliary muscle as the eyes focus on objects at differing distances. The lens has a tendency to lose its transparency in later years, a condition known as cataract. *(Woodburne, p 256)*

# CHAPTER 4

# The Back
## Questions

**DIRECTIONS (Questions 1 through 15): Each of the numbered items or incomplete statements in this section is followed by answers or by completions of the statement. Select the ONE lettered answer or completion that is BEST in each case.**

1.  Which of the following terms correctly applies to the fusion of the fifth lumbar vertebra into the sacrum?

    (A) lumbarization
    (B) sacralization
    (C) lumbago
    (D) lordosis
    (E) scoliosis

2.  Which of the following terms correctly applies to abnormal curvatures of the vertebral column?

    (A) sacralization
    (B) lumbarization
    (C) kyphosis
    (D) osteoporosis
    (E) osteomalacia

3.  Lordosis is characterized by an increased curve of the vertebral column that is

    (A) convex posteriorly
    (B) convex anteriorly
    (C) convex to the side
    (D) concave laterally
    (E) concave anteriorly

4.  The broad anterior longitudinal ligament tends to prevent which of the following?

    (A) hyperextension of the vertebral column
    (B) coccydynia
    (C) scoliosis
    (D) osteoporosis
    (E) hemivertebra

5.  A herniated or prolapsed disc usually occurs in which of the following directions?

    (A) anterior
    (B) posterior
    (C) posterolateral
    (D) anterolateral
    (E) inferiorly

6.  Clinicians often refer to which of the following articulations as "facet joints"?

    (A) opposing articular processes of adjacent vertebral arches
    (B) ribs and transverse ventral processes
    (C) vertebrae and adjacent intervertebral discs
    (D) adjacent vertebral bodies
    (E) adjacent vertebral spinous processes

7.  The broad, yellow, elastic bands that join the laminae of adjacent vertebral arches are known as the

    (A) interspinous ligaments
    (B) supraspinous ligaments
    (C) ligamentum nuchae
    (D) intertransverse ligaments
    (E) ligamenta flava

8.  Which of the following statements correctly applies to the tectorial membrane?

    (A) it is the upward continuation of the posterior longitudinal ligament
    (B) it extends from the first thoracic vertebral body to the occipital bone
    (C) it is covered by the alar ligament
    (D) it covers the transverse ligament
    (E) it attaches to the dens

9.  In severe neck flexion injuries, which of the following ligaments usually is torn?

    (A) anterior longitudinal
    (B) posterior longitudinal
    (C) apical
    (D) ligamentum nuchae
    (E) ligamentum flavum

10. The suboccipital triangle contains which of the following structures?

    (A) vertebral artery
    (B) lesser occipital nerve
    (C) spinal accessory nerve
    (D) occipital artery
    (E) posterior auricular artery

11. The spinal cord in adults usually ends opposite the intervertebral disc between which of the following vertebrae?

    (A) L4–5
    (B) L1–2
    (C) S2–3
    (D) S4–5
    (E) T10–11

12. The spinal cord is suspended in the dura mater by which of the following structures?

    (A) filum terminale
    (B) cauda equina
    (C) conus medullaris
    (D) denticulate ligament
    (E) alar ligament

13. The principal site of absorption of cerebrospinal fluid into the venous system is through which of the following structures?

    (A) diploic veins
    (B) arachnoid villa
    (C) pterygoid plexus
    (D) vertebral venous plexus
    (E) cavernous sinus

14. Which of the following statements correctly applies to the medial column of the erector spinae muscle?

    (A) it is known as the longissimus muscle
    (B) it arises from spinous processes
    (C) it inserts into transverse processes
    (D) it is a flexor of the vertebral column
    (E) it is a superficial muscle

15. Which of the following muscles is concerned with the maintenance of posture and movements of the vertebral column?

    (A) serratus posterior inferior
    (B) trapezius
    (C) latissimus dorsi
    (D) levator scapulae
    (E) longissimus

**DIRECTIONS (Questions 16 through 27): Each group of items in this section consists of lettered headings followed by a set of numbered words or phrases. For each numbered word or phrase, select the ONE lettered heading that is most closely associated with it. Each lettered heading may be selected once, more than once, or not at all.**

**Questions 16 through 19**

For each of the vertebrae listed below, select the statement that applies to that vertebra.

    (A) it is called the vertebra prominens
    (B) the largest of all movable vertebrae
    (C) the spinous process is long and slender
    (D) it has no spinous processes
    (E) its distinguishing feature is the blunt toothlike dens

16. first cervical

17. second cervical

18. fifth lumbar

19. third thoracic

**Questions 20 through 23**

For each of the ligaments listed below, select the statement that applies to that ligament.

    (A) it is the upward continuation of the posterior longitudinal ligament
    (B) it extends from the tip of the dens to the axis
    (C) it is an adult derivative of the notochord
    (D) it holds the dens of the axis against the anterior arch of the atlas
    (E) it extends from the dens to the lateral margins of the foramen magnum

20. transverse ligament of the atlas

21. alar ligaments

22. tectorial membrane

23. apical ligament

**Questions 24 through 27**

For each of the muscles listed below, select the statement that applies to that muscle.

    (A) it inserts into the mastoid process of the temporal bone
    (B) a suboccipital muscle
    (C) it is the lateral column of the erector spinae muscle

(D) an extrinsic back muscle

(E) it extends the vertebral column and rotates it toward the opposite side

24. trapezius

25. longissimus capitis

26. multifidus

27. iliocostalis

DIRECTIONS (Questions 28 through 39): Each group of items in this section consists of lettered headings followed by a set of numbered words or phrases. For each numbered word or phrase, select

A if the item is associated with (A) only,

B if the item is associated with (B) only,

C if the item is associated with both (A) and (B),

D if the item is associated with neither (A) nor (B).

Questions 28 through 31

(A) filum terminale

(B) denticulate ligament

(C) both

(D) neither

28. it is formed by the pia mater

29. it attaches the pia mater to the dura mater

30. it lies within the cauda equina

31. it has an extradural prolongation

Questions 32 through 35

(A) posterior longitudinal ligament

(B) anterior longitudinal ligament

(C) both

(D) neither

32. it covers the alar and transverse ligaments

33. it tends to prevent hyperflexion of the vertebral column

34. it tends to prevent hyperextension of the vertebral column

35. it extends inside the vertebral canal from the atlas to the sacrum

Questions 36 through 39

(A) intervertebral foramen

(B) transverse foramen

(C) both

(D) neither

36. the location of a dorsal root ganglion

37. located in the cervical, thoracic, and lumbar vertebrae

38. the vertebral arteries pass through these foramina

39. it is formed by the superior vertebral notch and the inferior vertebral notch of adjacent vertebrae

DIRECTIONS (Questions 40 through 50): For each of the items in this section, ONE or MORE of the numbered options is correct. Choose the answer

A if only 1, 2, and 3 are correct,

B if only 1 and 3 are correct,

C if only 2 and 4 are correct,

D if only 4 is correct,

E if all are correct.

40. Which of the following statements correctly apply to the dura mater?

(1) it forms dural sleeves that continue into the intervertebral foramina

(2) it becomes continuous with the epineurium

(3) it is continuous with the cranial dura mater

(4) it usually ends at the level of the lower border of S2 in adults

41. Which of the following conditions correctly apply to a patient with complete transection of the spinal cord between the cervical and lumbosacral enlargements?

(1) the patient is quadriplegic

(2) the patient is paraplegic

(3) the patient may die due to respiratory failure

(4) the abdominal and back muscles are also affected

42. Which of the following statements correctly apply to the intervertebral discs?

(1) these are fibrocartilaginous discs

(2) they are derivatives of the notochord

(3) the nucleus pulposus acts as a shock absorber

(4) the anulus fibrosus is composed of concentric lamellae

43. Which of the following conditions are usually associated with a "whiplash injury"?

    (1) scoliosis
    (2) rachischisis
    (3) spina bifida
    (4) hyperextension injury of the neck

44. Which of the following conditions are associated with a complete transection of the spinal cord above the C3 level?

    (1) the patient is quadriplegic
    (2) the patient does not lose all sensation below level of the lesion
    (3) the patient may die due to respiratory failure
    (4) the patient does not lose all voluntary movement below the lesion

45. Which of the following statements correctly apply to the vertebral venous plexus?

    (1) not clinically significant
    (2) it usually drains into the inferior vena cava
    (3) it usually empties into the renal veins
    (4) it may allow blood to return to the heart by way of the superior vena cava

46. The floor of the suboccipital triangle is formed by which of the following structures?

    (1) posterior atlantooccipital membrane
    (2) semispinalis muscle
    (3) posterior arch of the atlas
    (4) carotid artery

47. Which of the following statements correctly apply to the blood supply of the spinal cord?

    (1) the spinal cord is supplied by three longitudinal arteries and several ridicular arteries
    (2) the anterior spinal artery is formed by the union of two small branches of the carotid arteries
    (3) each posterior spinal artery arises as a small branch of the vertebral or the posterior inferior cerebellar
    (4) the ventral median fissure contains the posterior spinal arteries

48. Which of the following statements correctly apply to the layers of the meninges?

    (1) together the pia mater and the arachnoid are called the leptomeninges
    (2) the subarachnoid space is a potential space
    (3) the subarachnoid space contains cerebrospinal fluid
    (4) the subarachnoid space extending from L2 to S2 is known as the cerebellomedullary cistern

49. Which of the following statements correctly apply to the epidural space?

    (1) commonly used by the clinician to obtain cerebrospinal fluid
    (2) the epidural or extradural space is not a real space
    (3) the epidural space is filled with fibrocartilage
    (4) may be approached by the clinician through the sacral hiatus

50. Which of the following statements correctly apply to disc protrusions?

    (1) they most commonly occur posterolaterally
    (2) midline posterior protrusions of lumbar discs are unusual
    (3) the protruding nucleus pulposus often affects one or more spinal nerves
    (4) the posterior longitudinal ligament prevents anterolateral protrusions

# Answers and Explanations

1. **(B)** In some people the fifth lumbar vertebra is partly or completely incorporated into the sacrum (sacralization of the fifth lumbar vertebra). In others the first sacral vertebra is separated from the sacrum (lumbarization of the first sacral vertebra). *(Moore, p 607)*

2. **(C)** Kyphosis is characterized by an abnormal curve that is convex posteriorly (dorsal curvature of the vertebral column) and that usually occurs in the thoracic region (humpback). *(Moore, p 608)*

3. **(B)** Lordosis is characterized by an increased curve of the vertebral column that is convex anteriorly (backward bending). *(Moore, p 608)*

4. **(A)** The anterior longitudinal ligament is a strong, broad, fibrous band that runs longitudinally along the anterior surfaces of the intervertebral discs and the bodies of the vertebrae. Its fibers are firmly fixed to the intervertebral discs and to the periosteum of the vertebral bodies. The broad anterior longitudinal ligament tends to prevent hyperextension of the vertebral column. *(Moore, p 625)*

5. **(A)** A herniated or prolapsed disc usually occurs in a posterolateral direction, where the anulus fibrosus is weakest and poorly supported by the posterior longitudinal ligament. The protruding part of the nucleus pulposus may compress an adjacent spinal nerve root, causing leg or low back pain. *(Moore, pp 628–629)*

6. **(A)** True synovial joints of the plane variety, known as zygapophyseal joints, are formed by the opposing articular processes (zygapophyses) of adjacent vertebral arches. Because the contact surfaces of these articular processes are called articular facets, clinicians often refer to zygapophyseal joints as "facet joints." *(Moore, p 630)*

7. **(E)** The laminae of adjacent vertebral arches are joined by broad, yellow, elastic bands called ligamenta flava. Their fibers extend to the capsules of the zygapophyseal joints between the articular processes and contribute to the posterior boundary of the vertebral foramen. *(Moore, p 630)*

8. **(D)** The tectorial membrane is the upward continuation of the posterior longitudinal ligament. It runs from the body of the axis to the internal surface of the occipital bone and covers the alar ligaments and the transverse ligament. *(Moore, p 632)*

9. **(B)** In severe flexion injuries the posterior longitudinal and interspinous ligaments may be torn, and the vertebral arches may be dislocated and/or fractured along with crush fractures of the vertebral bodies. The anterior longitudinal ligament is usually not torn in flexion injuries. *(Moore, p 633)*

10. **(A)** The suboccipital triangle is important clinically because it contains the vertebral artery and the suboccipital nerve (dorsal ramus of the first cervical nerve). These structures lie in a groove on the superior surface of the posterior arch of the atlas. *(Moore, p 646)*

11. **(B)** The spinal cord in adults often ends opposite the intervertebral disc between L1 and L2, but it may terminate as high as T12 or as low as L3. Thus the spinal cord occupies only the upper two thirds of the vertebral canal. *(Moore, p 648)*

12. **(D)** The spinal cord is suspended in the dura mater by a saw-toothed denticulate ligament on each side. This ribbonlike ligament, composed of pia mater, is attached along the lateral surface of the spinal cord, midway between the dorsal and ventral nerve roots. There are 21 toothlike processes attached to the dura mater. *(Moore, p 657)*

13. **(B)** The principal site of absorption of cerebrospinal fluid into the venous blood is through the arachnoid villi projecting into the dural venous sinuses, particularly the superior sagittal sinus. Many arachnoid villi show hypertrophy in older persons and are called arachnoid granulations. *(Moore, p 670)*

14. **(B)** The spinalis muscle, the medial column of the erector spinae muscle, arises from spinous processes, and inserts into spinous processes. It is an extensor of the vertebral column. *(Moore, p 642)*

15. **(E)** The muscles of the deep or intrinsic group, the true back muscles, are concerned with the maintenance of posture and movements of the vertebral column (flexion, extension, lateral bending, rotation, and circumduction). The intermediate layer of intrinsic muscles includes the erector spinae muscle, which contains large bundles known as the spinalis, longissimus, and iliocostalis. *(Moore, pp 640–641)*

16. **(D)** The atlas has no spinous process or body; it consists of anterior and posterior arches, each of which bears a tubercle and a lateral mass. *(Moore, p 615)*

17. **(E)** The second cervical vertebra, known as the axis, has two flat-bearing surfaces, the superior articular facets, upon which the atlas rotates. Its distinguishing feature is the blunt toothlike dens. *(Moore, p 615)*

18. **(B)** The fifth lumbar vertebra, the largest of all movable vertebra, is characterized by stout transverse processes. *(Moore, p 617)*

19. **(C)** All 12 thoracic vertebrae articulate with ribs. Thus they are characterized by articular facets for them. The spinous processes tend to be long and slender. *(Moore, p 615)*

20. **(D)** The transverse ligament of the atlas is a strong band extending between the tubercles on the medial sides of the lateral masses of the atlas. It holds the dens of the axis against the anterior arch of the atlas, with a synovial joint between them. *(Moore, p 632)*

21. **(E)** The alar ligaments extend from the dens to the lateral margins of the foramen magnum. They check lateral rotation and side-to-side movements of the head and attach the skull to the axis. *(Moore, p 632)*

22. **(A)** The tectorial membrane is the upward continuation of the posterior longitudinal ligament. It runs from the axis to the internal aspect of the occipital bone and covers the alar and transverse ligaments. *(Moore, p 632)*

23. **(C)** The apical ligament is an adult derivative of the notochord in the embryo. It extends from the tip of the dens to the internal surface of the occipital bone. *(Moore, p 632)*

24. **(D)** The extrinsic muscles of the back include the trapezius, latissimus dorsi, levator scapulae, and rhomboid muscles. They connect the upper limb to the axial skeleton and are related to movements of the upper limbs. *(Moore, p 638)*

25. **(A)** The longissimus capitis inserts into the mastoid process of the temporal bone. *(Moore, p 642)*

26. **(E)** The multifidus muscle extends the vertebral column and rotates it toward the opposite side. Each bundle of muscle ascends obliquely from its origin and inserts two to five vertebrae superiorly. *(Moore p 642)*

27. **(C)** The iliocostalis muscle is the lateral column of muscles that form the erector spinae muscle. It arises from the iliac crest and inserts into the ribs. *(Moore p 641)*

28. **(C)** The filum terminale, which has no functional significance, consists of connective tissue and pia mater. The saw-toothed denticulate ligament, composed of pia mater, is attached along the lateral surface of the spinal cord, midway between the dorsal and ventral nerve roots. *(Moore, pp 651, 657)*

29. **(B)** The spinal cord is suspended in the dura mater by a saw-toothed denticulate ligament on each side. There are 21 toothlike processes attached to the dura mater between the foramen magnum and the level at which the dura is pierced by the nerve roots. *(Moore, p 657)*

30. **(A)** The spinal cord tapers rather abruptly into the conus medullaris, and from its inferior end a long slender filament, called the filum terminale, arises. It lies within the cauda equina. *(Moore, p 651)*

31. **(A)** The filum terminale has an extra dural prolongation, which inserts into the dorsum of the coccyx. *(Moore, p 651)*

32. **(D)** The anterior longitudinal ligament runs along the anterior surface of the intervertebral discs and the bodies of the vertebrae. The tectorial membrane is the upward continuation of the posterior longitudinal ligament. It runs from the body of the axis to the internal surface of the occipital bone and covers the alar and transverse ligaments. *(Moore, pp 625, 632)*

33. **(A)** The posterior longitudinal ligament tends to prevent hyperflexion of the vertebral column. *(Moore, p 625)*

34. **(B)** The broad anterior longitudinal ligament tends to prevent hyperextension of the vertebral column. *(Moore, p 625)*

35. **(A)** The posterior longitudinal ligament extends inside the vertebral canal from the atlas to the sacrum and is attached to the intervertebral discs and to the edges of the vertebral bodies. *(Moore, p 625)*

36. **(A)** The dorsal and ventral nerve roots are in the vertebral canal, and the dorsal root ganglia are in the invertebral foramina. These roots join each other at the outer margin of the invertebral foramen to form a spinal nerve. *(Moore, pp 613–614)*

37. **(A)** In all vertebrae there is a notch above each pedicle, the superior vertebral notch, which varies in size at different levels of the vertebral column. The notch below each pedicle, the inferior vertebral notch, is usually larger. When vertebrae are in articulation, these notches are adjacent to each other and form a bony ring, the intervertebral foramen. *(Moore, p 613)*

38. **(B)** The vertebral artery passes through the foramen transversarium, except in C7, which transmits only small, accessory, vertebral veins. *(Moore, p 615)*

39. **(A)** When two vertebrae are in articulation, the superior and inferior vertebral notches are adjacent to each other and form a complete ring of bone, the invertebral foramen. *(Moore, p 613)*

40. **(E)** The dura mater is evaginated along the dorsal and ventral nerve roots of the spinal nerves, forming dural sleeves. Beyond the spinal ganglia, these dural sleeves fuse to become continuous with the epineurium. The spinal dura is continuous with the cranial dura mater and ends downward at the level of the lower border of S2 in adults. *(Moore, p 657)*

41. **(C)** The patient is paraplegic (lower limbs paralyzed) if the transection is between the cervical and lumbosacral enlargements. The abdominal and back muscles are also affected. *(Moore, p 657)*

42. **(E)** The intervertebral discs are fibrocartilaginous discs that are derivatives of the notochord. Each disc contains an anulus fibrosus that is composed of concentric lamellae of collagenous fibers that surround the nucleus pulposus, which acts like a shock absorber. *(Moore, pp 625–627)*

43. **(D)** Although a hyperextension injury of the neck is popularly called a "whiplash injury," there is no well-defined clinical syndrome or fixed pathology associated with the injury. *(Moore, p 668)*

44. **(B)** Complete transection of the spinal cord results in loss of all sensation and voluntary movement be-

low the lesion. The patient is quadriplegic (upper and lower limbs paralyzed) if the cervical cord above C3 is transected, and the patient may die owing to respiratory failure. *(Moore, pp 656–657)*

45. **(C)** The vertebral venous plexus is important clinically because blood may return from the pelvis or the abdomen by way of it and reach the heart through the superior vena cava. It usually empties into the heart by way of the inferior vena cava. *(Moore, p 655)*

46. **(B)** The floor of the suboccipital triangle is formed by the posterior atlanto-occipital membrane and the posterior arch of the atlas. Its roof is formed by the semispinalis capitis muscle. *(Moore, p 645)*

47. **(B)** The spinal cord is supplied by three longitudinal vessels (an anterior spinal artery and two posterior spinal arteries) and anterior and posterior radicular arteries. The anterior spinal artery is formed by the union of two small branches of the vertebral arteries. It runs the length of the spinal cord in the ventral median fissure. Each posterior spinal artery arises from the vertebral or the posterior inferior cerebellar artery. *(Moore, pp 651–653)*

48. **(B)** The arachnoid mater is separated from the dura by a potential subdural space. The arachnoid is separated from the pia mater by an actual space, the subarachnoid space, which contains cerebrospinal fluid. The subarachnoid space extending from L2 to S2 is known as the lumbar cistern. *(Moore, p 658)*

49. **(C)** Local anesthetic solutions may be injected into the extradural (epidural) space for several reasons. First of all, it is not a real space, and it is filled with areolar tissue, fat, and veins. The sacral hiatus is used by the clinicians to administer epidural anesthesia. *(Moore, p 667)*

50. **(A)** Disc protrusions most commonly occur posterolaterally where the anulus fibrosus is thin. The posterior longitudinal ligament strengthens the middle portion of the posterior part of the anulus fibrosus. Consequently, midline posterior protrusions of lumbar discs are unusual. As the dorsal and ventral nerve roots cross this region, the protruding nucleus pulposus often affects one or more spinal nerve roots. *(Moore, pp 668–669)*

# CHAPTER 5

# The Thorax
## Questions

**DIRECTIONS (Questions 1 through 82): Each of the numbered items or incomplete statements in this section is followed by answers or by completions of the statement. Select the ONE lettered answer or completion that is BEST in each case.**

1. The mediastinum contains all the following structures EXCEPT the

   (A) heart
   (B) lungs
   (C) pulmonary arteries
   (D) trachea
   (E) esophagus

2. Anatomically components of the thoracic wall proper include the

   (A) pectoral muscles
   (B) serratus anterior
   (C) trapezius
   (D) intercostal muscles
   (E) latissimus dorsi

3. The superior thoracic aperture can be described correctly by which of these statements?

   (A) it is large and irregular
   (B) it is bounded by the third thoracic vertebra
   (C) its plane slopes downward and forward
   (D) it includes the costal arch
   (E) it involves costal cartilages of the tenth rib

4. The sternal angle is found at which of these locations?

   (A) jugular notch
   (B) xiphoid process
   (C) level with the fourth costal cartilage
   (D) level with the lower border of the sixth thoracic vertebra
   (E) manubriosternal joint

5. Ossification of the parts of the body of the sternum usually is complete by age (in years)

   (A) one
   (B) three
   (C) six
   (D) 15
   (E) 21

6. Which of the following defines true ribs?

   (A) upper seven pairs
   (B) all 12 pairs
   (C) lower five pairs
   (D) 10th and 11th pairs
   (E) 12th pair

7. All the following are correct statements about rib fractures EXCEPT

   (A) most frequently they occur as a result of compression forces on the thorax
   (B) most often they occur just anterior to the costal angle
   (C) splinting the chest wall is essential treatment
   (D) broken rib ends tend to spring outward
   (E) fractured bone ends may injure the lungs

8. The vertical length of the rib cage is increased by which of the following?

   (A) movement of ribs at the costovertebral joints
   (B) movement of the diaphragm
   (C) movement of ribs at the costotransverse joint
   (D) pump-handle movement of ribs
   (E) bucket-handle movement of ribs

9. Which of these statements correctly describes intercostal muscles?

   (A) external intercostals begin anteriorly
   (B) external intercostal membrane is posterior
   (C) fibers of external intercostals slant upward and backward
   (D) fibers of internal intercostals run upward and forward
   (E) innermost intercostals are the best developed of the intercostals

10. Innervation of the thoracic wall can be described correctly by all the following statements EXCEPT

    (A) it receives a nerve supply from spinal nerves T1–12
    (B) it receives no nerve supply from cervical nerves
    (C) cutaneous innervation of skin over paravertebral regions of the thorax is provided by dorsal rami of spinal nerves
    (D) ventral rami of T1–11 are called intercostal nerves
    (E) the ventral ramus of T12 is the subcostal nerve

11. Which of these statements is true in relation to intercostal nerves?

    (A) in the intercostal space, they run between the internal and external intercostal muscles
    (B) they are located in the costal groove above the artery and vein
    (C) they are all confined to the thorax
    (D) they are entirely cutaneous to the thoracic wall
    (E) the upper six nerves terminate as anterior cutaneous branches

12. Which of the following statements is correct regarding the blood vessels of the thoracic wall?

    (A) in the intercostal space, the vessels run just below the respective intercostal nerve
    (B) branches of the vessels vary widely from those of the intercostal nerves
    (C) superficial structures of the thorax are served by intercostal vessels
    (D) posterior intercostal arteries are branches of the internal thoracic artery
    (E) branches of the descending thoracic aorta become anterior intercostal arteries

13. All the following are true statements about the anterior thoracic artery EXCEPT

    (A) it is a branch of the arch of the aorta
    (B) it descends behind the subclavian vein
    (C) it divides into two terminal branches
    (D) it gives branches to the mediastinum
    (E) the musculophrenic artery is one of its terminal branches

14. The surface of the developing lung is covered by a layer of mesothelium designated as the

    (A) parietal pleura
    (B) visceral pleura
    (C) diaphragmatic pleura
    (D) costal pleura
    (E) mediastinal pleura

15. Innervation of pleura can be described correctly by all of these statements EXCEPT

    (A) costal pleura is supplied by intercostal nerves
    (B) the central portion of diaphragmatic pleura is supplied by the phrenic nerve
    (C) peripheral diaphragmatic pleura is supplied by intercostal nerves
    (D) the pain of pleurisy is mediated by autonomic nerves
    (E) mediastinal pleura is supplied by the phrenic nerve

16. Which of these items is true about inflammation of pleura?

    (A) usually lung infections do not involve the pleura
    (B) formation of pleural exudate may minimize mechanical irritation
    (C) inflammation affects motor nerves to the pleura
    (D) pleural pain is caused by irritation of autonomic nerves
    (E) usually pleural inflammation does not result in formation of pleural adhesions

17. Two thirds of the increase in thoracic capacity in respiratory movements is the result of

    (A) pump-handle movement of ribs
    (B) bucket-handle movement of ribs
    (C) diaphragmatic movement
    (D) accessory respiratory muscle contraction
    (E) elastic recoil of lungs and rib cage

18. All the following statements about the lungs are true EXCEPT

    (A) squamous epithelium forms alveoli
    (B) alveoli account for the greatest volume of the lung
    (C) alveoli lose all their air during expiration
    (D) lungs filled with air sound hollow to percussion
    (D) bronchial circulation supplies connective tissue of the lung

19. Which of the following statements is correct in relation to internal anatomy of the lung?

    (A) bronchopulmonary segments are the anatomic units
    (B) arteries and veins provide the major lung framework
    (C) bronchi branch symmetrically
    (D) bronchi are hollow tubes without particular wall support
    (E) pulmonary vessels show no particular relationship to bronchial branching

20. Which of these items is true regarding external anatomy of the lung?

   (A) the upper tapered end of the lung is its base
   (B) the root of the lung is located at its base
   (C) visceral pleura covers all lung surfaces
   (D) lobes are comparable to bronchopulmonary segments
   (E) each lung has three lobes

21. Which of these statements is correct regarding pulmonary circulation of the lung?

   (A) it is the main blood supply to the bronchi
   (B) it is the main blood supply to the connective tissue of the lung
   (C) the pulmonary trunk goes directly to the left lung
   (D) pulmonary veins enter the right atrium of the heart
   (E) the pulmonary trunk arises from the right ventricle

22. The ligamentum arteriosum is located correctly by which of the following?

   (A) between the left pulmonary artery and the aortic arch
   (B) between the ductus arteriosus and the right pulmonary artery
   (C) between the left pulmonary vein and the aorta
   (D) between the right pulmonary vein and the pulmonary trunk
   (E) between the left bronchial artery and the aortic arch

23. All the following are correct statements about pulmonary veins EXCEPT

   (A) two veins pass from the hilum of each lung
   (B) usually they enter the right atrium of the heart
   (C) they show more variation than do the pulmonary arteries
   (D) they are formed by confluence of capillaries in the lung
   (E) their primary tributaries are related to particular bronchopulmonary segments

24. The bronchial arteries may arise from all the following EXCEPT the

   (A) descending aorta
   (B) right intercostal artery
   (C) arch of the aorta
   (D) subclavian artery
   (E) anterior thoracic artery

25. Which of the following statements describes nerve supply of the lungs correctly?

   (A) sympathetic fibers in the pulmonary plexus are preganglionic fibers
   (B) vagal fibers in the pulmonary plexus are postganglionic fibers
   (C) visceral afferents from the lung have been demonstrated only in the vagus nerve
   (D) the vagus innervates the smooth muscle in walls of pulmonary vessels
   (E) sympathetic fibers control constriction of the bronchi

26. Establishment of internal gross anatomy of the lung takes place during which of these developmental periods?

   (A) primitive
   (B) canalicular
   (C) pseudoglandular
   (D) neonatal
   (E) terminal sac

27. The segment of the primitive heart tube giving rise to the aortic sac is the

   (A) bulbis cordis
   (B) primitive ventricle
   (C) truncus arteriosus
   (D) primitive atrium
   (E) sinus venosus

28. All of the following statements about the pericardial sac are true EXCEPT

   (A) the outer layer is fibrous
   (B) epicardium completely invests the heart
   (C) the pericardial sac and its content comprise the middle mediastinum
   (D) the fibrous pericardium lubricates the moving surfaces of the heart
   (E) the pericardial sac is fused to the central tendon of the diaphragm

29. Which of these items correctly describes the heart?

   (A) all the great veins enter its apex
   (B) its base is made largely of the left atrium and a portion of the right atrium
   (C) the apex points forward and toward the right
   (D) the diaphragmatic surface is formed largely by the right ventricle and atrium
   (E) the coronary sinus occupies the posterior interventricular sulcus

30. The right atrium includes all of these structures EXCEPT the

   (A) tricuspid valve
   (B) crista terminalis
   (C) musculi pectinati
   (D) fossa ovalis
   (E) trabeculae carneae

31. Which of these statements correctly describes the azygos venous system?

    (A) primarily it drains blood from the body wall
    (B) normally it drains into the inferior vena cava
    (C) it is located entirely on the right side of the vertebral column
    (D) it receives no blood from thoracic viscera
    (E) it has a number of valves

32. All of these items correctly describe the thoracic duct EXCEPT

    (A) it returns lymph from the greater part of the body to the venous system
    (B) it is the upward continuation of the cisterna chyli
    (C) it ends at the confluence of the right subclavian and brachiocephalic veins
    (D) in most of its course it lies behind the esophagus
    (E) it contains valves

33. Which of the following statements is correct in relation to thoracic splanchnic nerves?

    (A) their fibers relay in the sympathetic ganglion chain
    (B) they are part of the cardiac plexus
    (C) usually they are five in number
    (D) they consist of parasympathetic nerve fibers
    (E) they are composed predominantly of preganglionic fibers

34. Characteristics of thoracic vertebrae include all of the following EXCEPT

    (A) long vertical spinous processes of T5, T6, T7, and T8
    (B) a transverse foramen in each vertebra
    (C) a small circular vertebral foramen
    (D) progressively shorter transverse processes from T10–12
    (E) thoracic articular processes set on an arc to permit rotation

35. Correct description of structure of the sternum includes which of the following statements?

    (A) the jugular notch is located at the lower border of the manubrium
    (B) the upper border of the body is located at the level of the costal cartilage of the second rib
    (C) the sternal angle is located at the articulation of the body and the xiphoid process
    (D) the body consists of five fused sternebrae
    (E) the xiphoid process consists of bone thicker than that of the body

36. Ribs may be described correctly by all the following EXCEPT

    (A) every rib articulates with the vertebral column
    (B) the upper seven pairs of ribs are called vertebrosternal
    (C) ribs 8, 9, and 10 are called vertebrochondral ribs
    (D) floating ribs are the last two pairs
    (E) ribs one through 12 are called true ribs

37. Which of these facts about ribs is NOT true?

    (A) the typical rib takes an upward slope
    (B) the sternal end of each arch lies at a lower level than the vertebral end
    (C) ribs and cartilages increase in length progressively from first to seventh rib
    (D) the transverse diameter of the thorax increases progressively from first to eighth rib
    (E) the ninth is the most obliquely placed rib

38. All of these statements about intercostal arteries are correct EXCEPT

    (A) the upper two posterior intercostal arteries arise from the supreme intercostal artery
    (B) the lower nine posterior intercostal arteries arise from the aorta
    (C) the superior epigastric artery supplies anterior intercostal arteries
    (D) intercostal arteries run under the shelter of a costal groove
    (E) intercostal arteries accompany each intercostal nerve

39. Which of the following structures is NOT located in the mediastinum?

    (A) heart and pericardium
    (B) trachea
    (C) vessels proceeding to and from the heart
    (D) lungs
    (E) vagus nerves

40. All of the following are parts of the parietal pleura EXCEPT

    (A) costal
    (B) mediastinal
    (C) diaphragmatic
    (D) pulmonary
    (E) cervical

41. If the root of the left lung was completely severed, which of these structures would be spared?

    (A) pulmonary ligament
    (B) pulmonary veins
    (C) vagus nerve
    (D) pulmonary artery
    (E) bronchus

42. All of these statements describe the aortic arch correctly EXCEPT

   (A) it lies in the superior mediastinum
   (B) it begins at the level of the sternal angle
   (C) it gives off the left common carotid artery
   (D) it gives off the right subclavian artery
   (E) it gives off the right and left coronary arteries

43. Correct description of the recurrent laryngeal nerves includes which of the following?

   (A) they are branches of the phrenic nerve
   (B) they branch from the sympathetic trunk
   (C) their neuronal cell bodies are in the cervical spinal cord
   (D) the left nerve recurs around the ligamentum arteriosum
   (E) the right nerve recurs around the superior vena cava

44. Which of these statements does NOT describe the sympathetic trunk correctly?

   (A) ganglia are limited to those for the 12 thoracic nerves
   (B) the trunk lies a little wide of the mediastinum
   (C) intercostal arteries and veins cross the trunk posteriorly
   (D) inferior cervical and first thoracic ganglia form the stellate ganglion
   (E) postganglionic fibers pass from the upper five thoracic ganglia to the cardiac plexus

45. Each segmental bronchus together with the portion of lung it supplies is called

   (A) primary segment
   (B) bronchopulmonary segment
   (C) lobar segment
   (D) epiarterial segment
   (E) alveolar segment

46. Correct information about bronchial arteries includes all the following items EXCEPT

   (A) they arise by a stem from the aorta
   (B) they supply the pulmonary pleura
   (C) they supply the bronchi
   (D) the pressure within them is low
   (E) they run through the interlobar structures

47. Nerves of the lungs and pleura include all of these EXCEPT

   (A) branches of the vagus contribute to the pulmonary plexus
   (B) branches of the thoracic sympathetic ganglia 1 through 5 help to form the pulmonary plexus
   (C) sensory vagal fibers constitute the afferent limb of the respiratory reflex arc
   (D) efferent vagal fibers are secretomotor
   (E) visceral pleura has many afferent nerves sensitive to mechanical stimulation

48. The tracheal bifurcation can be seen at which of these levels?

   (A) T4–5 in the supine living subject
   (B) T8 in the erect subject
   (C) T6 during inspiration
   (D) T12 during expiration
   (E) T2–3 in the supine cadaver

49. If the phrenic nerve was cut close to its origin, which of these effects could be seen?

   (A) loss of bronchoconstriction
   (B) loss of sensation in the middle diaphragm
   (C) loss of power in intercostal muscles
   (D) difficulty in expiration
   (E) loss of the respiratory reflex arc

50. The brachiocephalic veins are formed from the union of the

   (A) external jugular and inferior vena cava
   (B) ductus arteriosum and superior vena cava
   (C) azygos vein and the axillary vein
   (D) internal jugular and subclavian
   (E) pulmonary vein and inferior vena cava

51. All the following structures empty into the right atrium of the heart EXCEPT the

   (A) coronary sinus
   (B) inferior vena cava
   (C) superior vena cava
   (D) anterior cardiac veins
   (E) pulmonary veins

52. Which of these items is an exception to the general typical arrangement of costovertebral joints?

   (A) costovertebral articulations consist of synovial joints
   (B) the head of the rib articulates with its own vertebral body
   (C) an articular capsule surrounds each joint
   (D) the tubercle of a rib articulates with the tip of a transverse process
   (E) the head of the rib articulates with an intervertebral disc

53. In the adult, which of the following structures is NOT prominent in the anterior mediastinum?

   (A) loose areolar tissue
   (B) lymph vessels and nodes
   (C) fat
   (D) thymus gland
   (E) sternopericardial ligaments

**54.** The thoracic aorta gives off all of these branches EXCEPT

(A) the first pair of posterior intercostal arteries
(B) bronchial arteries
(C) pericardial arteries
(D) diaphragmatic arteries
(E) one pair of subcostal arteries

**55.** Which of these statements correctly describes the position of the esophagus?

(A) it passes anterior to the left principal bronchus
(B) it begins at the level of the cricoid cartilage
(C) it descends on the left of the aortic arch
(D) it runs in the anterior mediastinum
(E) it passes through the hiatus formed by the median arcuate ligament of the diaphragm

**56.** Which of the following statements is true of the trachea?

(A) it descends behind the esophagus
(B) its posterior surface is convex
(C) it ends at the level of the sternal angle
(D) during inspiration, its bifurcation ascends
(E) it contains O-shaped bars of cartilage

**57.** All of the following statements correctly describe the left vagus nerve EXCEPT

(A) it gives off the left recurrent laryngeal nerve
(B) posterior to the left bronchus, it breaks up into the left pulmonary plexus
(C) it supplies the left side of the diaphragm
(D) it forms part of the esophageal plexus
(E) it descends from the neck posterior to the left common carotid artery

**58.** Which of the following correctly describes the aortic arch?

(A) it begins posterior to the sternal angle
(B) it passes to the right of the trachea
(C) it receives the ligamentum arteriosum from the right pulmonary artery
(D) it gives off the right subclavian artery
(E) it gives off the left brachiocephalic trunk

**59.** The superior vena cava returns blood from all of these structures EXCEPT the

(A) head
(B) neck
(C) upper limb
(D) lungs
(E) thoracic wall

**60.** All of the following statements correctly describe the brachiocephalic veins EXCEPT

(A) each is formed by the union of the internal jugular and the subclavian veins
(B) they contain valves to prevent backflow of blood
(C) they unite to form the superior vena cava
(D) each vein receives the internal thoracic vein
(E) they arise posterior to the medial ends of the clavicle

**61.** All of these structures occupy the superior mediastinum EXCEPT the

(A) heart and pericardium
(B) thymus
(C) aortic arch
(D) trachea
(E) esophagus

**62.** Which of the following statements is true regarding the coronary arteries?

(A) sharp lines of demarcation exist between their distribution to right and left ventricles
(B) most of the blood in these arteries returns to the left atrium
(C) they arise from the right and left aortic sinuses
(D) variations of these arteries are uncommon
(E) they are infrequent sites of arteriosclerosis

**63.** Which of these statements correctly describes sensation of the heart?

(A) referred pain from the heart rarely occurs
(B) heart ischemia rarely results in generation of painful stimuli
(C) afferent pain fibers from the heart run in the vagus nerve
(D) usually pain from the heart is felt only in the left chest
(E) the heart is insensitive to cold, heat, and touch

**64.** All of the following correctly describe innervation of the heart EXCEPT

(A) sympathetics increase rate of heart beat
(B) parasympathetics reduce rate and force of heart beat
(C) the impulse-conducting system of the heart is controlled by the autonomic nervous system
(D) the parasympathetic cardiac nerves supply three branches to each side
(E) vagal stimulation dilates coronary arteries

**65.** Initiation of the impulse for contraction of the heart is accomplished by the

(A) atrioventricular node (AV node)
(B) atrioventricular bundle (AV bundle)

(C) sinoatrial node (SA node)
(D) sympathetics
(E) bundle of His

66. Results of an abnormal aortic valve may include all the following EXCEPT
(A) hypertrophy of the right ventricle
(B) production of a heart murmur
(C) aortic regurgitation
(D) collapsing pulse
(E) valvular incompetence

67. Which of the following is NOT a characteristic of the left ventricle?
(A) its wall is much thicker than that of the right ventricle
(B) its interior is covered by trabeculae carneae
(C) the chordae tendinae of papillary muscles are distributed to cusps of the atrioventricular valve
(D) it forms the base of the heart
(E) the aorta arises from its anterior uppermost part

68. The heart valve most frequently diseased is the
(A) aortic
(B) mitral
(C) pulmonary
(D) tricuspid
(E) coronary sinus valve

69. Characteristics of the left atrium consist of all the following EXCEPT
(A) it forms most of the base of the heart
(B) it contains a few musculi pectinati
(C) it receives the pulmonary arteries
(D) much of this atrium lies posterior to the right atrium
(E) the auricle overlaps the root of the pulmonary trunk

70. Which of the following is characteristic of the right ventricle?
(A) it gives origin to the pulmonary trunk
(B) usually it has only two papillary muscles
(C) it receives blood through the mitral valve
(D) it has internal muscular ridges, the musculi pectinati
(E) it contains the fossa ovalis

71. Which of the following correctly describes chambers of the heart?
(A) the coronary sulcus separates the two ventricles
(B) the right ventricle forms the right border of the heart

(C) the valve of the superior vena cava directs blood downward
(D) the superior vena cava opens into the right atrium
(E) the interventricular septum contains the fossa ovalis

72. The heart may correctly be described by all the following EXCEPT
(A) an apex formed by the tip of the left ventricle
(B) a diaphragmatic surface formed by both ventricles
(C) a base formed by the atria
(D) a location in the middle mediastinum
(E) an anterior surface formed mainly by the left atrium

73. The epicardium receives its arterial blood supply from the
(A) pericardiophrenic
(B) coronary arteries
(C) musculophrenic
(D) superior phrenic
(E) bronchial

74. Which of these statements is true of the fibrous pericardium?
(A) it has no close relationship with the central tendon of the diaphragm
(B) it extends upward to the level of the sternal angle
(C) it moves freely within the thoracic cavity
(D) its base is pierced by the aorta
(E) it has no attachment to the sternum

75. Which of the following structures is NOT located in the mediastinum?
(A) heart
(B) thymus
(C) vertebral bodies
(D) great vessels
(E) vagus nerve

76. Which of these statements correctly describes lymphatic drainage of the lungs?
(A) usually both bronchomediastinal lymph trunks terminate in the thoracic duct
(B) only one lymphatic plexus is involved in this drainage
(C) little transfer of lymph drainage from side to side occurs
(D) no lymph vessels are located in the walls of the pulmonary alveoli
(E) rarely is this lymph drainage responsible for transfer of cancer cells to other organs

77. In what way does the root of the right lung differ from that of the left lung?

    (A) in numbers of pulmonary arteries
    (B) in numbers of pulmonary veins
    (C) in numbers of primary bronchi
    (D) in the presence of a pulmonary plexus of nerves
    (E) in numbers of bronchial veins

78. Characteristics of the left lung include which of the following?

    (A) it is heavier than the right lung
    (B) it is composed of three lobes
    (C) the azygos vein arches over its root
    (D) it has a horizontal fissure
    (E) the cardiac notch is found on its superior lobe

79. The central part of the parietal diaphragmatic pleura is supplied by which of these nerves?

    (A) intercostals
    (B) vagus
    (C) phrenic
    (D) sympathetics
    (E) parasympathetics

80. Which of the following items correctly describes the visceral pleura?

    (A) it receives its arterial supply from pulmonary arteries
    (B) its lymphatic vessels drain directly into the thoracic duct
    (C) its nerves accompany the pulmonary veins
    (D) it receives its nerve supply from autonomic nerves to the lung
    (E) it is intensely sensitive to pain

81. Which of the following statements correctly describes the pleural cavities?

    (A) they contain much lubricating fluid
    (B) each pleural cavity is a closed potential space
    (C) visceral and parietal pleura never become continuous
    (D) costal pleura covers the apex of the lung
    (E) at no point does the pleura extend below the costal margin

82. Which of these statements does NOT describe dermatomes correctly?

    (A) there is practically no overlapping of contiguous dermatomes
    (B) dermatomes are areas of skin supplied by the sensory root of a spinal nerve
    (C) areas of touch are more extensive than those for temperature
    (D) T10 nerve supplies the area of the umbilicus
    (E) The lower six intercostal nerves supply both thoracic and abdominal walls

**DIRECTIONS (Questions 83 through 102):** Each group of items in this section consists of lettered headings followed by a set of numbered words or phrases. For each numbered word or phrase, select the ONE lettered heading that is most closely associated with it. Each lettered heading may be selected once, more than once, or not at all.

**Questions 83 through 86**

    (A) cervical vertebrae
    (B) thoracic vertebrae
    (C) coccyx
    (D) sacrum
    (E) lumbar vertebrae

83. costal facets

84. wedge-shaped structure

85. foramen transversarium

86. large, kidney-shaped bodies

**Questions 87 through 90**

    (A) head of the rib
    (B) neck of the rib
    (C) body of the rib
    (D) tubercle of the rib
    (E) scalene tubercle

87. costal groove

88. two articular facets

89. located between head and tubercle

90. angle

**Questions 91 through 94**

    (A) sternum
    (B) costal margin
    (C) vertebral articulation with rib
    (D) jugular notch
    (E) cervical rib

91. synovial joint

92. upper seven costal cartilages

93. neurovascular symptoms

94. seventh through tenth costal cartilages

**Questions 95 through 98**

(A) increase in vertical diameter of the thorax
(B) increase in transverse diameter of the thorax
(C) increase in anteroposterior diameter of the thorax
(D) increase in pulmonary compression
(E) expiration

95. elastic recoil

96. diaphragm

97. pump-handle movement

98. bucket-handle movement

**Questions 99 through 102**

(A) fixes the arm in respiration
(B) inspiratory muscles
(C) keeps intercostal spaces rigid
(D) depresses ribs
(E) stabilizes scapula

99. intercostal muscles

100. pectoralis minor

101. serratus posterior muscles

102. subcostal muscles

**DIRECTIONS (Questions 103 through 122): Each group of items in this section consists of lettered headings followed by a set of numbered words or phrases. For each numbered word or phrase, select**

A if the item is associated with (A) only,
B if the item is associated with (B) only,
C if the item is associated with both (A) and (B),
D if the item is associated with neither (A) nor (B).

**Questions 103 through 106**

(A) thoracic aorta
(B) internal thoracic artery
(C) both
(D) neither

103. intercostal arteries for the first two intercostal spaces

104. intercostal arteries for the third through the sixth intercostal spaces

105. intercostal arteries for the seventh through the ninth intercostal spaces

106. intercostal arteries for the lower two intercostal spaces

**Questions 107 through 110**

(A) right lung
(B) left lung
(C) both
(D) neither

107. horizontal fissure

108. pulmonary ligament

109. pulmonary trunk

110. groove for the arch of the azygos vein

**Questions 111 through 114**

(A) right ventricle of the heart
(B) left ventricle of the heart
(C) both
(D) neither

111. musculi pectinati

112. coronary sinus

113. papillary muscles

114. pulmonary veins

**Questions 115 through 118**

(A) interatrial septum of the heart
(B) interventricular septum of the heart
(C) both
(D) neither

115. atrioventricular (AV) bundle

116. sinoatrial (SA) node

117. foramen ovale

118. AV node

**Questions 119 through 122**

(A) sympathetic nerves
(B) parasympathetic nerves
(C) both
(D) neither

119. constricts coronary arteries

120. carries afferent pain fibers from the heart

121.  slows the heart rate

122.  initiates heart beat

DIRECTIONS (Questions 123 through 173): For each of the items in this section, ONE or MORE of the numbered options is correct. Choose the answer

A if only 1, 2, and 3 are correct,
B if only 1 and 3 are correct,
C if only 2 and 4 are correct,
D if only 4 is correct,
E if all are correct.

123.  Correct description of the coronary arteries includes which of the following?

(1) blood in these vessels returns to the heart via the coronary sinus
(2) they arise from the right and left aortic sinuses
(3) the right artery gives off the atrioventricular (AV) nodal artery
(4) the circumflex branch supplies the right atrium

124.  Which of these statements is true regarding the veins of the heart?

(1) the middle cardiac vein runs in the coronary sulcus
(2) the coronary sinus is the main vein of the heart
(3) the anterior cardiac veins end in the left atrium
(4) the great cardiac vein ascends in the anterior interventricular sulcus

125.  Correct description of the thymus includes which of the following?

(1) location in the middle mediastinum
(2) a blood supply from the aorta
(3) a structure of one lobe
(4) involution by adulthood

126.  Which of the following is characteristic of thoracic vertebrae?

(1) heart-shaped bodies
(2) long, vertical, spinous processes for some
(3) progressively shorter transverse processes from the first to the twelfth vertebra
(4) triangular-shaped vertebral foramen

127.  Which of the following structures is closely related to costovertebral joints?

(1) intervertebral discs
(2) intra-articular ligaments
(3) costotransverse ligament
(4) tubercle of the rib

128.  Contents of an intercostal space include the

(1) anterior ramus of a spinal nerve
(2) internal intercostal membrane
(3) posterior intercostal artery
(4) posterior ramus of the spinal nerve

129.  Parts of the parietal pleura include the

(1) pulmonary pleura
(2) costal pleura
(3) visceral pleura
(4) mediastinal pleura

130.  The right aspect of the mediastinum displays which of the following structures?

(1) pericardial sac
(2) inferior vena cava
(3) vagus nerve
(4) jugular vein

131.  Structures found in the root of the lung include the

(1) pulmonary artery
(2) two pulmonary veins
(3) bronchus
(4) nerve plexuses

132.  The left vagus nerve can be described correctly by which of the following?

(1) it traverses the neck in the carotid sheath
(2) it accompanies the pericardiophrenic artery
(3) it gives off a recurrent laryngeal nerve
(4) it arises from the ventral ramus of spinal nerves C3, C4, and C5

133.  Characteristics of the sympathetic trunk include which of the following?

(1) it is found only in the midmediastinum
(2) intercostal arteries and veins cross the trunk anteriorly
(3) the cardiac plexus receives preganglionic fibers from the middle six thoracic ganglia
(4) typically a ganglion is found in front of each rib

134.  Splanchnic nerves are components of the sympathetic nervous system that

(1) consist of postganglionic fibers
(2) include a greater splanchnic nerve springing from the eighth, ninth, and tenth sympathetic ganglia
(3) include a lesser splanchnic nerve springing from the 12th sympathetic ganglion
(4) end in the celiac and renal ganglia

135.  The lungs can be described correctly by which of these statements?

(1) their apex rises to the neck of the first rib
(2) the cardiac impression is deeper on the left side than on the right
(3) a lobe of the azygos vein may be found on the right lung
(4) usually the left side shows ten segmental bronchi

136. Vessels of the lung include which of the following?

(1) two pulmonary veins on each side
(2) pulmonary arteries following the bronchial pattern
(3) bronchial arteries arising by a stem from the aorta
(4) lymph channels accompanying small blood vessels

137. Nerves of the lungs and pleura can be described correctly by which of the following statements?

(1) sensory vagal fibers constitute the afferent limb of the respiratory reflex arc
(2) branches of thoracic sympathetic ganglia 5 through 10 form the pulmonary plexus
(3) efferent sympathetic fibers are bronchodilators
(4) the visceral pleura is extremely sensitive to mechanical stimulation

138. A vertical line drawn from the neck to the abdomen less than finger breadth from the right margin of the sternum represents the right border of which of these structures?

(1) internal jugular vein
(2) superior vena cava
(3) inferior vena cava
(4) right ventricle

139. A continuation of the right ventricle of the heart is the

(1) ascending aorta
(2) right pulmonary artery
(3) transverse pericardial sinus
(4) pulmonary trunk

140. The pattern of coronary arteries can be described correctly by which of the following statements?

(1) the anterior interventricular artery is a branch of the left coronary artery
(2) the circumflex artery is a branch of the right coronary artery
(3) the posterior interventricular artery is a branch of the right coronary artery
(4) the transverse artery is a branch of the left coronary artery

141. In development of the heart, axial rotation results in which of the following?

(1) the right ventricle forming the right border of the heart
(2) the left ventricle becoming largely posterior
(3) the interatrial septum facing backward and to the left
(4) the left atrium being conspicuous posteriorly

142. Which of these statements correctly describes the right atrium of the heart?

(1) it develops from a merging of the right half of the sinus venarium and the primitive atrium
(2) the right atrium forms the right margin of the heart
(3) it contains musculi pectinati
(4) it receives the right and left pulmonary veins

143. Ventricles of the heart are described correctly by which of the following?

(1) they lie in front of the atria
(2) atrioventricular entrances are posterior
(3) their walls are lined with trabeculae carneae
(4) they contain papillary muscles

144. Which of the following statements is true in relation to the impulse-conducting system of the heart?

(1) the sinoatrial (SA) node is a muscular connection between the musculature of the atria and ventricles
(2) the SA node initiates the heart beat
(3) the atrioventricular bundle is composed of autonomic nerve fibers
(4) Atrioventricular (AV) node is situated in the interatrial septum

145. Sympathetic nerves of the heart conduct

(1) cardioaccelerator fibers
(2) vasodilator fibers
(3) sensory fibers
(4) fibers initiating heart beat

146. Which of the following items correctly describes the cardiac plexus?

(1) it contains only vagal fibers
(2) the vagal fibers are postganglionic when reaching the plexus
(3) the plexus is situated above the arch of the aorta
(4) its extensions are the right and left pulmonary plexuses

| | SUMMARY OF DIRECTIONS | | | |
|---|---|---|---|---|
| **A** | **B** | **C** | **D** | **E** |
| 1, 2, 3 only | 1, 3 only | 2, 4 only | 4 only | All are correct |

147. The bronchi have which of these characteristics?

    (1) they provide the framework of the lung parenchyma
    (2) normally they develop symmetrically
    (3) they contain incomplete rings of hyaline cartilage
    (4) they have little relationship to the pattern of arteries

148. The principal bronchi branching from the trachea demonstrate which of these characteristics?

    (1) a carina at their bifurcation
    (2) a right bronchus wider than the left
    (3) a left bronchus longer than the right
    (4) more foreign bodies being inhaled into the right bronchus than into the left

149. Which of the following statements accurately describes the blood vessels of the lung?

    (1) bronchial arteries are larger than pulmonary arteries
    (2) bronchial arteries carry blood to the connective tissue of the lung
    (3) no anastomoses are seen between the bronchial and pulmonary systems
    (4) the pulmonary veins return oxygenated blood to the left atrium

150. Nerve supply to the lung consists of which of the following?

    (1) visceral efferents by the vagus nerve
    (2) visceral efferents by thoracic sympathetics
    (3) visceral afferents by the vagus
    (4) afferents concerned with innervation of the bronchial mucosa

151. The pericardial sac is described correctly by which of these statements?

    (1) its base rests on the diaphragm
    (2) it has no attachment to the sternum
    (3) it and its contents comprise the middle mediastinum
    (4) it is stretched during ascent of the diaphragm in inspiration

152. After the heart has been removed, which severed connections of structures can be seen on the posterior wall of the pericardial sac?

    (1) superior vena cava
    (2) ascending aorta
    (3) right pulmonary veins
    (4) pulmonary trunk

153. Pericardial pain is mediated by which of these nerves?

    (1) sympathetic fibers
    (2) somatic afferents
    (3) vagus nerves
    (4) phrenic nerves

154. Correct relationships of surfaces of the heart include which of these?

    (1) a diaphragmatic surface formed largely by the right atrium
    (2) a sternocostal surface dominated by the left ventricle
    (3) a coronary sulcus separating the left atrium from the ventricle
    (4) the conus arteriosus continuing into the pulmonary trunk

155. The right border of the heart consists of the

    (1) superior vena cava
    (2) right atrium
    (3) inferior vena cava
    (4) pulmonary trunk

156. Correct description of the right atrium of the heart includes which of these?

    (1) it has an opening for the azygos vein
    (2) the interior is divided partially by the crista terminalis
    (3) the right atrioventricular ostium is guarded by the mitral valve
    (4) the coronary sinus opens into its lower, posterior corner

157. The left ventricle of the heart contains which of these structures?

    (1) trabeculae carneae
    (2) valvule of the foramen ovale
    (3) aortic vestibule
    (4) pulmonary valve

158. The great vessels include which of the following?

    (1) ascending aorta
    (2) pulmonary trunk
    (3) two venae cavae
    (4) four pulmonary veins

159. Which of these statements correctly describes audible heart sounds?

    (1) they are associated closely with opening of the valves

(2) audible areas for the sounds coincide with surface projections of each valve

(3) the first heart sound signals the beginning of diastole

(4) heart sounds of the tricuspid valve can be heard best over the anterior wall of the right ventricle

**160.** In congenital cardiac abnormalities, obstruction of blood flow to the lungs is found in

(1) tetralogy of Fallot
(2) pulmonary atresia
(3) tricuspid atresia
(4) anomalous pulmonary veins

**161.** Structural relationships of the trachea include which of the following?

(1) its bifurcation in the lower superior mediastinum
(2) its being crossed by the azygos vein above the root of the right lung
(3) its position behind the esophagus
(4) its being crossed by the arch of the aorta above the root of the left lung

**162.** Direct branches or connections of the arch of the aorta include which of these?

(1) right subclavian artery
(2) right common carotid artery
(3) inferior thyroid artery
(4) ligamentum arteriosum

**163.** Correct descriptions of the brachiocephalic veins include which of the following?

(1) they are formed by union of the subclavian and internal jugular vein
(2) the left vein runs in a vertical direction
(3) the left vein is considerably longer than the right vein
(4) confluence of the right and left veins forms the inferior vena cava

**164.** Nerves that enter the superior thoracic aperture consist of which of these?

(1) vagus nerves
(2) sympathetic trunks
(3) cardiac nerves
(4) phrenic nerves

**165.** Which of these items is correct regarding the phrenic nerve?

(1) they are accompanied by posterior intercostal nerves
(2) they carry motor fibers to the diaphragm
(3) they give off recurrent laryngeal nerves
(4) they convey pain sensation from the fibrous pericardium

**166.** Which of the following locations in the esophagus is a probable site for lodging of foreign bodies?

(1) the region of its contact with the aortic arch
(2) where it is crossed by the left bronchus
(3) the esophageal hiatus of the diaphragm
(4) the lower esophageal sphincter

**167.** Innervation of the muscle of the esophagus functions in which of the following ways?

(1) it causes spontaneous peristaltic contractions
(2) it causes contraction initiated by swallowing
(3) it governs contraction of the striated muscle through visceral efferent fibers of the sympathetics
(4) it governs contraction of the smooth muscle through visceral efferent fibers of the vagus

**168.** Which of these arteries is a branch of the descending thoracic aorta?

(1) bronchial
(2) coronary
(3) esophageal
(4) brachiocephalic

**169.** Correct description of the azygos venous system would include which of the following?

(1) its primary function is to drain blood from the body wall
(2) normally it drains upward into the superior vena cava
(3) the azygos vein arches over the root of the right lung
(4) the vessels of the two sides are asymmetrical

**170.** Which of these statements is true of the thoracic duct?

(1) it empties into the right subclavian vein
(2) it is the upward continuation of the cisterna chyli
(3) it enters the thorax through the esophageal hiatus of the diaphragm
(4) in most of its course it lies behind the esophagus

**171.** Aspects of the sympathetic nervous system can be described correctly by which of the following?

(1) all preganglionic fibers synapse in the sympathetic trunk
(2) Sympathetic trunks are present only in the thoracic region
(3) preganglionic neurons are located at all levels of the spinal cord
(4) postganglionic fibers from the sympathetic trunk enter the spinal nerves via gray rami communicants

172. Mediastinal groups of lymph nodes include which of the following?

    (1) parasternal
    (2) tracheobronchial
    (3) intercostal
    (4) diaphragmatic

173. If the vagus nerve was cut just before it enters the thorax, which of these functions would be impaired?

    (1) diaphragm contraction
    (2) secretion of bronchial glands
    (3) vasoconstriction of pulmonary blood vessels
    (4) bronchoconstriction

# Answers and Explanations

1. **(B)** The mediastinum is a broad median septum that partitions the thoracic cavity and separates the two hollow spaces filled by the right and left lungs. The bulk of the mediastinum is made up of the heart, enclosed in the pericardial sac, and of the major pulmonary and systemic arteries and veins. The trachea and esophagus enter the mediastinum from the neck through the superior thoracic aperture. *(Hollinshead, p 465)*

2. **(D)** The walls of the thoracic cavity are made up of layers of muscle and fascia supported by the thoracic skeleton. The thoracic cage supports the bones and muscles of the pectoral girdle. Much of the thoracic skeleton is covered by muscles that belong to the upper limb or vertebral column (e.g. pectoral muscles, serratus anterior, trapezius, and latissimus dorsi). These muscles, supplied by nerves other than those serving the true body wall, anatomically are not considered as components of the thoracic wall. Intercostal muscles are layers of the thoracic wall proper. *(Hollinshead, pp 466–467)*

3. **(C)** The superior thoracic aperture slopes downward and forward. This aperture is small, and its boundaries are the first thoracic vertebra, the first pair of ribs with their cartilages, and the superior margin of the manubrium of the sternum. The costal arch, made up of costal cartilages 7 through 10, forms part of the inferior thoracic aperture, not the superior aperture. *(Hollinshead, p 468)*

4. **(E)** The joint between the manubrium and body of the sternum is palpable as the sternal angle; this is a smooth ridge produced by the angulation of the bones and their slightly everted articular edges. The sternal angle lies at the level of the second costal cartilage and the lower border of the fourth thoracic vertebra. This angle serves as a reference point for counting the ribs, as it identifies the second rib. *(Hollinshead, pp 469,472)*

5. **(E)** The sternum ossifies from a number of centers that form independently of the ribs. Usually there is a single center for the manubrium and a single one for the uppermost part of the body. Each of the remaining three segments that contribute to the body may have a single or paired center of ossification. Fusion of the parts of the body begins below and extends upward, being completed at the 21st year. *(Hollinshead, p 469)*

6. **(A)** There are 12 pairs of ribs of which the upper seven are called true ribs because they form complete arches between the vertebrae and the sternum. The lower five ribs, which fail to reach the sternum, are regarded as false ribs. Ten of the 12 pairs of ribs form arches between respective vertebrae and the sternum, whereas the last two pairs of ribs float freely anteriorly (floating ribs). *(Hollinshead, pp 467–469,471)*

7. **(C)** Although ribs may fracture at any point under direct violence, the most frequent fractures are due to compression forces on the thorax. These forces occur most frequently just anterior to the costal angle, the weakest part of the rib. Usually rib fractures heal without any splinting. The broken rib ends tend to spring outward; however, a direct force may drive them inward, causing hemorrhage or injury to the lung. *(Hollinshead, p 471)*

8. **(B)** Vertical increase in the diameter of the thoracic cavity is the result of movement of the diaphragm. The anteroposterior and lateral increase depends on movement of the ribs. All of the items listed, except B, relate to thoracic cage increased measurement caused by movement of the ribs. *(Hollinshead, pp 472–474)*

9. **(D)** The fibers of internal intercostal muscles run upward and forward from one rib to the one above. The external intercostals begin posteriorly just lateral to the tubercle of the rib; anteriorly the muscle is represented in the upper intercostal spaces by the external intercostal membrane. Fibers of external intercostal muscles slant downward and forward from one rib to the next one below. Innermost intercostal muscles are less well developed than the

other two intercostal muscles. *(Hollinshead, pp 474–475)*

10. **(B)** All items are correct, except B, which states that the thoracic wall receives no innervation from cervical nerves. Anteriorly, as far down as the sternal angle, the skin is innervated by supraclavicular nerves, ventral rami of C4 and C5. Ventral rami of spinal nerves T1–12 supply the thoracic wall, those of T1–11 being called intercostal nerves and that of T12, the subcostal nerve. Over the paravertebral area of the back, the skin is supplied by dorsal rami of corresponding spinal nerves, which also innervate the erector spinae muscles. *(Hollinshead, p 476)*

11. **(E)** Intercostal nerves are examples of typical spinal nerves; the second through the sixth intercostals are confined to the thorax. The seventh to the 11th intercostals and subcostal nerve continue into the abdominal wall. The upper six intercostal nerves follow the curve of the ribs and costal cartilages toward the sternum and terminate as anterior cutaneous branches. The nerves run in the costal groove of the intercostal space between the internal and innermost intercostal muscles. *(Hollinshead, pp 477–478)*

12. **(C)** The thoracic wall is supplied chiefly by intercostal arteries and is drained by intercostal veins. These vessels are found in each intercostal space running in the costal groove just above the respective intercostal nerve. Their course and branches conform closely to those of the nerve. Superficial structures of the thorax are served by intercostal vessels and also by branches of the axillary and subclavian arteries and veins. Most anterior intercostal arteries are branches of the internal thoracic artery. Posterior intercostal arteries, of all but the first two intercostal spaces, are branches of the descending thoracic aorta. *(Hollinshead, pp 478–479)*

13. **(A)** The internal thoracic artery is given off by the subclavian artery at the base of the neck. This artery, like the intercostal vessels, runs in the neuromuscular plane of the thorax, that is, anterior to the transversus thoracis muscle and on the internal surface of the costal cartilage and the internal intercostal muscles. The other items listed are correct. *(Hollinshead, p 479)*

14. **(B)** The developing lungs are covered by a layer of mesothelium designated as the visceral or pulmonary pleura. As the lungs grow more and more into the pleural cavity, their pulmonary mesothelial surface contacts the parietal pleura, the outer wall of the sac that is draped over the thoracic wall and the fibrous pericardium. The costal pleura is the parietal pleura lining the thoracic wall. The mediastinal pleura is continuous anteriorly and posteriorly with the costal pleura. The diaphragmatic

pleura is an inferior reflection of the costal pleura over the diaphragm, thus forming the floor of the pleural and thoracic cavities. *(Hollinshead, pp 480–482)*

15. **(D)** All of the items are correct except D. The pain of pleurisy (pleural inflammation or irritation) is mediated along somatic afferent pathways of these nerves. By contrast, visceral pleura, being supplied by autonomic nerves of the lung, is insensitive to pain stimuli. *(Hollinshead, pp 484–485)*

16. **(B)** Inflammation resulting from infection, irritation, emboli, or neoplasms in segments of the lung causes hyperemia of the overlying pleura, which leads to exudate formation. Respiratory movements become painful when the parietal pleura primarily is involved because somatic pain afferents are excited. Sufficient exudate sooner or later separates the pleural layers and minimizes the mechanical irritation. Resolution of the inflammation often leaves adhesions between parietal and visceral pleura; usually these are painless. *(Hollinshead, p 485)*

17. **(C)** Ribs increase the anteroposterior and transverse diameters of the thorax by their pump-handle– and bucket-handle–type movements. It has been stated that two thirds of the increase in thoracic capacity is due to diaphragmatic movements. This is true even in those in whom costal breathing is quite prominent. In normal breathing expiration largely is passive, due to elastic recoil of lungs and rib cage and the relaxation of the diaphragm. *(Hollinshead, p 487)*

18. **(C)** Item C is incorrect because the alveoli are connected to the exterior by a branching system of tubes, the bronchial tree, and remain filled with air even during expiration. They account for the greatest volume of lung, by far. The lung has two circulations: the pulmonary and the bronchial. The latter is part of the systemic circulation and carries blood for nutrition of the bronchi and connective tissue of the lung. *(Hollinshead, pp 489,497)*

19. **(A)** The internal anatomy of the lungs conforms to a segmental pattern laid down during development by the budding of bronchi. The bronchopulmonary segments are the anatomic units of the lung. Clinical evaluation, as well as surgical resection, of diseased portions of the lungs, relies on the anatomy of bronchopulmonary segmentation. A bronchopulmonary segment is a pyramidal-shaped unit of the lung aerated by a segmental bronchus. Each segment is served principally by a segmental branch of the pulmonary artery and a segmental vein. The bronchopulmonary segments can be conceived of as the anatomic bronchovascular units of the lungs. *(Hollinshead, pp 490,493–494, and 506)*

**20.** **(C)** Each lung presents a tapered upper end, the apex, and a broad base. All lung surfaces are covered completely by visceral pleura that unites with mediastinal parietal pleura around the root of the lung. The root of the lung is a narrow pedicle that suspends the lung from the mediastinum in the pleural cavity and enters the lung at the pulmonary hilum. Although the right lung has three lobes and the left only two, the bronchopulmonary segments in right and left lungs correspond. *(Hollinshead, pp 490–492)*

**21.** **(E)** The lung has two circulations: the pulmonary and the bronchial. The pulmonary trunk arises from the right ventricle of the heart and divides into right and left pulmonary arteries. These arteries end in the capillary networks around the alveolar sacs of the lung. Venules arise from these capillaries and form the pulmonary veins, which enter the left atrium of the heart. The bronchial circulation, a part of the systemic circulation, carries blood to the bronchi and connective tissue of the lungs. *(Hollinshead, pp 497–499)*

**22.** **(A)** The left pulmonary artery is connected to the aortic arch by the ligamentum arteriosum, the fibrous remains of the ductus arteriosum. In the fetus this structure serves as a shunt between the two vessels. Following birth, the ductus closes and becomes the ligamentum arteriosum. *(Hollinshead, p 499)*

**23.** **(B)** The two pulmonary veins running from each lung enter the left, not the right, atrium of the heart. Their pattern shows more variation than that of the pulmonary arteries. Anomalous pulmonary veins may drain into systemic veins and prevent oxygenated blood from being delivered to the left side of the heart for circulation to the body. Such a condition also overloads the right side of the heart and, if widespread enough, may be incompatible with life. *(Hollinshead, p 500)*

**24.** **(E)** Descriptions of the anterior thoracic artery do not state that any bronchial arteries arise from it. All of the other listed sources of bronchial arteries are correct. Both the descending portion and arch of the aorta, along with the subclavian artery, provide sources of blood for the bronchi. *(Hollinshead, pp 479,500–501)*

**25.** **(C)** Each lung is supplied by visceral afferent and visceral efferent nerve fibers through the pulmonary plexus. Visceral efferents are contributed to the plexus by the vagus and thoracic sympathetic ganglia. The sympathetic fibers in the plexus are postganglionic, and the vagal fibers are preganglionic. Visceral afferents from the lung have been demonstrated only in the vagus. Apparently the vagus does not innervate smooth muscle in the walls of pulmonary vessels. The vagus, not the sympathetics, is responsible for constriction of the bronchi. *(Hollinshead, pp 501–503)*

**26.** **(C)** The development of the lung is divided into five periods: the embryonic, pseudoglandular, canalicular, terminal sac, and neonatal periods. Establishment of the internal gross anatomy of the lung takes place during the first two periods. The pseudoglandular period (eight to 16th week) is the principal growth time of the bronchi and of the future respiratory bronchioles and alveolar ducts. The first period, the embryonic period (fourth to seventh weeks) is the period of budding. *(Hollinshead, p 512)*

**27.** **(C)** In development of the primitive heart tube, differential growth defines five segments of the tube, which receive the names listed as items A through E in this question. Of these segments, at the rostral end is the truncus arteriosus, and the aortic sac appears as a dilatation on the truncus. Six pairs of aortic arches that arise from this sac skirt around the foregut and link the truncus to the bilateral dorsal aorta. *(Hollinshead, pp 516–517)*

**28.** **(D)** All the items listed are correct except D. The function of the fibrous pericardium is to retain the heart in position within the thoracic cavity and to limit distention of the heart. The function of the serous sac is to lubricate the moving surfaces of the heart. *(Hollinshead, pp 518–519)*

**29.** **(B)** The base of the heart largely is made up of the left atrium and a portion of the right atrium. All of the great veins enter the base of the heart and fix it posteriorly to the pericardial wall. The diaphragmatic surface is formed largely by the left ventricle and a narrower portion of the right ventricle. The coronary sulcus separates the right atrium from the right ventricle, and a large vein, the coronary sinus, is lodged in the sulcus. *(Hollinshead, pp 522–523)*

**30.** **(E)** All of the structures listed are found in the right atrium except trabeculae carneae. These structures are fleshy ridges found in the right ventricle, not the atrium. *(Hollinshead, pp 527–528)*

**31.** **(A)** The azygos venous system drains blood from the body wall into the superior vena cava. It does, however, receive venous blood from some of the thoracic viscera. The azygos system consists of two longitudinal venous channels, one on each side of the vertebral column. The channel on the right is called the azygos vein; on the left, usually it consists of two veins: the hemiazygos and the accessory hemiazygos veins. The two left veins are interconnected with each other, and both empty independently into the azygos vein through connecting transverse veins. *(Hollinshead, pp 570–571)*

**32.** **(C)** The incorrect statement is related to the ending of the thoracic duct. In the base of the neck the duct ends at the confluence of the left subclavian and the left internal jugular veins. The thoracic duct is the upward continuation of the cisterna chyli. It lies behind the esophagus in most of its course, and its beaded appearance is due to the presence of valves. *(Hollinshead, pp 90,572–573)*

**33.** **(E)** Preganglionic fibers leave the thoracic sympathetic trunks as thoracic splanchnic nerves destined for the innervation of abdominal or pelvic viscera. These nerves relay in collateral ganglia of the abdomen (celiac, superior, and inferior mesenteric ganglia). They also contain afferent fibers (pain sensation) from abdominal and pelvic viscera. These nerves usually are three in number, called the greater, lesser, and lowest splanchnic nerves. *(Hollinshead, pp 573–575)*

**34.** **(B)** All of the other items listed correctly describe thoracic vertebrae. The structural characteristic that thoracic vertebrae do not have is a transverse foramen. This foramen is characteristic of cervical vertebrae and functions to transmit the vertebral artery. *(Basmajian, pp 69–70,482–483)*

**35.** **(B)** The sternum consists of the manubrium, the body, and the xiphoid process. The upper border of the body of the bone articulates with the lower border of the manubrium, forming the sternal angle. This location is at the level of the costal cartilage of the second rib. *(Basmajian, pp 70–71)*

**36.** **(E)** The true ribs are the upper seven pairs of ribs that articulate directly with the sternum. They are otherwise known as the vertebrosternal ribs. The remaining five pairs are false ribs. The cartilages of ribs 8, 9, and 10 articulate with the cartilages immediately above them and are called vertebrochondral ribs. The last two pairs of ribs form floating or vertebral ribs. *(Basmajian, p 72)*

**37.** **(A)** The typical rib takes a downward, not an upward slope. The first arch slopes downward throughout. The ribs increase in obliquity progressively from the first to the ninth; the ninth rib is the most obliquely placed. Items B, C, and D are correct statements. *(Basmajian, pp 72–73)*

**38.** **(C)** The internal thoracic artery supplies most of the branches that are anterior intercostal arteries. The superior epigastric artery is one of the two terminal branches of the anterior thoracic artery; it descends behind the seventh costal cartilage and rectus abdominis muscle to anastomose with the inferior epigastric branch of the external iliac artery, thereby bringing the great vessels of the upper and lower limbs into communication. The other items are correct. *(Basmajian, p 76)*

**39.** **(D)** The mediastinum lies between the right and left pleural cavities. These are closed potential spaces within thin-walled sacs of serous membrane. The lung, in development, invaginates the pleural cavity. The other structures listed are contents of the mediastinum. *(Basmajian, pp 77–79,79–84)*

**40.** **(D)** The parietal layer of pleura lines each pleural cavity. This cavity possesses two walls (costal and mediastinal), a base, and an apex. Therefore its parts are (1) the costal pleura lining the rib cage, (2) the mediastinal pleura applied to the mediastinum, (3) the diaphragmatic pleura, and (4) the cupola or cervical pleura that rises into the neck. The pulmonary or visceral pleura coats the lung. *(Basmajian, p 77)*

**41.** **(C)** The root or connecting portion of the mediastinal and pulmonary pleura is a tube or sleeve of pleura in whose upper half lie all the structures that pass to and from the lung. Its lower half is collapsed and is known as the pulmonary ligament. If the root was sectioned, it would sever this ligament as well as all structures listed except the vagus nerve. The left vagus nerve passes in the mediastinum behind, not through, the root of the lung; it therefore would not be severed. *(Basmajian, pp 77,81–83)*

**42.** **(D)** The left subclavian artery is given off directly by the aortic arch. The right subclavian and the right common carotid are branches of the brachiocephalic trunk that arises from the aortic arch. The right and left coronary arteries are the first two branches of the arch. *(Basmajian, pp 82–83,115–116)*

**43.** **(D)** The left recurrent laryngeal nerve springs from the vagus, where the latter crosses the aortic arch. Embryologically the left recurrent laryngeal nerve recurs around the ligamentum arteriosum rather than the aorta, as the ligamentum arteriosum is the obliterated half of the primitive VI left aortic arch. The right recurrent nerve recurs around the right subclavian artery. None of the other items is correct. *(Basmajian, pp 83,116)*

**44.** **(A)** The sympathetic trunks are paired, each consisting of a series of ganglia connected by intervening fibers. Each trunk extends from the level of the first cervical vertebra to the tip of the coccyx, where the two trunks may meet. Item A thus is erroneous in stating that ganglia are limited to the thoracic segments. *(Basmajian, pp 84–85; Hollinshead, p 60)*

**45.** **(B)** The trachea bifurcates into a right and left primary bronchus for supply of the respective lungs. Primary bronchi give off secondary bronchi (lobar), three on the right side and two on the left, for the corresponding lobes of the lung. The secondary bronchi divide into tertiary segmental bronchi.

Each segmental bronchus, together with the portion of the lung it supplies, is called a bronchopulmonary segment. *(Basmajian, pp 88–89)*

46. **(D)** All of the statements about bronchial arteries are correct except that the pressure in them is low. The pressure in these vessels is high; in the pulmonary arteries it is low. Nevertheless, anastomoses between bronchial and pulmonary vessels occur at capillaries of the respiratory bronchioles and even earlier. *(Basmajian, p 90)*

47. **(E)** Branches of the vagus and of the thoracic sympathetic ganglia form pulmonary plexuses, and these supply the lungs. Vagal fibers form the afferent limb of the respiratory reflex arc. Efferent vagal fibers are secretomotor. The visceral pleura is insensitive to mechanical stimulation. *(Basmajian, pp 91–92)*

48. **(A)** The tracheal bifurcation both in the cadaver and supine living subject lies at about the level of T4–5. When the subject is erect, it usually lies at T6 or lower. It descends during inspiration. The domes of the diaphragm fall 2 to 3 cm when the erect posture is assumed. *(Basmajian, p 92)*

49. **(B)** If the phrenic nerve was cut at its origin, the only result listed would be B. This is because the phrenic nerve not only is motor to its own half of the diaphragm but also is sensory to the central part of the diaphragmatic pleura. *(Basmajian, pp 91–92)*

50. **(D)** On each side the internal jugular and the subclavian veins unite to form the brachiocephalic vein. This occurs behind the sternal end of the clavicle. The left brachiocephalic vein passes obliquely behind the upper half of the manubrium and joins the right brachiocephalic vein to form the superior vena cava. *(Basmajian, p 96)*

51. **(E)** The pulmonary veins (4) carrying oxygenated blood from the lungs, enter the left, not the right, atrium. The other vessels listed enter the right atrium. In addition to the tributaries of the coronary sinus, there are several small anterior cardiac veins that arise on the surface of the right ventricle and pass across the coronary sulcus to penetrate directly the anterior wall of the right atrium. *(Basmajian, p 104; Hollinshead, pp 536–537)*

52. **(B)** In the typical arrangement of costovertebral joints, the head of each rib articulates with the demifacets of two adjacent vertebrae and the intervertebral disc between them. There are some exceptions to this general arrangement. The heads of the first and of the last three ribs articulate only with the bodies of their own vertebra. The other items listed are correct. *(Moore, p 101)*

53. **(D)** In infants and children the anterior mediastinum may contain the lower part of the thymus gland. During childhood the thymus begins to diminish in relative size (undergoes involution). By adulthood often it is scarcely recognizable. *(Moore, pp 89,101)*

54. **(A)** All of the arteries listed are branches of the thoracic aorta, except the first pair of posterior intercostal arteries. These are given off by the superior intercostal artery that is a branch of the costocervical trunk of the subclavian artery. *(Moore, pp 31,99)*

55. **(B)** The esophagus extends from the lower end of the pharynx, at the level of the cricoid cartilage, to the stomach. The esophagus enters the superior mediastinum between the trachea and the vertebral column and passes posterior, not anterior, to the left principal bronchus. It descends behind and to the right, not the left, of the aortic arch. It occupies the posterior, not the anterior, mediastinum. It passes through the esophageal hiatus, not the hiatus formed by the median arcuate ligament of the diaphragm (aortic hiatus). *(Moore, pp 97–98,252)*

56. **(C)** The trachea begins in the neck as the continuation of the lower end of the larynx. It ends at the level of the sternal angle by dividing into right and left principal bronchi. During inspiration the tracheal bifurcation descends. The trachea is kept open by a series of C-shaped bars of cartilage. It descends in front of the esophagus, and its posterior side is flat where it is applied to the esophagus. *(Moore, p 97)*

57. **(C)** All of the statements about the left vagus nerve are correct except that it supplies the left diaphragm. The phrenic nerves are the sole motor supply to the diaphragm. They arise from ventral rami of the third, fourth, and fifth cervical nerves. *(Moore, p 96)*

58. **(A)** The aortic arch, the continuation of the ascending aorta, begins posterior to the right half of the sternal angle. It passes to the left of the trachea. The ligamentum arteriosum passes from the root of the left pulmonary artery to the inferior concave surface of the aortic arch. The aortic arch gives off the left subclavian artery and the brachiocephalic trunk (on the right). *(Moore, pp 91–94)*

59. **(D)** The superior vena cava enters the right atrium vertically from above. It returns blood from everything above the diaphragm (parts listed) except from the lungs. The blood from the lungs enters the left atrium via the pulmonary veins. *(Moore, p 91)*

60. **(B)** The brachiocephalic veins have no valves. Each is formed by union of the internal jugular and sub-

clavian veins. Each receives the internal thoracic, vertebral, inferior thyroid, and highest intercostal veins. The brachiocephalic veins unite to form the superior vena cava. *(Moore, p 89)*

61. **(A)** All of the structures listed are found in the superior mediastinum, except the heart and pericardium. The middle mediastinum contains the heart and pericardium, with the adjacent phrenic nerves and the roots of the great vessels passing to and from the heart. *(Moore, pp 55,89)*

62. **(C)** The coronary arteries arise from the right and left aortic sinuses, respectively, at the root of the aorta just after it leaves the heart. They are the first vessels given off by the aorta. Most of the blood in these vessels returns to the chambers of the heart via the coronary sinus. Some small venous channels, however, empty directly into its chambers. The coronary sinus drains all the venous blood from the heart except that carried by the anterior cardiac veins and the venae cordis minimae, which open directly into the right atrium. *(Moore, pp 83–87)*

63. **(E)** The heart is insensitive to touch, cutting, cold, and heat, but ischemia and resulting accumulation of metabolic products stimulate pain endings in the myocardium. The pain of angina pectoris and of myocardial infarction commonly radiates to the left shoulder and the medial aspect of the arm (referred pain). The afferent pain fibers run centrally in the middle and inferior cervical branches and thoracic cardiac branches of the sympathetic trunk. *(Moore, pp 81–82)*

64. **(E)** All of the statements regarding innervation of the heart are true except that vagal stimulation dilates the coronary arteries. The vagus, the parasympathetic cardiac nerve, constricts the coronary arteries. It also slows the rate and reduces the force of the heart beat. The intrinsic impulse-conducting system of the heart is under control of the cardiac nerves of the autonomic nervous system; this enables the heart to respond to changing physiological needs of the body. *(Moore, p 81)*

65. **(C)** The SA node initiates the impulse for contraction, which is conducted rapidly to the cardiac muscle cells of the atria, causing them to contract. The impulse enters the AV node and is transmitted through the AV bundle and its branches to the papillary muscles and then throughout the walls of the ventricles. The bundle of His is the old name for the AV bundle. *(Moore, pp 79–80)*

66. **(A)** In aortic valve stenosis, the edges of the valve are unusually fused together. This valvular stenosis causes extra work for the left, not the right, ventricle, resulting in ventricular hypertrophy. A heart murmur also is produced. If the aortic valve is damaged by disease, the valvular incompetence results in aortic regurgitation that produces a heart murmur and a collapsing pulse. *(Moore, p 77)*

67. **(D)** The point of the left ventricle forms the apex, not the base of the heart. All of the other items listed are correct. *(Moore, pp 65,74–75)*

68. **(B)** The mitral valve is the most frequently diseased of the heart valves. Rheumatic fever formerly was a common cause of this type of heart disease. Nodules on the valve cusps roughen them, resulting in irregular blood flow and a heart murmur. Later the diseased cusps undergo scarring and shortening, resulting in a condition called valvular incompetence. With this, blood in the left ventricle regurgitates into the left atrium, producing a murmur when the ventricles contract. *(Moore, p 74)*

69. **(C)** All of the items are correct except C. The left atrium receives the four pulmonary veins, not the pulmonary arteries. The pulmonary veins carry oxygenated blood returning from the lungs. *(Moore, p 73)*

70. **(A)** The right ventricle gives origin to the pulmonary trunk. This vessel leaves the right ventricle and divides into the right and left pulmonary arteries that take blood to the lungs. There are three papillary muscles in the right ventricle. Their chordae tendinae are inserted into the free edges and ventricular surfaces of the right atrioventricular (tricuspid) valve. *(Moore, pp 70–71)*

71. **(D)** The coronary sulcus separates both atria from both ventricles. The right atrium, not the right ventricle, forms the right border of the heart. The superior vena cava opening into the right atrium does not have a valve. The fossa ovalis is a prominent feature of the interatrial, not the interventricular, septum. *(Moore, p 70)*

72. **(E)** The heart is described correctly by all the items except E. The anterior (sternocostal) surface of the heart is formed mainly by the right ventricle and the right atrium, not by the left atrium, as stated in E. The left ventricle and left atrium lie more posteriorly and form only a small strip on the sternocostal surface. *(Moore, pp 62–65)*

73. **(B)** The epicardium of the heart is supplied by the coronary arteries. The blood supply to the pericardium is derived from the internal thoracic artery via its branches and from pericardial branches of the bronchial, esophageal, and superior phrenic arteries. The veins of the pericardium are tributaries of the azygos system. *(Moore, p 61)*

74. **(B)** The ascending aorta carries the pericardium upward beyond the heart to the level of the sternal angle. The base of the fibrous pericardium is fused to the central tendon of the diaphragm. Owing to its many connections, the pericardium is anchored firmly, not freely, within the thoracic cavity. Its truncated apex, not its base, is pierced by the aorta. In the anterior median line, the pericardium is attached to the posterior surface of the sternum by the superior and inferior sternopericardial ligaments. *(Moore, pp 56–57)*

75. **(C)** The median region between the two pleural sacs is called the mediastinum. It extends from the sternum and costal cartilages in front to the anterior surfaces of the 12 thoracic vertebrae behind. The vertebral bodies, however, are not in the mediastinum. All the other structures listed are contents of the mediastinum. *(Moore, pp 54–55)*

76. **(D)** No lymph vessels are located in the walls of the pulmonary alveoli. The bronchomediastinal lymph trunks usually terminate on each side at the junction of the subclavian and internal jugular veins; the left trunk may terminate in the thoracic duct. There are two lymphatic plexuses: a superfocial and a deep plexus. Cancer cells may spread to the opposite side through the lymphatics because some lymph from the left lung also drains into nodes on the right. Lymph from the lungs drains into the venous system; hence lymph from the lungs may carry cancer cells into the blood. Common sites of spread of cancer cells via the blood from a bronchogenic carcinoma are the brain, bones, and adrenal glands. *(Moore, pp 52–54)*

77. **(A)** The roots of the left and the right lungs are similar, except that there is only one pulmonary artery in the root of the left lung and two pulmonary arteries on the right. Each side has two pulmonary veins, a bronchus, a pulmonary plexus of nerves, lymph vessels, and lymph nodes enclosed in connective tissue. *(Moore, pp 50–51)*

78. **(E)** The left lung is divided into inferior and superior lobes by a long, deep, oblique fissure. The superior lobe has a wide cardiac notch on its anterior border where the lung is deficient owing to the bulge of the heart. The right lung is larger and heavier than the left lung. The azygos vein arches over the root of the right, not the left, lung. A horizontal fissure is found in the right, not the left, lung. *(Moore, pp 40–41,44)*

79. **(C)** The central part of the parietal diaphragmatic pleura is supplied by the phrenic nerve; it also supplies the mediastinal pleurae. The costal pleura and the pleura on the peripheral part of the diaphragm are supplied by the intercostal nerves. *(Moore, p 39)*

80. **(D)** The visceral pleura receives nerve supply from autonomic nerves to the lung. It receives its arterial supply from the bronchial arteries. Its numerous lymphatics drain into nodes at the hila of the lungs. The visceral pleura is insensitive to pain: the parietal pleura is very sensitive to pain. *(Moore, pp 35, 39–40)*

81. **(B)** The only correct item is B, which states that each pleural cavity is a closed potential space. Normally the pleural cavity contains only a capillary layer of serous, lubricating fluid. This substance lubricates the surface and reduces friction between parietal and visceral layers of pleura. Visceral and parietal pleura become continuous at the root of the lung. The apex of the lung is covered by cervical pleura (cupola of the pleura). The pleurae descend below the costal margin in three regions, where an abdominal incision might enter a pleural sac. These regions are (1) the right xiphisternal angle, (2) the right costovertebral angle, and (3) the left costovertebral angle. *(Moore, pp 32–35)*

82. **(A)** A dermatome is an area of skin supplied by a dorsal (sensory) root of a spinal nerve. There is considerable overlapping of contiguous dermatomes. The area around the umbilicus is a helpful landmark for identification of innervation level because it is supplied by T10. *(Moore, p 30)*

83. **(B)**

84. **(D)**

85. **(A)**

86. **(E)**

83–86. The thoracic vertebrae are characterized by having articular facets for articulation with ribs. There is one or more facets on each side of the vertebral body for articulation with the head of a rib and one on each transverse process of the upper ten vertebrae for the tubercle of a rib. The sacrum is a large, triangular, wedge-shaped bone that usually is composed of five fused sacral vertebrae in the adult. On the pelvic and dorsal surfaces of the sacrum there typically are four pairs of foramina for exit of the anterior and posterior primary divisions of the sacral nerves. The cervical vertebrae form the bony axis of the neck. Their distinctive feature is the large, oval-shaped foramen transversarium (transverse foramen) in each transverse process. The vertebral arteries pass through these foramina, except in C7, which transmit only small accessory veins. Lumbar vertebrae may be distinguished by their relatively large bodies, as compared with thoracic and cervical vertebrae and by the absence of costal facets; their vertebral bodies, seen from above, are kidney shaped, and their vertebral foramina are oval to triangular. *(Moore, pp 615–618)*

**87.  (C)**

**88.  (A)**

**89.  (B)**

**90.  (C)**

**87–90.** The body of a rib is thin, flat, long, and curved. The sharp inferior border of the rib projects downward and is located external to the costal groove on the internal surface of the body, near its inferior border. The point of greatest change in curvature is called the angle of the rib. The head of the rib is wedge shaped and presents two articular facets for articulation with the numerically corresponding vertebra and the vertebra superior to it. These facets are separated by the crest of the head, which is joined to the intervertebral disc by an intra-articular ligament. The neck of the rib is the stout, flattened part located between the head and the tubercle; the neck lies anterior to the transverse process of the corresponding vertebra. Its upper border, called the crest of the neck, is sharp, and its lower border is rounded. *(Moore, pp 5–7)*

**91.  (C)**

**92.  (A)**

**93.  (E)**

**94.  (B)**

**91–94.** Synovial joints connect the rib with the facets of the vertebral bodies and transverse processes of the vertebrae. They are the plane type of synovial joint that allows gliding movement. The head of each typical rib articulates with the demifacets of two adjacent vertebrae and the intervertebral disc between them. The first seven pairs of ribs are called true ribs or vertebrosternal ribs because they articulate with the sternum through their costal cartilages. The true ribs increase in length from above downward. The seventh through the tenth costal cartilages meet on each side, and their inferior edges form the costal margin. The diverging costal margins form an infrasternal (subcostal) angle below the xiphisternal joint. Cervical ribs are in about 0.5% to 1% of people and articulate with the seventh cervical vertebra. Usually these ribs produce no symptoms, but in some cases the subclavian artery and the lower trunk of the brachial plexus are kinked where they pass over the cer-

vical rib. Compression of these structures between the extra rib and the anterior scalene muscle produces symptoms of nerve and arterial compression, called the neurovascular compression syndrome. *(Moore, pp 1,9–11,101,701)*

**95.  (E)**

**96.  (A)**

**97.  (C)**

**98.  (B)**

**95–98.** During normal expiration the elastic recoil of the lungs produces a subatmospheric pressure in the pleural cavities. This and the weight of the thoracic walls cause the lateral and anteroposterior diameters of the thorax to return to normal following inspiration. Movements of the thorax primarily are concerned with increasing and decreasing the intrathoracic volume. The resulting changes in pressure result in air being alternately drawn into and expelled from the lungs. The vertical diameter of the thorax is increased primarily when the diaphragm contracts, lowering it. The transverse diameter of the thorax is increased by the ribs swinging outward during the so-called "bucket-handle" movement; this elevates ribs 2 through 10, approximately, and everts their lower borders. This pulls the lateral portions of the ribs away from the midline, thereby increasing the transverse diameter of the thorax. The anteroposterior diameter of the thorax also is increased by raising the ribs. Movement of the costovertebral joints through the long axes of the necks of the ribs results in raising and lowering their sternal ends, the "pump-handle" movement. *(Moore, pp 18–19,20–21)*

**99.  (C)**

**100.  (E)**

**101.  (B)**

**102.  (D)**

**99–102.** The exact role of the intercostal muscles in the widening of intercostal spaces on inspiration is controversial. It is known that all three layers of intercostal muscles act together in keeping the intercostal spaces rigid. This action prevents the spaces from bulging out during expiration and from being drawn in during inspiration. The serratus posterior superior and serratus posterior inferior muscles run from the vertebrae to the ribs. Both of these muscles are concerned with inspiration. The pectoralis minor muscle arises from the third, fourth, and fifth ribs and inserts on the coracoid of the scapula. It stabilizes the scapula by drawing it forward and downward and elevates the ribs from which it arises; thus it is an accessory respiratory muscle. The subcostal muscles are thin muscular slips, variable in size and shape. They extend from the inside of the angle of one rib to the internal surface of the rib below, crossing one or two intercostal spaces. They run in the same direction as the internal intercostal muscles and lie internal to them. These muscles probably depress the ribs. *(Moore, pp 21–28)*

**103. (B)**

**104. (C)**

**105. (A)**

**106. (A)**

**103–106.** Three arteries, a large posterior intercostal artery and two small, paired, anterior intercostal arteries, supply the intercostal spaces. The upper six intercostal spaces are supplied by (1) paired anterior intercostal arteries derived from the internal thoracic artery, (2) posterior intercostal arteries supplied by the superior intercostal artery (a branch of the costocervical trunk) to the first two intercostal spaces. Therefore the thoracic aorta is not involved as intercostal arteries to these first two spaces, but the internal thoracic artery is involved. Intercostal arteries for the third through the sixth intercostal spaces have posterior intercostal arteries arising from the back of the thoracic aorta and anterior intercostal arteries from the anterior thoracic arteries. Intercostal spaces 7 through 9 receive their anterior arteries from the musculophrenic artery, a branch of the internal thoracic. The posterior intercostal arteries for these spaces come from the thoracic aorta. For the lower two intercostal spaces, there are no anterior intercostal arteries. Only the posterior intercostals from the thoracic aorta supply these spaces. *(Moore, p 31)*

**107. (A)**

**108. (C)**

**109. (D)**

**110. (A)**

**107–110.** The right lung is divided into superior, middle, and inferior lobes by the oblique and horizontal fissures. The horizontal fissure separates the superior and middle lobes. The oblique fissure separates the inferior lobe from the middle and superior lobes. The hilum of the lung is where the bronchi, pulmonary vessels, bronchial vessels, lymph vessels, and nerves enter and leave the lung. The hilum is surrounded by a sleeve of pleura that is reflected off the lung onto the mediastinum; the pulmonary ligament is the downward extension of this pleural sleeve. The structures enter and leave the lung from the root of the lung, which is attached to the hilum. The azygos vein forms a groove as it arches over the root of the right lung. This vein drains blood from the thoracic wall. The upper part of the right ventricle of the heart tapers into a cone-shaped structure (conus arteriosus) that gives origin to the pulmonary trunk. The pulmonary trunk bifurcates into right and left pulmonary arteries that proceed to the lungs. The pulmonary trunk ends with the division into pulmonary arteries. *(Moore, pp 34,40,41,44,71, and 101)*

**111. (D)**

**112. (D)**

**113. (C)**

**114. (D)**

**111–114.** Musculi pectinati are muscular ridges resembling the coarse teeth of a comb. They are found in both atria, not in the ventricles. Also related to the atria are the coronary sinus and the pulmonary veins. The heart is drained mainly by veins that empty into the coronary sinus. It is the main vein of the heart and runs from left to right in the coronary sulcus. It opens into the right atrium. The four pulmonary veins enter the sides of the posterior half of the left atrium. They bring oxygenated blood from the lungs back to the heart. Papillary muscles are found in both ventricles. In the right ventricle there are usually three papillary muscles. These are irregular muscle bundles on the ventricular walls that have a number of slender fibrous threads arising from their apices. The chordae are inserted into the free edges and ventricular surfaces of the cusps of the right atrioventricular (AV) (tricuspid) valve. The chordae tendinae prevent the cusps of the valve from being driven into the right atrium as the ventricular pressure rises. There are two large papillary muscles in the left ventricle. They are larger than those in the right ventricle, and their chordae tendinae are thicker but less numerous. The chordae tendinae of each muscle are distributed to the contiguous halves of the two cusps of the left AV valve. *(Moore, pp 70–71,74–75,73,87)*

**115. (C)**

**116. (D)**

**117. (A)**

**118. (A)**

**115–118.** The AV bundle arises from the AV node located in the posterior-inferior part of the interatrial septum. It runs forward to the membranous part of the interventricular septum. Thus both the atrial and ventricular septa are involved. The SA node is a small mass of specialized cardiac muscle fibers located in the wall of the right atrium. This structure initiates the impulse for heart contraction; the impulse then spreads through the cardiac cells of both atria, causing them to contract. The fossa ovalis is a shallow, translucent depression in the interatrial septum. The oval fossa is a remnant of the fetal foramen ovale, through which oxygenated blood from the placenta passes from the inferior vena cava through the right atrium to the left atrium. The AV node receives the impulse for contraction from the SA node and transmits it to the AV bundle and its branches. From there the impulse goes first to the papillary muscles and then throughout the walls of the ventricles. The papillary muscles tighten the chordae tendinae, drawing the cusps of the atrioventricular valves together. Next the contraction of the ventricle occurs. *(Moore, pp 70,78–79)*

**119. (B)**

**120. (A)**

**121. (B)**

**122. (D)**

**119–122.** Stimulation through the parasympathetics (vagus nerve) slows the heart rate, reduces the force of the heart beat, and constricts the coronary arteries. The cardiac plexus contains small ganglia near the SA node, which belong chiefly to the parasympathetic system. The afferent pain fibers run centrally in the middle and inferior cervical ganglion branches and thoracic cardiac branches of the sympathetic trunk. The axons of these primary sensory neurons enter spinal cord segments T1 to T4 or T5 on the left side. The heart beat is not initiated by the nerves but by specialized cardiac muscle fibers that initiate the normal heart beat (SA node) and coordinate the contraction of the heart chambers. Both atria contract together, as do both ventricles; but atrial contraction occurs first. This system gives the heart its automatic rhythmic beat. The sinoatrial node initiates the impulse for contraction. It is called the pacemaker of the heart. The rate at which the node produces impulses can be altered by nervous stimulation: sympathetic stimulation speeds it up and vagal stimulation slows it down. *(Moore, pp 78–79,81)*

**123. (A)** The first three items are correct, while item 4 is erroneous. The circumflex artery is a division of the left coronary artery. It follows the coronary sulcus around the left border of the heart to the posterior surface. It terminates to the left of the posterior interventricular sulcus by giving branches to the left ventricle and the left, not the right, atrium. The right and left coronary arteries arise from the aortic sinuses. The blood in these vessels returns to the heart via the coronary sinus. The right coronary artery gives off the AV nodal artery. *(Moore, pp 83–84)*

**124. (C)** The heart is drained mainly by veins that empty into the coronary sinus and partly by small veins that open directly into the chambers of the heart, principally those on the right side. The coronary sinus is the main vein of the heart and drains most of the venous blood from it except that carried by the anterior cardiac veins and the vena cordis minimae. The coronary sinus opens into the right atrium immediately to the left of the inferior vena cava and posterior to the atrioventricular orifice. The great cardiac vein ascends in the anterior interventricular sulcus. The middle cardiac vein runs in the posterior interventricular sulcus to enter the coronary sinus. The anterior cardiac veins enter the right, not the left, atrium. The left atrium receives blood returning from the lungs but receives none of the veins carrying blood from the heart itself. *(Moore, p 87)*

**125. (D)** In early childhood, the thymus is a prominent feature of the superior mediastinum. It is a bilobed structure that has a pink lobulated appearance during early life. During childhood, the thymus begins to diminish in relative size (undergoes involution). The blood supply of the thymus is the inferior thyroid and internal thoracic arteries. *(Moore, p 89)*

**126. (A)** Typical features of thoracic vertebrae include items 1, 2, 3. The bodies in the middle of the series are heart shaped. The transverse processes are stout and conform to the backward sweep of the ribs, having on their tips facets for the tubercles of the ribs. The transverse processes become progressively shorter from the first to the 12th vertebra. Triangular-shaped vertebral foramina are found in lumbar vertebrae, while thoracic vertebral foramina are small and circular. *(Basmajian, pp 69–70,186)*

127.  **(E)** At costovertebral joints, ribs are attached to intervertebral discs by transversely placed intra-articular ligaments. Costotransverse ligaments bind the tubercle of the rib to a transverse process of the vertebra. *(Basmajian, p 73)*

128.  **(A)** Spinal nerves divide into a ventral and a dorsal ramus. Roughly, the dorsal rami supply muscles and skin of the back; the ventral rami supply muscles and skin of the anterior three quarters of the body wall. The anterior ramus becomes the intercostal nerve that runs in the intercostal space; the dorsal ramus proceeds toward the back. The internal intercostal membrane (the posterior portion of the internal intercostal muscle) and the posterior intercostal artery also are contents of the intercostal space. *(Basmajian, pp 75,336)*

129.  **(C)** Each pleura has three parts: parietal, visceral, and connecting. The parts of the parietal layers are (1) costal pleura, (2) mediastinal pleura, (3) diaphragmatic pleura, and (4) cervical pleura (the cupola). Pulmonary and visceral pleura are both names for the pleural layer that coats the lung. *(Basmajian, p 77)*

130.  **(E)** All of the structures listed are seen on the right side of the mediastinum. The pericardial sac is separated from the right half of the body of the sternum by the thickness of the anterior border of the right lung and pleura. The superior and inferior venae cavae are like two rivers that flow due south and north to empty into the right atrium. The right vagus nerve passes obliquely downward to the back of the root of the lung where it takes part in the posterior pulmonary plexus. *(Basmajian, p 80)*

131.  **(E)** The three chief structures in the root of the lung are (1) the pulmonary artery that brings blood charged with carbon dioxide from the heart to the lungs; (2) the pulmonary veins, two on each side, that return oxygenated blood to the heart; (3) the bronchus or air passage. Also in the root of the lung are nerve plexuses, bronchial vessels, and lymphatics. *(Basmajian, pp 81–82)*

132.  **(B)** The left vagus nerve is a cranial nerve and therefore does not arise from the spinal cord as listed in item 4. This nerve traverses the neck in the carotid sheath. The left recurrent laryngeal nerve springs from the vagus, where the latter crosses the left side of the aortic arch. The recurrent nerve recurs around the ligamentum arteriosum. *(Basmajian, p 83)*

133.  **(D)** The true statement is that typically a ganglion of the sympathetic trunk is found in front of each rib. Since ganglia are segmental structures, there is one, developmentally, for each of the 31 spinal nerves. The sympathetic trunk is found lying a little wide of the mediastinum; it is not limited to the thoracic region but extends the length of the vertebral column. Intercostal arteries and veins cross the trunk posteriorly, not anteriorly. From the upper five thoracic sympathetic ganglia, postganglionic fibers pass the cardiac plexus, posterior pulmonary plexus, and upper parts of the esophagus and aorta. *(Basmajian, pp 84–85)*

134.  **(C)** There are three, paired, splanchnic nerves in the thorax consisting of preganglionic fibers making "nonstop journeys" through sympathetic ganglia. They end in the celiac and renal ganglia, where they are relayed as unmyelinated postganglionic fibers. The three, paired, splanchnic nerves are the greater, lesser, and least. Typically the greater splanchnic springs from the eighth, ninth, and tenth sympathetic ganglia. The lesser splanchnic nerve springs from the tenth, 11th, and perhaps the 12th ganglia; the lowest splanchnic, when present, springs from either the 11th or 12th ganglion or from both ganglia. *(Basmajian, p 86)*

135.  **(A)** The apex of the lung rises to the neck of the first rib. In front of the hilus of the lung is an excavation, the cardiac impression, that is deeper on the left side than on the right because two thirds of the heart lies to the left of the median plane. A lobe of the azygos vein results when the apex of a developing right lung encounters the arch of the azygos and is cleft by it. The vein is suspended by a pleural "mesentery," and it may cast a shadow by x-ray. Usually, there are eight, not ten, segmental bronchi on the left side. *(Basmajian, pp 87–89)*

136.  **(E)** The pulmonary arterial tree, in general, follows the bronchial pattern. On each side, two pulmonary veins, an upper and a lower, enter the left atrium. The bronchial arteries arise from the aorta, but the right may arise from an intercostal artery. Each bronchial artery supplies the bronchi and related structures and runs through interlobar septa to supply the pulmonary pleura. Lymph channels, accompanying small blood vessels, occur adjacent to the alveolar walls but not in the interalveolar partitions. Lymph drainage is into the bronchopulmonary nodes in the hilus of the lung. *(Basmajian, p 90)*

137.  **(B)** Branches of the vagus and of the thoracic sympathetic ganglia 1 through 5 form the pulmonary plexus, and these supply the lungs. Sensory vagal fibers constitute the afferent limb of the respiratory reflex arc. Efferent vagal fibers are bronchoconstrictor and secretomotor. Efferent sympathetic

fibers are bronchodilator. The visceral pleura is insensitive to mechanical stimulation. *(Basmajian, pp 91–92)*

**138.** **(A)** If a vertical line is drawn, in the manner described, from above downward, the right borders of (1) the right internal jugular vein, (2) the right brachiocephalic vein, (3) the superior vena cava, (4) the right atrium, and (5) the inferior vena cava will be represented. The right ventricle is not in a position to have its right border represented by this line. *(Basmajian, pp 95–96)*

**139.** **(D)** The pulmonary trunk is the continuation of the right ventricle. It passes upward and to the left between the right and left auricles, which embrace it. It lies in front of, and conceals, the root of the aorta. Below the aortic arch it divides into right and left pulmonary arteries. The stems of the pulmonary trunk and ascending aorta, lying within a single sleeve of serous pericardium, form the anterior boundary of a potential space, the transverse pericardial sinus. *(Basmajian, pp 97–98)*

**140.** **(B)** The left coronary artery divides into an interventricular branch and a circumflex branch, which run in the coronary sulcus around the left margin of the heart. The right coronary artery divides into a posterior interventricular branch and a transverse branch, which, continuing in the coronary sulcus, meet the circumflex branch of the left coronary artery. *(Basmajian, p 100)*

**141.** **(D)** The heart undergoes a slight rotation to the left on its long axis. This results in (1) the right atrium being conspicuous at the right border of the heart, (2) the left ventricle being largely inferior and slightly in front, (3) the interatrial and ventricular septa facing forward and to the right, and (4) the left atrium being conspicuous posteriorly. *(Basmajian, p 103)*

**142.** **(A)** The first three items are correct. The pulmonary veins enter the left, not the right atrium. The right atrium developed from the sinus venarum and the primitive atrium, which merged to form the single chamber. The rotation of the heart left the right atrium forming the right margin of the heart. Musculi pectinati are structural characteristics of the right atrium. *(Basmajian, pp 104–105)*

**143.** **(E)** All of the items are correct. Both of the ventricles lie in front of their atria. The atrioventricular (AV) orifices are posterior; the orifices of the aorta and pulmonary trunk are superior, so the blood pursues a V-shaped course within the ventricle. The ventricular walls are lined with muscular bundles, the trabeculae carneae, some of which form projections, the papillary muscles. From the apices of the papillary muscles, fibrous cords, the chordae ten-

dinae, pass to the cusps of the AV valves. *(Basmajian, pp 105–106)*

**144.** **(C)** The SA node initiates the heart beat. The AV node is situated in the interatrial septum beside the mouth of the coronary sinus. The AV bundle is the sole muscular connection between the musculature of the atria and the musculature of the ventricles. This structure is composed of muscle and not of nerve fibers as stated. *(Basmajian, p 109)*

**145.** **(A)** The sympathetic nerves are cardioaccelerator and vasodilator and carry sensory fibers. They do not initiate the heart beat accomplished by the sinoatrial (SA) node. The vagal fibers are cardioinhibitory. *(Basmajian, p 109)*

**146.** **(D)** The cardiac plexus, composed of both sympathetic and vagal fibers, is situated in front of the bifurcation of the trachea, above the bifurcation of the pulmonary trunk, and therefore below the arch of the aorta. The vagal efferent fibers are preganglionic until they synapse with the postganglionic fibers in the ganglia of the cardiac plexus and in the intrinsic cardiac ganglia (practically confined to the atria, interatrial septum, and areas near the roots of the great vessels). The extensions of the cardiac plexuses are the right and left pulmonary plexuses and the plexus on the thoracic aorta. *(Basmajian, p 114)*

**147.** **(B)** Budding of the lung diverticulum does not progress in a strictly dichotomous fashion; therefore the bronchi do not develop symmetrically. They do provide the framework of the lung parenchyma. The major branches of the pulmonary veins and arteries conform to the bronchial tree in a definable manner. The bronchi are hollow tubes kept open by incomplete rings or plates of hyaline cartilage. *(Hollinshead, p 493)*

**148.** **(E)** The bifurcation of the trachea into principal bronchi lies, in the cadaver, at about the level of the sternal angle. It is marked by an internal ridge, the carina, that separates the openings of the two principal bronchi. The right principal bronchus is wider than the left, but the left is almost twice as long as the right. A foreign body inhaled into the trachea is much more likely to lodge in the right bronchus because this is more directly in line with the trachea and because it is larger than the left main bronchus. *(Hollinshead, pp 494–495)*

**149.** **(C)** The bronchial arteries are insignificant in size by comparison to the pulmonary arteries. Bronchial arteries deliver blood for nutrition of bronchi and connective tissue of the lung. The four pulmonary veins return oxygenated blood to the left atrium. Anastomoses do exist between the bronchial and pulmonary systems at the capillary level and also between some larger precapillary branches of the bronchial and pulmonary arteries. *(Hollinshead, pp 497–499)*

150. **(E)** Visceral efferents are contributed to the pulmonary plexus by the vagus and thoracic sympathetic ganglia. The sympathetic fibers in the plexus are postganglionic, and the vagal fibers are preganglionic. Visceral afferents from the lung have been demonstrated only in the vagus, their cell bodies being in the superior and inferior vagal ganglion. Afferents from the lung are concerned with innervation of the bronchial mucosa; with sensing stretch in the alveoli, the interalveolar septa, and the pleura; with monitoring pressure in the pulmonary veins; and with mediating pain sensation. All these impulses travel in the vagus and are concerned with the afferent limb of such reflexes as the cough reflex and the stretch reflex, which regulates respiration. *(Hollinshead, pp 502–503)*

151. **(B)** The fibrous pericardium rests with its base on the diaphragm and thus is fixed to the central tendon of the diaphragm. The pericardium and its contents comprise the middle mediastinum. The anterior surface of the pericardium is attached to the sternum by two poorly defined sternopericardial ligaments. Ascent of the diaphragm during expiration, not inspiration, relaxes, not stretches, the pericardium; this permits the sac to bulge laterally and the heart to become more horizontal. *(Hollinshead, pp 518–520)*

152. **(E)** All of the severed structures listed can be seen on the posterior wall of the pericardium after the heart has been removed. In addition to the vessels listed, also seen are the inferior vena cava, the left pulmonary veins, and the cut pericardial sleeve around the aorta and pulmonary trunk. *(Hollinshead, p 521)*

153. **(C)** Pericardial pain is mediated by somatic afferents distributed to the fibrous and parietal lamina of the pericardium by the phrenic nerve. The epicardium, supplied by autonomic nerves from the coronary plexus, is insensitive to pain. Pericardial pain is felt behind the sternum and usually is due to inflammation of the pericardium resulting from viral or bacterial infections or by neoplastic involvement. *(Hollinshead, p 522)*

154. **(D)** The diaphragmatic surface of the heart is formed largely by the left ventricle and a narrow portion of the right ventricle. The sternocostal surface faces anteriorly and is dominated by the right ventricle. The coronary sulcus separates the right atrium from the ventricle and runs from the root of the aorta toward the inferior vena cava. The right ventricle tapers toward the origin of the pulmonary

trunk, and this funnel-shaped portion forms the conus arteriosus or infundibulum. The infundibulum continues into the pulmonary trunk. *(Hollinshead, pp 524–525)*

155. **(A)** The right border of the heart consists of the first three structures listed. The right ventricle gives origin to the pulmonary trunk, which is not a part of the right border. *(Hollinshead, p 525)*

156. **(C)** The interior of the right atrium is divided partially by the crista terminalis, a smooth muscular ridge that commences on the roof of the atrium and extends to the anterior lip of the inferior vena cava. The coronary sinus opens into the posterior corner of the atrium. The right atrioventricular ostium is guarded by the tricuspid, not the mitral, valve. The azygos vein does not open directly into the right atrium but into the superior vena cava. *(Hollinshead, pp 527,531,571)*

157. **(B)** The trabeculae carneae in the left ventricle are fine and delicate in contrast to the coarse trabeculae in the right ventricle. The outflow tract of the left ventricle sometimes is designated as the aortic vestibule; it corresponds to the conus arteriosus on the right, which lies directly anterior to it. The valvule of the foramen ovale is found on the septum of the left atrium, visible in the right atrium as the floor of the fossa ovalis. The pulmonary valve closes the orifice of the pulmonary trunk that opens from the right ventricle. *(Hollinshead, pp 528,531)*

158. **(E)** Although the pulmonary trunk conveys venous blood from the heart toward the lungs, structurally its wall is similar to that of the aorta. The pulmonary trunk begins at the pulmonary orifice of the right ventricle. The aorta begins at the aortic orifice of the left ventricle. The pulmonary veins enter the posterior half of the left atrium. *(Hollinshead, pp 541,530)*

159. **(D)** No doubt, audible heart sounds are connected closely with closure of the valves; opening of the valves in the normal heart is inaudible. There are specific auscultation areas where sounds made by individual valves can be heard to the best advantage. For all but one valve, these areas do not coincide with the surface projection of the valves; sounds of the tricuspid valve are heard over the anterior wall of the right ventricle, which happens to superimpose on the projection of the valve itself. The first heart sound signals the beginning of ventricular systole, not diastole. *(Hollinshead, p 544)*

160. **(A)** A decrease in the volume of blood reaching the lungs is one cause of cyanosis. Other reasons for cyanosis result from mixing venous and arterial blood or transposition of the great arteries. Struc-

tural abnormalities resulting in a decrease in the volume of blood reaching the lungs are the tetralogy of Fallot, one characteristic of which is a narrowing of the right ventricular outflow tract. Other causes are pulmonary atresia and tricuspid atresia. (Hollinshead, pp 550–552)

161. **(C)** Both the trachea and the esophagus fill the superior thoracic aperture in the median plane. They descend through the superior mediastinum; the trachea bifurcates into the right and left principal bronchi near the lower boundary of the superior mediastinum, while the esophagus continues into the posterior mediastinum. The esophagus lies behind the trachea on the anterior surface of the vertebral bodies. Above the root of the lung, both trachea and esophagus are crossed by the azygos vein on the right side and by the arch of the aorta on the left side. (Hollinshead, pp 558–559)

162. **(D)** The ligamentum arteriosum connects the inferior surface of the aortic arch to the left pulmonary artery. The arch of the aorta gives off the brachiocephalic artery that divides into the right subclavian and right common carotid arteries. The inferior thyroid artery is a branch of the thyrocervical trunk, which arises from the subclavian artery. (Hollinshead, pp 561,840)

163. **(B)** The brachiocephalic veins are formed by the confluence of the subclavian and internal jugular veins on each side of the root of the neck. The right brachiocephalic vein is vertical and is close to the right border of the manubrium of the sternum. The left vein is much longer than the right, and its course is oblique as it runs in almost a transverse direction toward the right. Behind the sternal end of the right first rib, the left vein unites with the right brachiocephalic vein and their confluence forms the superior vena cava, not the inferior vena cava. (Hollinshead, p 563)

164. **(E)** The main nerves that enter the superior thoracic aperture are the vagus, the phrenic, and the sympathetic trunks. Cervical cardiac branches of the vagus and sympathetic ganglia also enter the superior thoracic aperture. The phrenic nerves are not autonomic but are spinal nerves derived mainly from the fourth, and to some extent, from the third and fifth cervical segments of the spinal cord. (Hollinshead, p 565)

165. **(C)** The phrenic nerves carry motor innervation to the diaphragm. They also convey somatic afferents (pain sensation) from the fibrous and parietal serous pericardium, from the mediastinal and diaphragmatic portions of the parietal pleura, and from parietal peritoneum on the inferior surface of the diaphragm. These nerves are accompanied by the pericardiophrenic vessels (branches of the in-

ternal thoracics). The recurrent laryngeal nerves are branches of the vagus. (Hollinshead, pp 566–567)

166. **(A)** There are four places of narrowing along the esophagus where foreign bodies are prone to lodge and where swallowed corrosives produce the greatest injury. Three of these are listed as items 1, 2, and 3 of this question. The other site is at the beginning of the esophagus, in the neck, where it is surrounded by the upper esophageal sphincter. The lower esophageal sphincter is not an anatomical but a "physiological sphincter," the most important function of which is to prevent regurgitation of gastric contents into the esophagus. (Hollinshead, pp 567–568)

167. **(B)** The esophagus is innervated by branches of the vagus and the sympathetic trunks. Vagal branches to the striated muscle (branchial efferents) are comprised of the fibers of the accessory (11th cranial) nerve. Branches to the smooth muscle are preganglionic, visceral efferents of the vagus nerve. Effects of the sympathetic visceral innervation on esophageal muscle are unknown. Unlike other portions of the alimentary tract, the esophagus does not have spontaneous peristaltic contractions; its waves of contraction are initiated by swallowing. (Hollinshead, p 569)

168. **(B)** Several small visceral branches are given off from the anterior surface of the descending thoracic aorta. These include two or more bronchial and esophageal arteries. The coronary arteries arise from the aorta just above its origin, and the brachiocephalic artery springs from the arch of the aorta in the midline. (Hollinshead, pp 570,561,533)

169. **(E)** The primary function of the azygos venous system is to drain blood from the body wall. The system, however, does receive venous blood from some of the thoracic viscera. All of the other statements relating to this system of veins are true. (Hollinshead, pp 570–571)

170. **(C)** The thoracic duct returns lymph, draining from the greater part of the body, to the venous system. This duct is the upward continuation of the cisterna chyli. In most of its course it lies behind the esophagus and ends at the confluence of the left subclavian and left internal jugular veins. It enters the thorax behind the aorta through the aortic, not the esophageal, hiatus. (Hollinshead, p 572)

171. **(D)** The only correct answer is item 4. Preganglionic sympathetic fibers enter the sympathetic trunk and synapse with neurons in the trunk ganglia. The postganglionic fibers return to the spinal nerve via gray rami communicantes. Each sympathetic trunk extends from the level of the first cervical vertebra to the tip of the sacrum. The

preganglionic fibers of the sympathetic system usually leave the spinal cord only through the 12 thoracic and the first two lumbar nerves. The preganglionic neurons are not found at all spinal cord levels but only at the thoracic and the first two lumbar levels, as stated above. Many of the preganglionic fibers do not terminate in the sympathetic trunk ganglia but merely pass through it and proceed toward the viscera. See reference for further details. *(Hollinshead, pp 574,59–61)*

172. **(E)** All of these listed groups of lymph nodes are located in the mediastinum. The posterior diaphragmatic, the intercostal, and the posterior mediastinal lymph nodes of the left side send their efferent vessels directly into the thoracic duct. The efferents of the parasternal and tracheobronchial nodes form the right and left bronchomediastinal lymph trunks. Before its termination, the thoracic duct usually receives the left internal jugular and subclavian lymph trunks. *(Hollinshead, p 572)*

173. **(D)** The only function listed for which the vagus nerve is responsible is bronchoconstriction. Diaphragmatic contraction is the function of the phrenic nerve. The sympathetics furnish secretomotor fibers to the glands of the bronchial tree. Sympathetics also produce vasoconstriction of pulmonary blood vessels. *(Hollinshead, pp 502–503)*

---

# CHAPTER 6

# Pelvis and Perineum
## Questions

**DIRECTIONS (Questions 1 through 100): Each of the numbered items or incomplete statements in this section is followed by answers or by completions of the statement. Select the ONE lettered answer or completion that is BEST in each case.**

1. Which of the following types of pelves predominates in the male?

   (A) anthropoid and android
   (B) gynecoid and android
   (C) platypelloid and anthropoid
   (D) gynecoid and platypelloid
   (E) platypelloid and android

2. Which of these statements does NOT describe differences between male and female pelves accurately?

   (A) the female pelvis is more cylindrical than the male
   (B) the female sacrum is shorter and wider than the male
   (C) the female pubic tubercles are farther apart than the male
   (D) the female pubic arch makes a more acute angle than in the male
   (E) the female anterolateral pelvic wall is relatively wider than in the male

3. The pelvic diaphragm is composed of all the following muscles EXCEPT the

   (A) coccygeus
   (B) iliococcygeus
   (C) piriformis
   (D) pubococcygeus
   (E) puborectalis

4. Visceral branches of the internal iliac artery include all EXCEPT which one of the following?

   (A) umbilical
   (B) inferior vesical
   (C) middle rectal
   (D) uterine
   (E) iliolumbar

5. Which of the following statements about the sacral plexus is true?

   (A) it takes form on the anterior wall of the pelvis
   (B) its major part lies on the obturator internus muscle
   (C) it gives off the pudendal nerve
   (D) most of its branches pass above the piriformis to appear in the buttock
   (D) it supplies the medial cutaneous area below the knee

6. Which of these nerves is formed by posterior divisions of the sacral plexus?

   (A) common peroneal
   (B) anococcygeal
   (C) pudendal
   (D) nerve to the quadratus femoris
   (E) nerve to the obturator internus and superior gemellus

7. Which of these nerves is formed by anterior divisions of the sacral plexus?

   (A) superior gluteal
   (B) tibial
   (C) inferior gluteal
   (D) nerve to the piriformis
   (E) lateral part of the posterior femoral cutaneous nerve

8. These statements about pelvic splanchnic nerves are correct EXCEPT for which of the following?

   (A) they contain preganglionic parasympathetic fibers
   (B) they convey visceral afferents from the pelvic plexus to sacral segments of the spinal cord
   (C) they contribute to formation of the inferior hypogastric plexus
   (D) they are branches of the sciatic nerve
   (E) they are known as nervi erigentes

9. Which of these statements correctly describes the obturator nerve?

 (A) it arises from the sacral plexus
 (B) it passes along the lateral pelvic wall
 (C) it supplies abductor muscles of the thigh
 (D) it arises from posterior rami of sacral nerves
 (E) it gives off pelvic splanchnic nerves

10. In relation to autonomic plexuses of the pelvis, all these items are correct EXCEPT which one of the following?

 (A) the superior rectal plexus consists chiefly of sympathetic fibers
 (B) the ovarian plexus consists chiefly of sympathetic fibers
 (C) parasympathetic components predominate in the inferior hypogastric plexus
 (D) the superior hypogastric plexus is known as the pelvic plexus
 (E) hypogastric nerves help to form the inferior hypogastric plexus

11. Components of the superior hypogastric plexus include all but which one of the following?

 (A) Preganglionic sympathetic fibers
 (B) Postganglionic sympathetic fibers
 (C) Visceral afferents for pain
 (D) Visceral efferents from T10–L2
 (E) Parasympathetic fibers

12. Which of these statements does NOT describe correctly the pelvic splanchnic nerves?

 (A) they are known as nervi erigentes
 (B) they are major pathways for visceral afferents from pelvic viscera
 (C) afferents from the uterus follow a route through these nerves
 (D) their visceral efferent fibers have cell bodies at spinal cord levels S2–4
 (E) their visceral afferents ascend to a dorsal root ganglion cell body of S2–4 spinal nerves

13. Correct description of the rectum includes which of the following statements?

 (A) it begins at the rectosigmoid junction
 (B) it structurally is straight, as its name implies
 (C) the rectal ampulla is at its upper end
 (D) the rectourethral muscle is found only in the female
 (E) its mucosa is identical to that of the sigmoid colon

14. The rectum receives its blood supply primarily by the branches of the

 (A) superior mesenteric artery
 (B) inferior mesenteric artery
 (C) inferior rectal arteries
 (D) branches of the external iliac artery
 (E) femoral artery

15. Which one of the following statements is incorrect about innervation of the rectum?

 (A) its motor fibers include sympathetics
 (B) its motor fibers are conveyed in the middle rectal plexus
 (C) its afferent supply of nerves belongs to the parasympathetic system
 (D) pelvic splanchnic nerves are involved in its innervation
 (E) the superior rectal plexus may supply rectal blood vessels

16. All of these statements are true about the urinary bladder EXCEPT that

 (A) its base is the superior posterior surface
 (B) the median umbilical ligament is attached to its apex
 (C) its neck leads into the urethra
 (D) when distended, it becomes lower in position
 (E) its body shows superior and inferolateral surfaces

17. Which of these structures is NOT a component of the bladder?

 (A) vesical trigone
 (B) detrusor muscle
 (C) pubovesical muscle
 (D) ureteric ostia
 (E) vesical sphincter

18. Correct relationships of the female ureters include which of the following?

 (A) the ureters pass lateral to the vagina
 (B) the ureters are not in close proximity to the vesical nerve plexus
 (C) at the base of the broad ligament the uterine artery crosses above and in front of the ureters
 (D) the ureters enter the anterior aspect of the bladder
 (E) at the pelvic brim the ovarian vessels cross medial to the ureters

19. The male urethra traverses all of the following structures EXCEPT the

 (A) prostate gland
 (B) ejaculatory duct
 (C) urogenital diaphragm
 (D) sphincter urethra
 (E) internal urethral orifice

20. Which of these statements correctly describes the seminal vesicles?

    (A) they store sperm
    (B) they secrete a seminal fluid component
    (C) they lie medial to the ductus deferens
    (D) they form the ampulla of the ductus deferens
    (E) they empty directly, and alone, into the prostate

21. Which one of these structures is NOT a part of the female internal genital system?

    (A) ovary
    (B) perineal body
    (C) mesovarium
    (D) infundibulum
    (E) vagina

22. The ovarian artery is a branch of the

    (A) internal iliac
    (B) external iliac
    (C) inferior epigastric
    (D) abdominal aorta
    (E) uterine artery

23. Which of these nerves is NOT included in the utero-vaginal nerve plexus?

    (A) sympathetics
    (B) parasympathetics
    (C) somatic
    (D) afferent
    (E) vasomotor

24. Lymphatic drainage of the uterus includes all the following EXCEPT the

    (A) external iliac nodes
    (B) inferior mesenteric nodes
    (C) internal iliac nodes
    (D) lumbar nodes
    (E) superficial inguinal nodes

25. Which of these is characteristic of the vagina?

    (A) its anterior fornix is deeper than the posterior fornix
    (B) it terminates at the urogenital diaphragm
    (C) it fuses around the cervix of the uterus
    (D) normally it is flattened laterally
    (E) it allows little distention

26. The superior hypogastric plexus contains all these types of fibers EXCEPT

    (A) preganglionic sympathetic fibers
    (B) postganglionic sympathetic fibers
    (C) motor fibers to striated muscle
    (D) small ganglia
    (E) visceral afferent fibers

27. The greater part of the pelvic diaphragm is formed from the

    (A) obturator internus muscle
    (B) pelvic fascia
    (C) perineal membrane
    (D) levator ani
    (E) coccygeus

28. Visceral branches of the internal iliac artery include which of these?

    (A) middle rectal
    (B) obturator
    (C) iliolumbar
    (D) lateral sacral
    (E) internal pudendal

29. Branches of the internal iliac artery to the pelvic wall and lower limb include which of the following?

    (A) inferior vesical
    (B) superior vesical
    (C) umbilical
    (D) uterine
    (E) superior gluteal

30. Which of the following statements correctly describes the urogenital region of the perineum?

    (A) it contains the rectum
    (B) it contains the ischiorectal fossa
    (C) it includes the coccygeal triangle
    (D) it is pierced by the urethra
    (E) its floor is the obturator internus muscle

31. All of the following are correct statements about the pelvic splanchnic nerves EXCEPT which of these items?

    (A) they are parasympathetic nerves
    (B) they spring from cord levels S2, S3, S4
    (C) they are the sole innervation of prostatic musculature
    (D) they contribute to the formation of the pelvic plexus
    (E) they are the sole motor fibers to the rectum

32. Which of the following arteries is the chief blood supply of the perineum?

    (A) obturator
    (B) superior gluteal
    (C) iliolumbar
    (D) external iliac
    (E) internal pudendal

33. The major innervation of the perineum is which of these nerves?

    (A) sciatic
    (B) pudendal
    (C) femoral
    (D) obturator
    (E) inferior gluteal

34. The morphological homologue in the female of the penis in the male is the

    (A) clitoris
    (B) labia majora
    (C) labia minora
    (D) vestibular bulbs
    (E) vulva

35. Most perineal structures send their lymphatics to the

    (A) internal iliac nodes
    (B) nodes of the lumbar chain
    (C) superficial inguinal nodes
    (D) external iliac nodes
    (E) inferior mesenteric nodes

36. Which one of these statements about the ischiorectal fossa is incorrect?

    (A) they are spaces on each side of the anal canal
    (B) they contain the pudendal nerve and internal pudendal vessels
    (C) they contain connective tissue and fat
    (D) their inferior walls are formed by the levator ani muscle
    (E) they are bordered laterally by the obturator internus muscle

37. The inferior pelvic aperture is bordered by all the following EXCEPT the

    (a) inferior margin of the pubic symphysis
    (B) rami of the pubis and ischium
    (C) coccyx
    (D) sacrotuberous ligament
    (E) iliac crest

38. The urogenital diaphragm includes which of these structures?

    (A) a continuous sheet of smooth muscle
    (B) the anal opening
    (C) superior and inferior fascias
    (D) the coccygeus
    (E) the obturator internus muscle

39. The deep perineal space is described correctly by which of the following statements?

    (A) it is filled completely by the deep perineal muscles

    (B) in the male it surrounds the scrotum and penis
    (C) it is limited by the superficial fascia
    (D) its fascia is a continuation of abdominal fascia
    (E) it communicates directly with the ischiorectal fossa

40. The superficial perineal space includes which one of these structures?

    (A) deep transverse perineal muscle
    (B) puborectalis
    (C) sphincter urethrae
    (D) superficial perineal fascia
    (E) perineal body

41. Which of these statements correctly describes the course of the ureter?

    (A) the right ureter is related to the base of the sigmoid colon
    (B) after crossing the pelvic brim, the ureter lies lateral to the internal iliac artery
    (C) in the male, as it nears the bladder, it passes above the ductus deferens
    (D) the ureter enters the superior aspect of the bladder
    (E) in the female the ureter crosses behind and below the uterine artery

42. Nerves required for erection of the penis are the

    (A) pudendals
    (B) inferior rectals
    (C) pelvic splanchnics
    (D) posterior scrotals
    (E) sympathetics

43. The ejaculatory duct opens into the

    (A) prostatic urethra
    (B) bulbourethral gland
    (C) ureter
    (D) penis
    (E) ductus deferens

44. Which of the following structures does NOT attach into the perineal body?

    (A) perineal membrane
    (B) ischiocavernosus
    (C) sphincter ani externus
    (D) transversus perinei superficialis
    (E) bulbospongiosus

45. Which of these statements correctly describes the bulbourethral glands?

    (A) they are found only in the female
    (B) they open into the membranous urethra

(C) they are located in the deep perineal pouch

(D) they are located close to the sides of the anus

(E) their ducts open into the scrotum

46. Territories supplied by the pudendal nerve include all the following EXCEPT the

(A) pelvic surface of the levator ani

(B) anal triangle

(C) urogenital triangle

(D) labium majus

(E) scrotum

47. Which of the following is a branch of the internal pudendal artery?

(A) superior gluteal

(B) uterine

(C) inferior vesical

(D) umbilical

(E) inferior rectal

48. Correct statements regarding the middle rectal vein include which of these?

(A) it begins in the anal columns

(B) it becomes the inferior mesenteric vein

(C) it has no valves

(D) it is a more important vessel than the middle rectal artery

(E) it drains the rectum below the internal sphincter

49. The superior gluteal artery is a branch of the

(A) obturator

(B) internal iliac

(C) internal pudendal

(D) iliolumbar

(E) middle rectal

50. Branches of the sacral plexus include which of these nerves?

(A) obturator

(B) femoral

(C) sciatic

(D) iliohypogastric

(E) genitofemoral

51. Which of the following items correctly describes the pelvic splanchnic nerves?

(A) they send motor branches to the coccygeus

(B) they contract the sphincter ani internus

(C) they are sensory to the perineum

(D) they cause contraction of the arteries of the penis

(E) they arise from spinal levels S2, S3, S4

52. Which of the following nerves passes through the greater sciatic foramen and largely innervates the perineum?

(A) superior gluteal

(B) pudendal

(C) inferior gluteal

(D) sciatic

(E) posterior cutaneous nerve of the thigh

53. Which of these structures is attached to the posterior aspect of the broad ligament?

(A) ligament of the ovary

(B) uterine tube

(C) round ligament of the uterus

(D) uterine artery

(E) ureter

54. All of the following statements correctly describe the ovary EXCEPT

(A) it is covered with cuboidal epithelium

(B) its anterior border is attached to the broad ligament by the mesovarium

(C) the suspensory ligament of the ovary suspends its tubal pole

(D) it has a smooth surface

(E) the ligament of the ovary attaches it to the lateral margin of the uterus

55. The abdominal orifice of the uterine tube is located at its

(A) ampulla

(B) isthmus

(C) uterine part

(D) fundus

(E) infundibulum

56. The part of the uterus that rises above the uterine tubes is the

(A) external os

(B) body

(C) fundus

(D) cervix

(E) fornix

57. Correct relations of the female ureters include which of the following?

(A) they are equidistant from each side of the cervix

(B) they cross the lateral fornix of the vagina

(C) they cross above the broad ligament

(D) they enter the bladder behind the vagina

(E) they cross above the uterine artery

58. These statements regarding innervation of the uterus are correct EXCEPT for which one of the following?

    (A) it includes many parasympathetic efferents
    (B) the uterovaginal plexus holds many sympathetic efferent fibers
    (C) afferents from the body of the uterus travel with the sympathetic system
    (D) pain fibers from the uterus enter the spinal cord through the last two thoracic nerves
    (E) innervation is not necessary to the function of this organ

59. Which of the following structures is most important for support of the female pelvic viscera?

    (A) the supporting fasciae
    (B) the perivascular stalk as a suspensory structure
    (C) the uterosacral ligament
    (D) the levator ani muscle
    (E) the round ligament of the uterus

60. Which of these statements describes the superior hypogastric plexus correctly?

    (A) it lies above the bifurcation of the aorta
    (B) it contains mostly parasympathetic nerves
    (C) it is a downward prolongation of the preaortic plexus
    (D) it is joined by the first and second lumbar splanchnic nerves
    (E) its large branches do not contribute to the pelvic plexus

61. Functions of the pelvic splanchnic nerves include which of these?

    (A) to contract arteries of the erectile tissue
    (B) to carry motor fibers to striated muscle
    (C) to supply muscle of the prostate
    (D) to carry sensory fibers from the body of the uterus
    (E) to control muscular walls of the bladder

62. Pelvic lymph nodes include all the following EXCEPT the

    (A) common iliac
    (B) internal iliac
    (C) lateral sacral
    (D) superior hypogastric
    (E) median sacral

63. Which of these statements describes the extent of the perineum?

    (A) from the symphysis pubis to the central perineal tendon
    (B) from the symphysis pubis to the tip of the coccyx

    (C) from the central perineal tendon to the tip of the coccyx
    (D) from the ischial tuberosities to the symphysis pubis
    (E) from the symphysis pubis to the anus

64. The anterior half of the inferior pelvic aperture is closed by which of these structures?

    (A) the levator ani muscle
    (B) the coccygeus muscle
    (C) the urogenital diaphragm
    (D) the arcuate pubic ligament
    (E) the urethral sphincter muscle

65. All of the following structures are found in the area of the urogenital triangle EXCEPT the

    (A) external anal sphincter
    (B) sphincter urethrae muscle
    (C) arcuate pubic ligament
    (D) deep transversus perinei muscle
    (E) perineal membrane

66. The wedge-shaped mass of fibrous tissue located at the center of the perineum is the

    (A) urorectal septum
    (B) perineal membrane
    (C) ischiorectal tendon
    (D) sacrotuberous ligament
    (E) perineal body

67. In the male, the superficial perineal space contains all EXCEPT which one of these structures?

    (A) the root of the penis
    (B) branches of internal pudendal vessels
    (C) the proximal part of the spongy urethra
    (D) the membranous urethra
    (E) the pudendal nerve

68. In the female, the deep perineal space contains all EXCEPT which one of the following structures?

    (A) the greater vestibular glands
    (B) part of the urethra
    (C) the inferior part of the vagina
    (D) the deep transverse perineal muscles
    (E) the sphincter urethrae

69. The ischiorectal fossae are described correctly by which of these statements?

    (A) there is no communication between the two sides
    (B) they are located on each side of the rectum
    (C) they are filled with watery fluid
    (D) anteriorly they continue inferior to the urogenital diaphragm
    (E) they have no relation to rectal function

70. Which of the following nerves does NOT innervate the scrotum?

    (A) the ilioinguinal
    (B) medial branches of the perineal
    (C) branches of the pudendal
    (D) inferior gluteal
    (E) posterior femoral cutaneous

71. Infection may reach the ischiorectal fossa from all EXCEPT which one of the following?

    (A) inflammation of the anal sinuses
    (B) downward extension of a perirectal abscess
    (C) following a tear in the anal mucous membrane
    (D) a penetrating wound in the anal region
    (E) a ruptured urethra

72. Most of the innervation of the perineum is supplied by which of these nerves?

    (A) obturator
    (B) pudendal
    (C) femoral
    (D) sciatic
    (E) sympathetics

73. The deep perineal space of the male contains which of these structures?

    (A) bulbourethral glands
    (B) bulbospongiosus muscle
    (C) ischiocavernosus muscle
    (D) the spongy urethra
    (E) the root of the penis

74. The blood supply of the scrotum includes which of these arteries?

    (A) inferior vesical
    (B) umbilical
    (C) external pudendal
    (D) inferior gluteal
    (E) superior rectal

75. Which of the following structures is NOT located in the female superficial perineal space?

    (A) superficial transverse perineal muscle
    (B) ischiocavernosus muscle
    (C) bulbospongiosus muscle
    (D) greater vestibular glands
    (E) dorsal nerve of the clitoris

76. Embryologically, which of these structures in the female is homologous to the scrotum in the male?

    (A) labia majora
    (B) labia minora
    (C) mons pubis
    (D) vestibule of the vagina
    (E) hymen

77. The clitoris of the female is similar to the penis of the male EXCEPT that it

    (A) is composed of erectile tissue
    (B) is composed of two crura
    (C) has two corpora cavernosa
    (D) has a corpus spongiosum
    (E) is suspended by a suspensory ligament

78. Some obstetricians are reluctant to perform a median episiotomy because it may

    (A) injure the baby's head
    (B) involve the external anal sphincter
    (C) enlarge the distal end of the birth canal
    (D) injure the urethral sphincter
    (E) lacerate the labia minora

79. Which of the following statements correctly compares the female pelvis with the male pelvis?

    (A) the hip bones are closer together than in the male
    (B) the sacrum is narrower than in the male
    (C) there are fewer prominent bony markings than in the male
    (D) the ischial tuberosities are closer together than in the male
    (E) the subpubic angle of the pubic arch is less than in the male

80. The diagonal conjugate diameter of the pelvis is a measurement of the superior pelvic aperture between which of the following points?

    (A) the midpoint of the inferior border of the symphysis pubis to the midpoint of the sacral promontory
    (B) the midpoint of the superior border of the symphysis pubis to the midpoint of the sacral promontory
    (C) transversely from the linea terminalis on one side to this line on the opposite side
    (D) from one iliopubic eminence to the opposite sacroiliac joint
    (E) from one ischial spine to the opposite ischial spine

81. A pelvis in which the anteroposterior (AP) diameter of the superior pelvic aperture is short and the transverse diameter is long is called

    (A) gynecoid
    (B) anthropoid
    (C) platypelloid
    (D) android
    (E) piriform

82. Which of these muscles crosses the pelvic brim?

    (A) piriformis
    (B) obturator internus
    (C) levator ani
    (D) coccygeus
    (E) none

83. Which of these muscles is a part of the pelvic diaphragm?

    (A) obturator internus
    (B) piriformis
    (C) puborectalis
    (D) quadratus femoris
    (E) deep transverse perineal muscle

84. Which of these nerves does NOT arise from the sacral plexus?

    (A) sciatic
    (B) obturator
    (C) pudendal
    (D) superior gluteal
    (E) inferior gluteal

85. The nerve most likely to be injured during removal of cancerous lymph nodes from the side wall of the pelvis is the

    (A) femoral
    (B) lumbosacral trunk
    (C) sciatic
    (D) pudendal
    (E) obturator

86. All of the following arteries enter the pelvis minor EXCEPT the

    (A) internal iliac (paired)
    (B) median sacral
    (C) femoral
    (D) superior rectal
    (E) ovarian (paired)

87. The superior rectal artery is a branch of which of these arteries?

    (A) inferior mesenteric
    (B) umbilical
    (C) inferior vesical
    (D) internal pudendal
    (E) middle rectal

88. Which of the following statements about the umbilical vein is correct?

    (A) it closes immediately at the time of birth
    (B) the obliterated vein becomes the ligamentum teres of the liver
    (C) the obliterated vein becomes the falciform ligament
    (D) the obliterated vein becomes the median umbilical ligament
    (E) it drains into the common iliac vein

89. In the female, which of these statements about the route of the pelvic part of the ureter is correct?

    (A) it descends on the anterior wall of the pelvis
    (B) it forms the anterior boundary of the ovarian fossa
    (C) it passes lateral to the origin of the uterine artery
    (D) at the level of the ischial spine, it is crossed superiorly by the uterine artery
    (E) it enters the anterior angle of the bladder

90. In the male, obstruction of the ureter by ureteric stones occurs most often at which of the following portions of this structure?

    (A) nearest the kidney
    (B) on the side wall of the pelvis
    (C) just superior to the ischial spine
    (D) just lateral to the ductus deferens
    (E) where it crosses the external iliac artery and the brim of the pelvis

91. The base of the urinary bladder is its

    (A) posterior surface
    (B) anterior end
    (C) inferolateral surface
    (D) neck
    (E) superior surface

92. Arteries of the urinary bladder include all EXCEPT which one of the following?

    (A) superior vesical
    (B) inferior vesical
    (C) external iliac
    (D) inferior gluteal
    (E) obturator

93. Which statement about the male membranous urethra is true?

    (A) it is the longest portion of the urethra
    (B) it is the widest portion of the urethra
    (C) it traverses the hymen
    (D) it is the least dilatable part of the urethra
    (E) it has on either side a greater vestibular gland

94. The ejaculatory ducts open into which of the following structures?

    (A) external urethral orifice
    (B) urethral crest
    (C) ureter
    (D) prostatic urethra
    (E) urinary bladder

**95.** All of these are true statements about the female urethra EXCEPT which one of the following?

(A) it corresponds to the prostatic and membranous parts of the male urethra
(B) it is longer than the male urethra
(C) it passes through the pelvic diaphragm
(D) inferiorly it is associated intimately with the vagina
(E) it passes through the urogenital diaphragm

**96.** Which of these structures is a male accessory genital gland?

(A) testes
(B) deferent ducts
(C) seminal vesicles
(D) ejaculatory ducts
(E) prostate

**97.** The ductus deferens ends in which of the following?

(A) testes
(B) seminal vesicles
(C) ejaculatory duct
(D) prostate
(E) bulbourethral gland

**98.** Correct statements about the seminal vesicles include which of the following?

(A) their ducts contribute to formation of the ejaculatory duct
(B) their secretion probably is not valuable
(C) they are situated posterior to the rectum
(D) they store sperm
(E) they lie in front of the ureters

**99.** Which of these characteristics is NOT true of benign, nodular hyperplasia of the prostate?

(A) common in older males
(B) presence of urethral obstruction
(C) projection of the prostate into the urinary bladder
(D) impedance of urinary flow
(E) commonly, a poor reaction to surgery

**100.** Descriptions of the position of the uterus include all but which one of these?

(A) anteversion
(B) anteflexion
(C) vertical proflexion
(D) retroversion
(E) retroflexion

**DIRECTIONS (Questions 101 through 112):** Each group of items in this section consists of lettered headings followed by a set of numbered words or phrases. For each numbered word or phrase, select the ONE lettered heading that is most closely associated with it. Each lettered heading may be selected once, more than once, or not at all.

**Questions 101 through 104**

(A) lower third of the uterus
(B) expanded upper two thirds of the uterus
(C) rounded upper part of the body of the uterus
(D) region where the uterine tubes enter the uterus
(E) vaginal recess

**101.** fundus

**102.** cervix

**103.** cornu

**104.** fornix

**Questions 105 through 108**

(A) principal support of the uterus
(B) structure conveying the dividing zygote to the uterine cavity
(C) structure suspended by the mesovarium
(D) separation between the uterus and the bladder
(E) structure holding the uterus in a relatively normal position

**105.** broad ligament

**106.** uterine tube

**107.** ovary

**108.** levator ani

**Questions 109 through 112**

(A) distal end of the uterine tube
(B) widest and longest part of the uterine tube
(C) part of the uterine tube that joins the horn of the uterus
(D) fringed folds at the distal end of the uterine tube
(E) uterine part of the uterine tube

**109.** ampulla

**110.** fimbria

111.  infundibulum

112.  isthmus

**DIRECTIONS (Questions 113 through 124): Each group of items in this section consists of lettered headings followed by a set of numbered words or phrases. For each numbered word or phrase, select**

A if the item is associated with (A) only,
B if the item is associated with (B) only,
C if the item is associated with both (A) and (B),
D if the item is associated with neither (A) nor (B).

**Questions 113 through 116**

(A) rectum
(B) anal canal
(C) both
(D) neither

113.  puborectalis

114.  sigmoid colon

115.  rectouterine pouch

116.  pectinate line

**Questions 117 through 120**

(A) vagina
(B) hemorrhoids
(C) both
(D) neither

117.  anal canal

118.  rectal veins

119.  rectocele

120.  portal system of veins

**Questions 121 through 124**

(A) sympathetics
(B) parasympathetics
(C) both
(D) neither

121.  superior hypogastric plexus

122.  pudendal nerve

123.  pelvic splanchnic nerves

124.  inferior hypogastric plexus

**DIRECTIONS (Questions 125 through 184): For each of the items in this section, ONE or MORE of the numbered options is correct. Choose the answer**

A if only 1, 2, and 3 are correct,
B if only 1 and 3 are correct,
C if only 2 and 4 are correct,
D if only 4 is correct,
E if all are correct.

125.  The boundaries of the perineum include the

(1) symphysis pubis
(2) inferior pubic rami
(3) ischial rami
(4) sacrotuberous ligament

126.  The urogenital triangle contains the

(1) urethra
(2) root of the scrotum
(3) penis
(4) anus

127.  The anal region contains the

(1) perineal membrane
(2) deep transverse perineal muscle
(3) ischiocavernosus muscle
(4) ischiorectal fossa

128.  Muscles found in the superficial perineal space include which of the following?

(1) sphincter urethra
(2) deep transverse perineal muscle
(3) levator ani
(4) bulbospongiosus

129.  The arterial blood of the scrotum is provided by the

(1) external pudendal arteries
(2) internal pudendal arteries
(3) testicular arteries
(4) inferior vesical arteries

130.  The scrotum is supplied by which of the following nerves?

(1) ilioinguinal
(2) lateral femoral cutaneous
(3) branches of the perineal
(4) obturator

131.  Which of the following statements correctly describes the labia minora in the female perineum?

(1) they are folds of fat-free, hairless skin
(2) they lie between the labia majora
(3) they enclose the vestibule of the vagina
(4) they are united posteriorly by the frenulum of the labia minora

132. In the female, the external urethral orifice has which of these characteristics?

    (1) it is located 2 to 3 cm anterior to the clitoris
    (2) it has paraurethral glands opening on each side of it
    (3) it is surrounded by the hymen
    (4) usually it is a median slit

133. Which of these statements correctly describes the greater vestibular glands of the female?

    (1) they are four in number
    (2) they are located in the deep perineal space
    (3) they are homologous with the prostate of the male
    (4) they are located on each side just posterior to the bulb of the vestibule

134. Correct statements about the hip bones include which of the following?

    (1) fusion of the bony components is complete at about 5 years of age
    (2) they are connected to the sacrum on each side
    (3) the pubic bones are joined by a synovial joint
    (4) three bones make up the acetabulum

135. Boundaries of the pelvis major include which of these?

    (1) abdominal wall
    (2) sacrum
    (3) iliac fossa
    (4) symphysis pubis

136. Characteristics of a gynecoid pelvis include which of the following?

    (1) a circular superior aperture
    (2) a wide subpubic arch
    (3) widely spaced ischial spines
    (4) resemblance to a shallow, flat bowl

137. The iliolumbar ligaments are especially important because they

    (1) hold the sacrum in position
    (2) limit axial rotation of L5 vertebra on the sacrum
    (3) give origin to several muscles
    (4) assist the vertebral articular processes in preventing forward gliding of L5 vertebra on the sacrum

138. Which of these are functions of the sacrotuberous ligament?

    (1) it binds the sacrum to the ischium
    (2) it permits some movement of the sacrum
    (3) it resists backward rotation of the inferior end of the sacrum
    (4) it provides an origin for the piriformis muscle

139. True statements about the obturator internus muscle include which of the following?

    (1) it covers the side wall of the pelvis minor
    (2) it originates on the obturator membrane
    (3) it inserts into the greater trochanter of the femur
    (4) it leaves the pelvis via the greater sciatic foramen

140. The pelvic diaphragm includes which of these muscles?

    (1) piriformis
    (2) deep transverse perineal
    (3) obturator internus
    (4) levator ani

141. The coccygeus muscle is important because it

    (1) supports the fetal head during childbirth
    (2) forms the posterior part of the pelvic diaphragm
    (3) forms a U-shaped sling around the rectum
    (4) pulls the coccyx forward

142. Which of these nerves are branches of the sacral plexus?

    (1) pudendal
    (2) sciatic
    (3) inferior gluteal
    (4) obturator

143. Anterior branches of the internal iliac artery include which of the following?

    (1) umbilical
    (2) median sacral
    (3) obturator
    (4) superior rectal

144. Which of the following correctly describes the uterine artery?

    (1) it passes through the lesser sciatic foramen
    (2) usually it arises separately from the internal iliac
    (3) it is a direct continuation of the inferior mesenteric artery
    (4) it passes above the ureter near the lateral fornix of the vagina

145. Which of these correctly describes the internal pudendal artery?

    (1) it passes through the greater sciatic foramen
    (2) it is a branch of the internal iliac artery (posterior division)
    (3) it passes through the lesser sciatic foramen to enter the ischiorectal fossa
    (4) it anastomoses with the uterine artery

| SUMMARY OF DIRECTIONS | | | | |
|---|---|---|---|---|
| A | B | C | D | E |
| 1, 2, 3 only | 1, 3 only | 2, 4 only | 4 only | All are correct |

146. Pelvic venous plexuses are important clinically because

    (1) the rectal venous plexuses drain into the superior, middle, and inferior rectal veins
    (2) cancer cells may metastasize from the prostatic venous plexus
    (3) veins from the internal rectal plexus may become varicose
    (4) frequently portal obstruction results in rectal varicosities

147. The ureters receive their main blood supply from which of these sources?

    (1) renal artery
    (2) common iliac
    (3) vesical artery
    (4) middle rectal artery

148. Correct aspects of the urinary bladder include which of these?

    (1) its wall has three muscular layers
    (2) its superior surface receives openings of the ureters
    (3) the ureters pass obliquely through the bladder wall
    (4) sympathetic fibers are motor to the detrusor muscle

149. Which of the following correctly describes the male urethra?

    (1) the membranous part is its longest portion
    (2) the prostatic part begins at the urethral orifice of the bladder
    (3) the ejaculatory ducts open into the membranous part
    (4) ducts of the bulbourethral glands open into the spongy part

150. Features of the female urethra consist of which of these?

    (1) it is not easily distended
    (2) often it has urethral stricture
    (3) commonly it shows congenital hypospadias
    (4) compared to the male, it is relatively short

151. Which of the following statements correctly describes the ductus deferens?

    (1) it ends in the ejaculatory duct
    (2) it has few autonomic nerve fibers

    (3) its main artery is a branch of the umbilical
    (4) it is a thin-walled tube

152. Functions of the seminal vesicles include which of these?

    (1) secretion of a thick fluid
    (2) storage of sperm
    (3) promotion of sperm activation
    (4) promotion of ejaculatory duct opening

153. Which of these correctly describe vessels of the prostate?

    (1) it receives arteries from the internal pudendal artery
    (2) veins of the prostatic plexus drain into the external iliac vein
    (3) it receives branches from the middle rectal arteries
    (4) it receives arteries from the renal vessels

154. Benign prostatic hypertrophy results in which of these symptoms?

    (1) nocturia
    (2) dysuria
    (3) urgency
    (4) extravasation of urine into the superficial perineal pouch

155. Important relations of the vagina consist of which of the following?

    (1) its anterior wall is in contact with the cervix
    (2) its anterior wall is in contact with the base of the bladder
    (3) it is connected intimately with the urethra
    (4) usually its superior limit is covered by peritoneum

156. The wall of the uterus consists of which of these?

    (1) perimetrium
    (2) myometrium
    (3) endometrium
    (4) parametrium

157. Uterine tubes receive their blood from which of these arteries?

    (1) uterine
    (2) vesical
    (3) ovarian
    (4) umbilical

158. Which of the following characterize the ovarian arteries?

    (1) they run medial to the ovary on the uterine tube
    (2) they arise from the abdominal aorta

(3) they anastomose with the internal pudendal arteries

(4) they send branches to the ovary via the mesovarium

159. Correct anatomical features of the rectum include which of these statements?

(1) the iliococcygeus muscle surrounds it as a sling
(2) in the male it lies immediately anterior to the prostate
(3) it is covered entirely by the peritoneum
(4) superiorly it is continuous with the sigmoid colon

160. Anterior relations of the rectum consist of which of these?

(1) base of the bladder
(2) terminal parts of the ureters
(3) deferent ducts
(4) seminal vesicles

161. Which of these statements correctly describes the rectal vessels?

(1) rectal arteries anastomose freely with each other
(2) middle rectal arteries supply the terminal part of the sigmoid colon
(3) the superior rectal artery is a continuation of the inferior mesenteric artery
(4) inferior rectal arteries branch from the internal iliac branches

162. Which of the following structures can be palpated through the rectum?

(1) ischial spines
(2) enlarged internal iliac lymph nodes
(3) ischiorectal abscess
(4) prostate

163. Which of these structures relate to the anal canal?

(1) septum
(2) columns
(3) transverse folds
(4) valves

164. Identify the artery or arteries NOT supplying the anal canal.

(1) superior rectal
(2) median sacral
(3) middle rectal
(4) iliolumbar

165. The lymph vessels from the anal canal superior to the pectinate line drain into which of the following?

(1) internal iliac nodes

(2) common iliac nodes
(3) aortic lymph nodes
(4) superficial inguinal nodes

166. The nerve supply of the anal canal includes which of these?

(1) sympathetic nerves
(2) pudendal nerves
(3) parasympathetic nerves
(4) obturator nerves

167. In which of these ways can hemorrhoids be described correctly?

(1) they have no relation to portal hypertension
(2) internal hemorrhoids are varicosities of the superior rectal vein
(3) thrombus formation is commoner in internal than in external hemorrhoids
(4) external hemorrhoids are varicosities of the inferior rectal veins

168. Which of these statements is true regarding the pelvic autonomic nerves?

(1) sympathetic trunks unite in the ganglion impar
(2) sacral sympathetic trunks send gray rami to each ventral ramus of sacral and coccygeal nerves
(3) the superior hypogastric plexus lies just below the bifurcation of the aorta
(4) hypogastric nerves are branches of the superior hypogastric plexus

169. The inferior hypogastric plexus contains which of these nerves?

(1) hypogastric nerves
(2) direct prolongation of the intermesenteric plexus
(3) pelvic splanchnic nerves
(4) branches from the ganglion impar

170. Which of the following ligaments unites L5 transverse processes to the pelvis?

(1) supraspinous
(2) ligamenta flava
(3) interspinous
(4) iliolumbar

171. Which of these statements correctly describes the sacroiliac joints?

(1) they have considerable range of movement
(2) they are cartilagenous joints
(3) they contain a joint disc
(4) they have strong interosseous ligaments

**172.** The bones of the symphysis pubis are connected by which of these?

(1) hyaline cartilage
(2) interpubic disc
(3) superior pubic ligament
(4) arcuate pubic ligament

**173.** The structure or structures that usually rupture during the first stage of labor include the

(1) placenta
(2) amniotic sac
(3) perineal membrane
(4) chorionic sac

**174.** The umbilical artery usually supplies which of these structures?

(1) upper rectum
(2) uterus
(3) ovary
(4) upper part of the bladder

**175.** Branches of the internal iliac artery that supply the pelvis and lower limb include which of the following?

(1) iliolumbar
(2) lateral sacral
(3) obturator
(4) inferior gluteal

**176.** Somatic nerves entering the pelvis include which of these?

(1) femoral
(2) lumbosacral trunk
(3) hypogastric
(4) obturator

**177.** Which of these nerves branch from the sacral plexus?

(1) iliohypogastric
(2) genitofemoral
(3) obturator
(4) sciatic

**178.** Which of the following statements correctly describes the pelvic splanchnic nerves?

(1) they are considered as part of the sacral plexus
(2) they are given off by the ventral rami of S3 and S4

(3) they contain preganglionic sympathetic fibers
(4) they contribute to the inferior hypogastric plexus

**179.** The inferior hypogastric plexus is correctly described by which of these statements?

(1) all pelvic viscera receive efferent and afferent nerves through this plexus
(2) the superior rectal plexus nerves do not pass through this plexus
(3) parasympathetic components predominate in the plexus
(4) the plexus receives substantial contributions from the sympathetics

**180.** The superior hypogastric plexus consists of which of the following components?

(1) pelvic splanchnic nerves
(2) preganglionic and postganglionic sympathetic fibers
(3) visceral efferents from L3 and L4 spinal segments
(4) small ganglia and visceral afferents

**181.** Components of pelvic splanchnic nerves can be described accurately by which of these statements?

(1) their visceral efferents pass through ganglia of the sympathetic trunk
(2) their postganglionic neuronal cell bodies are found in the sympathetic trunk
(3) their visceral afferents mediate pain from the uterus
(4) their preganglionic cell bodies are found at spinal cord levels S2–4

**182.** Which of the following items correctly describes the rectum?

(1) it is a continuation of the sigmoid colon
(2) it presents an anteroposterior (AP) sacral flexure
(3) it has three lateral curves
(4) it shows sacculations, called haustra

**183.** Innervation of the rectum includes which of these?

(1) motor fibers are entirely parasympathetic
(2) motor fibers are conveyed in the middle rectal plexus
(3) afferent fibers concerned with pain are transmitted with the parasympathetics
(4) the superior rectal plexus is not important to general physiology of the rectum

**184.** Arteries of the prostate are derived from the

(1) inferior vesical
(2) middle rectal
(3) internal pudendal
(4) inferior gluteal

# Answers and Explanations

1. **(A)** Four major types of pelves are recognized: anthropoid, android, gynecoid, and platypelloid. In the first two, the conjugate diameter is longer than the transverse; in the latter two, the reverse is the case. Anthropoid and android pelves predominate in the male; most women have gynecoid or android types. Platypelloid pelvis is rare and shows a pronounced anteroposterior flattening; an anthropoid pelvis is flattened from side to side. *(Hollinshead, p 739)*

2. **(D)** In the male, the pubic arch makes a more acute angle than in the female. As compared to the male, the female pelvis is more cylindrical, shorter, and wider; the pubic tubercles are farther apart; and the anterolateral wall of the pubis is relatively wider. *(Hollinshead, p 740)*

3. **(C)** The pelvic diaphragm consists of the coccygeus muscle posteriorly and the more extensive levator ani. The named parts of the levator ani are the pubococcygeus, puborectalis, and iliococcygeus. The coccygeus is a small muscle arising from the ischial spine and expanding to insert into the lateral borders of the lower two sacral and upper two coccygeal segments. The piriformis is not a part of the pelvic diaphragm but is a muscle of the lower limb that lines part of the pelvic cavity. *(Hollinshead, pp 740–742)*

4. **(E)** The iliolumbar artery is a branch of the posterior trunk of the internal iliac artery; it branches to the pelvic wall, not to the viscera of the pelvis. The umbilical artery gives off superior vesical arteries to the bladder; the inferior vesical artery is present in the male and reaches the bladder and prostate. The middle rectal artery enters the rectum. The uterine artery, in addition to supplying the uterus, gives rise to the vaginal arteries and terminates in its tubal branch. *(Hollinshead, pp 745–747)*

5. **(C)** The sacral plexus gives off the chief somatic nerve of the perineum, the pudendal nerve. This plexus also sends branches to the pelvic diaphragm. The sacral plexus takes form on the posterior, not the anterior, wall of the pelvis; its major part lies on the anterior surface of the piriformis, not the obturator internus, muscle. All of its larger branches pass through the greater sciatic foramen, most of them below the piriformis, to appear in the buttock. The medial cutaneous area below the knee is supplied by the saphenous nerve, a branch of the femoral nerve, derived from the lumbar, not the sacral plexus. *(Hollinshead, p 749)*

6. **(A)** The common peroneal nerve is the only nerve listed to be formed by posterior divisions of the sacral plexus. The anococcygeal nerve, not considered to be a part of the sacral plexus, contributes to innervation of skin between the anus and tip of the coccyx. The other nerves listed are formed by anterior divisions of the plexus. *(Hollinshead, p 750)*

7. **(B)** The tibial nerve is the only nerve listed formed by anterior divisions of the sacral plexus. All the other nerves cited are formed by posterior divisions of the plexus. *(Hollinshead, p 750)*

8. **(D)** Pelvic splanchnic nerves have no particular relationship to the sciatic nerve; they are not considered to be part of the sacral plexus. The pelvic splanchnics are given off at spinal cord levels S3 and S4 (sometimes S2) and contribute to the formation of the inferior hypogastric or pelvic plexus. Their fibers are preganglionic parasympathetic fibers. These nerves also convey visceral afferents from the pelvic plexus to sacral segments of the spinal cord. They represent parasympathetic outflow and are known as the nervi erigentes because they are the nerves capable of causing erection of the penis or clitoris. *(Hollinshead, pp 751,754)*

9. **(B)** The obturator nerve arises from the anterior divisions of the lumbar plexus (L2, L3, L4). It passes along the lateral pelvic wall to the obturator canal. It supplies adductor, not abductor, muscles of the thigh. Pelvic splanchnic nerves are parasympathetic nerves, not branches of the obturator nerve. *(Hollinshead, pp 385–386,752)*

10. **(D)** The inferior hypogastric plexus is formed by lateral extensions of the superior hypogastric plexus, known as hypogastric nerves, and the pelvic splanchnic nerves. The inferior, not the superior, hypogastric plexus, is known as the pelvic plexus. Both the superior rectal and the ovarian plexuses consist chiefly of sympathetic fibers. Parasympathetic components predominate in the inferior hypogastric plexus. *(Hollinshead, pp 752–753)*

11. **(E)** The hypogastric nerves and the superior hypogastric plexus apparently do not contain any parasympathetic fibers. The superior hypogastric plexus contains a mixture of preganglionic and postganglionic sympathetic fibers, small ganglia, and visceral afferents that mediate pain sensation from the uterus. The sympathetic visceral efferents are from T10 to L2 segments; the afferents terminate in these same segments of the spinal cord. *(Hollinshead, pp 753–754)*

12. **(C)** Visceral afferents of the pelvic splanchnic nerves (parasympathetics) seem to mediate not only general visceral sensation but also pain from all pelvic organs. (This is noteworthy because pain sensation from thoracic and abdominal viscera, in general, follows the sympathetic pathway). There is evidence that the uterus is an exception to the rule. Afferent innervation from the body of the uterus travels the inferior hypogastric plexus with the sympathetic system through the superior hypogastric and aortic plexuses. Pain fibers from the body of the uterus enter the spinal cord through the last two thoracic nerves. *(Hollinshead, pp 754,787)*

13. **(A)** The rectum begins at the rectosigmoid junction in front of the third sacral vertebra. Contrary to what its name implies, the rectum is not straight but presents the anteroposterior sacral flexure and three lateral curves. The part of the rectum in the region of the middle and lower curves is called the rectal ampulla. As the rectum reaches the pelvic diaphragm, most of its longitudinal muscle fibers continue downward along the anal canal, but a few of the fibers reflect from it. In the male the anterior fibers are known as the rectourethral muscle; the slips that pass backward to the coccyx form the rectococcygeus muscle. The fluffy, rugose mucosa of the colon becomes smooth mucosa in the rectum. *(Hollinshead, pp 760–761)*

14. **(B)** The rectum is supplied primarily by the inferior mesenteric artery through the branches of its continuation, the superior rectal artery. The lower part of the rectum also receives the middle rectal arteries, branched from the internal iliac artery. They anastomose freely with the superior rectal arteries. The inferior rectal arteries supply the anal canal rather than the rectum. *(Hollinshead, pp 761–762)*

15. **(A)** The motor fibers to the rectum appear to be entirely parasympathetic. They are conveyed in the middle rectal plexus, derived from the inferior hypogastric plexus. The afferent supply of rectal nerves, both for pain and presence of feces or gas, belongs to the parasympathetic system. Thus the rectum receives both its afferent and efferent innervation through the pelvic splanchnic nerves in the rectal plexus. *(Hollinshead, p 763)*

16. **(D)** All the statements are true except D. As the bladder fills it rises above the pelvic brim. If fully distended it may rise as high as the umbilicus. Although the distended bladder rises as a dome, the empty bladder is flat. In children the dimensions are small, and even the empty bladder is largely above the pelvic brim. *(Hollinshead, p 764)*

17. **(E)** Several surveys have verified that there is no anatomically demonstrable vesical sphincter at the junction of the bladder and urethra. The "internal sphincter of the bladder" is a functional entity that prevents urine from entering the urethra and the ejaculate from entering the bladder; its mechanism is not understood completely. The smooth triangular area outlined by the two ureteric ostia and the internal urethral orifice is called the vesical trigone. The musculature of the bladder as a whole is referred to as the detrusor muscle. *(Hollinshead, pp 765–766)*

18. **(C)** At the base of the broad ligament the ureter passes forward and medial. In so doing it is crossed above and in front by the uterine artery. After crossing behind and below the uterine artery, the ureter passes to the front of the vagina, surrounded by the upper part of the vesical nerve plexus. It enters the posterolateral, not the anterior, aspect of the bladder. The ovarian vessels often cross the pelvic brim just lateral, not medial, to the ureters. *(Hollinshead, pp 767–768)*

19. **(B)** The male urethra commences at the internal urethral orifice. It traverses all the structures named except the ejaculatory duct. The ejaculatory ducts enter the prostatic urethra. They are formed by the union of the ampulla of the ductus deferens with the duct of the seminal vesicle. *(Hollinshead, pp 769–770,774,796)*

20. **(B)** The seminal vesicles secrete a fluid that adds fructose to the seminal plasma for maintaining the motility of the spermatozoa. Sperm are stored in the ampulla, the ductus, and the epididymus but not in the seminal vesicle. The seminal vesicles lie below and lateral, not medial, to the ampulla of the ductus. The distal end of the seminal vesicle becomes narrow and, together with the ductus deferens, forms the ejaculatory duct that opens in the prostatic urethra. *(Hollinshead, p 774)*

21. **(B)** All of the structures listed are parts of the internal female genital system except the perineal body. This structure, called the central tendon of the perineum, is a mass of fibromuscular tissue located between the anal canal and the vagina or bulb of the penis. A number of perineal muscles terminate in it. The perineal body is larger in the female than in the male and it is of considerable importance in obstetrics and gynecology. *(Hollinshead, pp 744–784,784,796)*

22. **(D)** The paired ovarian arteries usually arise in the abdomen from the front of the aorta, below the origin of the renal arteries. Their origin and much of their abdominal course is similar for the testicular arteries. Throughout most of its course, each artery is accompanied by the corresponding vein. *(Hollinshead, pp 686,785)*

23. **(C)** The uterovaginal autonomic plexus of nerves primarily consists of visceral afferents and sympathetic efferent fibers. It contains only a few, if any, parasympathetic efferents. It includes visceral but not somatic fibers. Innervation is not necessary to functions of the uterus. *(Hollinshead, p 787)*

24. **(B)** The inferior mesenteric lymph nodes are not described as being a part of the distribution of lymphatics from the uterus. Some lymphatics from the fundus and upper part of the body of the uterus drain into superficial inguinal nodes; many from these areas drain upward into nodes of the lumbar chain. Many lymphatics from the lower uterus and cervix end in the internal iliac nodes; others empty into external iliac and into sacral nodes. *(Hollinshead, p 786)*

25. **(C)** The vagina terminates by fusing around the cervix of the uterus. At the upper end, its lumen forms recesses (fornices)—the posterior fornix being deeper than the anterior and lateral fornices. Normally the vagina is flattened anteroposteriorly, not laterally. It is greatly distensible. *(Hollinshead, p 784)*

26. **(C)** No somatic motor fibers appear in the superior hypogastric nerve plexus, an autonomic plexus. It is a direct continuation of the aortic plexus below the aortic bifurcation. The visceral afferents found in the plexus mediate pain from the uterus and follow the course of sympathetic efferents. *(Hollinshead, p 753)*

27. **(D)** The pelvic diaphragm consists of the coccygeus muscle posteriorly and the more extensive and complex levator ani anterolaterally. The levator ani muscle originates along a semicircular line that skirts the pelvic walls from the pelvic surface of the body of the pubis to the ischial spine. In between these bony points, the levator ani is attached to a bandlike reinforcement in the obturator fascia, the arcus tendineus. *(Hollinshead, pp 741–742)*

28. **(A)** The internal iliac artery, a terminal branch of the common iliac, supplies the pelvic viscera and the perineum and proximal parts of the lower limb. In front of the greater sciatic foramen, the artery breaks up into a number of branches, many of which supply the pelvic viscera. The only visceral branch listed here is the middle rectal artery; all others named go to pelvic walls or perineum. *(Hollinshead, p 747)*

29. **(E)** The superior gluteal is the only branch of the internal iliac artery listed here that extends to the pelvic wall and lower limb. All other branches named are visceral branches to the bladder, ureter, ductus deferens, seminal vesicles, and uterus. The superior gluteal artery supplies areas of the hip (gluteus medius, gluteus minimus, and tensor fasciae latae). *(Hollinshead, pp 745–747)*

30. **(D)** The perineal area can be divided into two triangular regions by a line connecting the two ischial tuberosities: anteriorly, the urogenital region and posteriorly the anal region. The anal region contains the rectum, the anal canal, and the ischiorectal fossae on each side of the anal canal. In the urogenital region, a muscular shelf stretches between the conjoint ischiopubic rami of the two sides; this is the urogenital diaphragm. It is pierced by the urethra and, in the female, also by the vagina. The diaphragm serves for attachment of the external genitalia. The obturator internus muscle is not in the urogenital region. *(Hollinshead, p 789)*

31. **(C)** The pelvic splanchnic nerves are given off by S2–4 ventral rami of spinal nerves and contribute to the formation of the pelvic plexus. They take into that plexus preganglionic parasympathetic fibers and convey visceral afferents from the pelvic plexus to sacral segments of the spinal cord. The motor fibers to the rectum appear to be entirely parasympathetic and are conveyed in the middle rectal plexus derived from the inferior hypogastric plexus. The prostatic musculature apparently is innervated by the sympathetics and not by parasympathetic efferents. *(Hollinshead, pp 763,751,773)*

32. **(E)** The chief blood supply of the perineum is provided by the internal pudendal artery. It is a large branch of the anterior trunk of the internal iliac artery. It leaves the pelvis between the piriformis and coccygeus muscles and descends vertically on the exterior of the levator ani into the perineum. *(Hollinshead, pp 747,790)*

33. **(B)** The pudendal nerve is the sole somatic motor nerve of the perineum; also, it is sensory to most of the perineal skin. The pudendal nerve conducts sen-

sations from the prepuce and penis; from the glans penis or clitoris; from the vestibule of the vagina; from parts of the anal canal; and from the perineal skin, as well as from posterior parts of the labia majora and the scrotum. The areas on the periphery of the perineum are supplied by other cutaneous nerves. *(Hollinshead, pp 810–812)*

34. **(A)** The clitoris is the homologue of the penis, but it consists of only two erectile bodies, the corpora cavernosa clitoridis, and is not traversed by the urethra. The labia majora are homologues of the two halves of the scrotal sac. The labia minora are homologues of the skin that covers part of the penis. The female external genitalia collectively are referred to as the vulva. Partly in the substance of each labium minus is located an oval-shaped mass of erectile tissue called the bulb of the vestibule. Each bulb is a homologue of half of the bulb of the penis and the posterior part of the corpus spongiosum. *(Hollinshead, pp 803–804)*

35. **(C)** Most perineal structures send their lymphatics along the branches of the external pudendal vessels to the superficial inguinal nodes. Lymphatics of the lower anal canal, the perineal skin, the spongy urethra, and the entire vulva drain to the inguinal nodes. *(Hollinshead, p 810)*

36. **(D)** The levator ani forms, not the inferior walls of the ischiorectal fossae, but their sloping superomedial walls. These fossae are spaces on each side of the anal canal. They contain fat and connective tissue and also the pudendal canal. This canal is a fascial sheath in which the pudendal nerve and the internal pudendal artery reach the perineum and the internal pudendal veins leave it. The more or less vertical wall of each ischiorectal fossa is formed by the obturator internus muscle. *(Hollinshead, pp 793–794,738)*

37. **(E)** The iliac crest is not a boundary of the inferior pelvic aperture. This aperture forms the osseoligamentous frame of the perineum. It is formed anteriorly by the pubic arch, made up of the inferior margin of the pubic symphysis and the conjoint rami of the pubis and ischium. Posteriorly the aperture is bordered by the sacrotuberous ligament. The coccyx juts forward into the inferior aperture. *(Hollinshead, pp 739,789)*

38. **(C)** The urogenital region is the anterior portion of the perineum. The urogenital diaphragm is a muscular shelf stretched between the conjoint ischiopubic rami of the two sides and lying below the anterior part of the pelvic diaphragm (levator ani). Fascias cover the superior and inferior surfaces of the muscular shelf and are parts of the urogenital diaphragm; none of the other items listed are parts of the diaphragm. *(Hollinshead, pp 741–742,789,795)*

39. **(A)** The superior and inferior fascias of the urogenital diaphragm enclose what is called the deep perineal space or pouch. In fact, it is not a space at all, not even a partial one, since it is filled completely by the deep perineal muscles. The deep perineal space is closed completely, and it does not communicate with the ischiorectal fossa or other perineal or pelvic spaces. Items (B), (C), and (D) are characteristic of the superficial, not the deep, perineal space. *(Hollinshead, pp 797–798)*

40. **(D)** The superficial perineal fascia is a continuation of the superficial fascia from the abdominal wall into the perineum. It limits a potential space surrounding the external genitalia, called the superficial perineal space. Two named parts of this fascia are the superficial penile fascia and the tunica dartos. *(Hollinshead, p 798)*

41. **(E)** The only correct item states that in the female the ureter crosses behind and below the uterine artery. The crossing of the ureter by the uterine artery occurs close to the uterus but can vary markedly when pathological conditions have distorted relationships. This crossing is oblique and is a common site of surgical injury to the ureter. The left ureter, not the right, is related to the base of the sigmoid colon. After crossing the pelvic brim, the ureter lies medial, not lateral, to the internal iliac artery. In the male it passes below, not above, the ductus. It enters the posterolateral aspect, not the superior aspect, of the bladder and courses obliquely through its wall. *(Hollinshead, pp 768–769,774)*

42. **(C)** The visceral efferents required for erection of the penis are derived from the pelvic splanchnic nerves, which represent the sacral parasympathetic outflow. They are known also as nervi erigentes because they are capable of causing erection. Arteries of the penis, in response to excitation of the cavernous nerves (derived from the nervi erigentes), pour arterial blood into the cavernae at a faster rate than it can leave through the cavernous veins. Erection is thus achieved by tumescence of the corpora with arterial blood. *(Hollinshead, pp 754,800,812)*

43. **(A)** The ejaculatory duct is formed by union of the distal end of the duct of the seminal vesicle together with the ductus deferens. The ejaculatory ducts converge toward each other as they run through the prostate. They open into the prostatic urethra close together on the colliculus seminalis just lateral to the utriculus. *(Hollinshead, pp 770–771,774)*

44. **(B)** The perineal body is a small fibrous mass at the center of the perineum. Attached here are the base of the perineal membrane and several muscles. The

ischiocavernosus muscle surrounds the free surface of each crus of the penis. It is attached to the ischiopubic ramus but not to the perineeal body. *(Basmajian, pp 198,202; Hollinshead, p 803)*

**45.** **(C)** The bulbourethral glands are found in the male. They are two small glands, each the size of a pea, that lie deep to the urogenital diaphragm in the deep perineal pouch. Their long ducts travel in the wall of the urethra for 2 to 3 cm before opening into the spongy urethra. *(Basmajian, p 203)*

**46.** **(A)** All of the areas listed are supplied by the pudendal nerve, except the pelvic surface of the levator ani. The levator ani is supplied by branches of S2, S3, and S4 nerves. Its perineal surface is supplied by twigs from the perineal nerve. *(Basmajian, pp 203,212)*

**47.** **(E)** The inferior rectal artery (inferior hemorrhoidal) is a branch of the internal pudendal artery that supplies the anal triangle. The internal pudendal artery travels through the pudendal canal and deep perineal pouch to become the dorsal artery of the penis. Other arteries listed are branches of the internal iliac artery. *(Basmajian, pp 203–204)*

**48.** **(D)** The middle rectal vein is a much more important vessel than the corresponding artery. It is the chief link between the portal and caval systems and ends in the internal iliac vein. It drains the rectum above, not below, the internal sphincter. The superior, not the middle, rectal vein begins in the anal columns. The middle rectal vein belongs to the caval system and has valves. *(Basmajian, p 222)*

**49.** **(B)** The superior gluteal artery is the largest branch of the internal iliac artery. It makes a U turn around the angle of the greater sciatic notch into the gluteal region. Its vein and the superior gluteal nerve accompany it. *(Basmajian, p 223)*

**50.** **(C)** The sciatic is the only nerve listed that is a branch of the sacral plexus. The sacral plexus has many collateral branches and ends as two terminal branches: the sciatic and the pudendal nerves. Branches from roots of the plexus include muscular branches to the piriformis, levator ani, and coccygeus. Other nerves listed belong to the lumbar plexus, not the sacral. *(Basmajian, pp 225,189)*

**51.** **(E)** The pelvic splanchnic nerves arise from S2, S3, and S4. They are "mixed" parasympathetic nerves. They supply the involuntary sphincters of the rectum and bladder, causing them to relax while the organs are contracting; they are also sensory to them. They cause dilatation of the arteries of the erectile tissue of the penis or clitoris and thereby produce erection. They do not send motor fibers to

the coccygeus, as that is the function of somatic nerves. *(Basmajian, p 226)*

**52.** **(B)** The pudendal nerve largely innervates the perineum. This nerve escapes between the piriformis and coccygeus just medial to the sciatic nerve. It has three divisions: the rectal nerve in the anal triangle; the perineal nerve that sends branches to perineal pouches; and the dorsal nerve of the penis to the crus of the penis and clitoris. All of the nerves listed pass through the greater sciatic foramen. *(Basmajian, pp 203,258)*

**53.** **(A)** The ligament of the ovary stands out in relief from the back of the broad ligament. It joins the lower pole of the ovary to the angle between the side of the uterus and the uterine tube. A similar cord, the round ligament of the uterus, stands out from the front of the broad ligament. It passes from the angle between the uterus and tube across the pelvic brim to the deep inguinal ring. *(Basmajian, p 229)*

**54.** **(D)** The ovary does not have a smooth surface. It has pits and scars on its surface that mark the sites of the absorbed corpora lutea (corpus luteum). These scars occur monthly from shedding of the ova. All other statements are correct. *(Basmajian, p 229)*

**55.** **(E)** The abdominal orifice of the uterine tube lies at the bottom of a funnel-shaped depression, the infundibulum. Fringes or fimbria lined with ciliated epithelium project from the infundibulum and encourage ova, when shed, into the tube. One fimbria is attached to the ovary. Listed items (A), (B) and (C) are other parts of the uterine tube. *(Basmajian, p 230)*

**56.** **(C)** The fundus of the uterus is the part that rises above the uterine tubes. The fundus and the body form the upper 5 cm, the cervix the lower 3 cm of the uterus. The uterine tubes enter at its widest part. *(Basmajian, p 230)*

**57.** **(B)** In the female the ureter crosses the lateral fornix of the vagina. Because of the obliquity of the uterine axis, the ureter lies closer to the cervix on one side (generally the left). The ureter crosses below, not above, the broad ligament and the uterine artery. It enters the bladder in front of the vagina. *(Basmajian, pp 233–234)*

**58.** **(A)** The uterovaginal plexus contains few, if any, parasympathetic efferents to the uterus. This plexus consists primarily of visceral afferents and sympathetic efferent fibers. Afferents from the body of the uterus travel with the sympathetics; pain fibers enter the spinal cord through the last two thoracic nerves. The physiology of the motor supply to the uterus is not understood; innervation is not necessary to the function of this organ. *(Hollinshead, p 787)*

59.  **(D)** The thick pubic parts of the levator ani form a puborectal sling for the rectum, drawing it forward until it forms a sloping shelf. Upon this shelf the vagina rests, and on the vagina rests the bladder. This is the essential support. The pubic parts of the levator ani also insert into the perineal body and thus act as a sling for the posterior wall of the vagina. The urogenital diaphragm and its fascias blend with the lower third of the vagina and assist the levator ani to support it. *(Basmajian, p 234)*

60.  **(C)** The superior hypogastric plexus represents sympathetic input into the pelvis. It lies below the bifurcation of the aorta. This plexus is a downward prolongation of the preaortic plexus. It is joined by the 3rd and 4th lumbar splanchnic nerves. Its large branches, the right and left hypogastric nerves, contribute to the pelvic (inferior hypogastric) plexus. *(Basmajian, p 238, Hollinshead, p 624)*

61.  **(E)** Pelvic splanchnic nerves are generally believed to have exclusive control of the muscular walls of the bladder, urethra, and rectum. They cause relaxation, not constriction, of the arteries to erectile tissue, producing erection of the penis or clitoris. These nerves also carry many afferent fibers from the pelvis, but sympathetics carry sensory fibers from the body and fundus of the uterus. The pelvic splanchnics do not carry motor fibers to striated muscle. *(Basmajian, p 238)*

62.  **(D)** Superior hypogastric are not lymph nodes but the downward projection of the preaortic plexus of nerves. The pelvic lymph nodes include the common iliac nodes near the pelvic brim and the internal iliac, lateral sacral, and medial sacral nodes within the pelvic cavity. *(Basmajian, p 239)*

63.  **(B)** The perineum extends from the symphysis pubis to the tip of the coccyx. When the thighs are abducted, the perineum is a diamond-shaped region with its extent as stated. The boundaries are the symphysis pubis, the inferior pubic ramus, the ischial rami, the ischial tuberosities, the sacrotuberous ligament, and the coccyx. *(Moore, p 293)*

64.  **(C)** The anterior half of the inferior pelvic aperture is closed by the urogenital diaphragm; the posterior half is closed by the levator ani muscle. The urogenital diaphragm is a thin sheet of striated muscle stretching between the two sides of the pubic arch, which is formed by the converging ischiopubic rami. *(Moore, pp 293–294)*

65.  **(A)** The external anal sphincter surrounds the anus in the anal, not the urogenital, triangle. All the other structures are found in the urogenital triangle. The sphincter urethrae muscle encircles the membranous urethra in the male and the superior half of the urethra in the female. The arcuate pubic ligament lies just posterior to the pubic symphysis. The transverse muscle fibers posterior to the urethra are called the deep transversus perinei muscles. The deep fascia on the inferior surface of the urogenital diaphragm forms the dense perineal membrane. *(Moore, pp 294–295)*

66.  **(E)** The perineal body or central perineal tendon is a fibromuscular node located at the center of the perineum between the anal canal and the bulb of the penis or vagina. The central perineal tendon indicates where the urorectal septum divided the cloacal membrane in the embryo. It is the landmark of the perineum that gives attachment to the transverse perineal muscles, the bulbospongiosus, some fibers of the external anal sphincter, and the levator ani muscles of both sides. *(Moore, p 297)*

67.  **(D)** In the male the membranous urethra is found in the deep, not the superficial, perineal pouch (space). The superficial perineal space is the fascial space between the superficial perineal fascia and the perineal membrane. In the male this space contains the root of the penis and the muscles associated with it; the proximal part of the spongy urethra; branches of the internal pudendal vessels; and the pudendal nerves. *(Moore, pp 299,307–308)*

68.  **(A)** In the female the greater vestibular glands are found in the superficial, not the deep, perineal space. The deep perineal space is the fascial space enclosed by the superior and inferior fascias of the urogenital diaphragm. In the female this space is occupied by part of the urethra, the sphincter urethrae, the inferior part of the vagina, and the deep transverse perineal muscles. The deep perineal space in both sexes also contains the blood vessels and nerves associated with the structures within it. *(Moore, pp 299,324–325)*

69.  **(B)** The ischiorectal fossae are large, wedge-shaped, fascia-lined spaces on each side of the anal canal. They contain fat and loose connective tissue, not fluid, and allow the rectum and anal canal to distend. The fossae of the two sides do communicate with each other over the anococcygeal ligament. Posteriorly each fossa is continuous with the lesser sciatic foramen, superior to the sacrotuberous ligament. *(Moore, p 301)*

70.  **(D)** The scrotum is not innervated by the inferior gluteal nerve. Its anterior part is supplied by the ilioinguinal nerve; the posterior part by medial and lateral scrotal branches of the perineal nerve. It is also supplied by the perineal branch of the posterior femoral cutaneous nerve. *(Moore p 311)*

71. **(E)** Infection may reach the ischiorectal fossae from any of the sources listed except from a ruptured urethra. A ruptured urethra into the superficial perineal space allows urine to pass into the areolar tissue in the scrotum, around the penis, and upward into the anterior abdominal wall. *(Moore, p 303)*

72. **(B)** The pudendal nerve supplies most of the innervation of the perineum. Toward the distal end of the pudendal canal, the pudendal nerve splits to form the dorsal nerve of the penis (or clitoris) and the perineal nerve. The perineal nerve gives off the scrotal or labial branches and continues to supply the muscles of the urogenital diaphragm. *(Moore, p 306)*

73. **(A)** Of the structures listed, only the bulbourethral glands are found in the deep perineal space of the male. These glands lie posterolateral to the membranous urethra. Their relatively long ducts pass through the inferior fascia of the urogenital diaphragm and through the bulb of the penis to open into the proximal part of the spongy urethra. *(Moore, pp 308–309; 381)*

74. **(C)** Blood supply to the anterior and posterior scrotum is supplied by external and internal pudendal arteries respectively. None of the other arteries listed supply the scrotum; other sources of arterial supply to the scrotum are branches of the testicular and cremasteric arteries. *(Moore, pp 310–311)*

75. **(E)** The dorsal nerve of the clitoris is a branch of the pudendal nerve. It is found in the deep, not the superficial, perineal space. This nerve is homologous to the dorsal nerve of the penis in the male. *(Moore, pp 324–325)*

76. **(A)** Embryologically the labia majora are homologous to the scrotum of the male. The labia majora are two large folds of skin, filled largely with subcutaneous fat, that run downward and backward from the mons pubis, a rounded eminence lying anterior to the symphysis pubis. The labia minora are two thin, delicate folds of skin lying between the labia majora. The space between the labia minora is the vestibule of the vagina. The hymen is a thin, incomplete fold of mucous membrane surrounding the vaginal opening. *(Moore, pp 325–327)*

77. **(D)** The clitoris has no corpus spongiosum and is entirely separate from the urethra. It is composed of erectile tissue and is capable of enlargement as a result of engorgement with blood. It consists of two crura, two corpora cavernosa, and a glans. It is suspended by a suspensory ligament, as in the male. *(Moore, p. 329)*

78. **(B)** Some obstetricians are reluctant to perform a median episiotomy, fearing it may tear or extend posteriorly and involve the external anal sphincter or the rectum. In this type of episiotomy, the cut is in the midline of the perineum, beginning at the frenulum of the labia minora and passing posteriorly through the skin, the vaginal mucosa, and the central perineal tendon. The incision stops well short of the external anal sphincter. *(Moore, pp 331–333)*

79. **(C)** The general structure of the male pelvis is heavy and thick, and it has more prominent bony markings than does the female pelvis. Generally, the female pelvis is wider and shallower and has larger superior and inferior pelvic apertures than does the male. *(Moore, p 335)*

80. **(A)** The diagonal conjugate diameter of the pelvis is the measurement from the midpoint of the inferior border of the symphysis pubis to the midpoint of the sacral promontory. Item (B) measures the anteroposterior diameter of the superior pelvic aperture. Item (C) describes the transverse diameter of the superior pelvic aperture, its greatest width. Item (D) measures the oblique diameter of the superior pelvic aperture. Item (E), the midplane (interspinous) diameter of the pelvis between the ischial spines, cannot be measured but may be estimated by palpating the sacrospinous ligament during a vaginal examination. *(Moore, pp 337–339)*

81. **(C)** The platypelloid pelvis is a flattened type of pelvis that is present in about 2.5% of females. It resembles a shallow, flat bowl. In patients with this type of pelvis, there may be difficulty with the fetal head engaging in the superior pelvic aperture, which may necessitate a cesarean section. Other items listed are various shapes of pelves found in the female. Piriform is not a type of pelvic shape. *(Moore, pp 341–343)*

82. **(E)** The walls of the pelvic cavity are lined, in part, with muscles; however, no muscles cross the pelvic brim. The piriformis muscle is located partly within the pelvis minor and partly posterior to the hip joint. The obturator internus covers most of the side wall of the pelvis minor. The two levator ani muscles and the two coccygeus muscles form the pelvic diaphragm. *(Moore, pp 346–349)*

83. **(C)** The puborectalis is part of the levator ani muscle. The fibers from the two sides of the puborectalis loop around the posterior surfaces of the anorectal junction forming a U-shaped rectal sling. It thereby offers special support to this region of the pelvic diaphragm. *(Moore, p 350)*

84. **(B)** The obturator nerve arises from the lumbar plexus. It is formed from the anterior divisions of

the ventral rami of the second, third, and fourth lumbar nerves. It is, therefore, not a branch of the sacral plexus. The sacral plexus is formed by the lumbosacral trunk, the ventral rami of the first three sacral nerves, and the descending part of the fourth sacral nerves. All the other nerves listed arise from this plexus. *(Moore, pp 353–354,356)*

85.  **(E)** The obturator nerve is the only nerve supplying the lower limb, which lies on the side wall of the pelvis in the extraperitoneal fat. Here it is vulnerable to injury during removal of cancerous lymph nodes from the side wall of the pelvis. Injury to this nerve results in deficient adduction power of the thigh on the affected side. *(Moore, p 357)*

86.  **(C)** The femoral artery, the chief arterial supply to the lower limb, is the continuation of the external iliac artery. It enters the femoral triangle deep to the midpoint of the inguinal ligament. It does not enter the pelvis minor. The other arteries listed enter the pelvis minor. *(Moore, pp 357,454)*

87.  **(A)** The superior rectal artery is the direct continuation of the inferior mesenteric artery. It descends into the pelvis minor within the sigmoid mesocolon and sends a branch downward on each side of the rectum. The other arteries listed are branches of the internal iliac artery. *(Moore, pp 365,357–358)*

88.  **(B)** The umbilical vein drains into the left branch of the portal vein. Before birth it carries oxygenated blood from the placenta to the fetus. A fold of peritoneum, called the falciform ligament, contains the ligamentum teres of the liver, the obliterated umbilical vein. Usually this vessel is patent for some time after birth. The median umbilical ligament is the remnant of the urachus that developed from the intra-abdominal part of the allantois. *(Moore, pp 140,365)*

89.  **(D)** At the level of the ischial spine, the ureter is crossed superiorly by the uterine artery. This relationship is important to the surgeon for identification of these two structures; correct identification of the ureter can prevent its injury during surgery. *(Moore, p 369)*

90.  **(E)** Ureteric stones may cause complete or intermittent obstruction of urinary flow anywhere along the ureter. It occurs most often, however, at either of two points: (1) where the ureter crosses the external iliac artery and the brim of the pelvis and (2) where it passes obliquely through the wall of the urinary bladder. *(Moore, p 369)*

91.  **(A)** The posterior surface of the bladder is referred to as its base. Its anterior end is the apex. The inferior part of the organ, where the base and inferolateral surfaces converge, is called the neck of the bladder. The superior surface faces upward. *(Moore, pp 370–371)*

92.  **(C)** The main arteries supplying the bladder are branches of the internal, not the external, iliac artery. The superior vesical arteries, branches of the umbilical, supply anterosuperior parts of the bladder. The inferior vesical arteries supply the base of the bladder. The obturator and inferior gluteal arteries also supply small branches to the bladder. *(Moore, p 371)*

93.  **(D)** The membranous (second) part of the male urethra is its shortest portion. It is the least, not the most, dilatable part. Except for the external urethral orifice, the membranous urethra is the narrowest urethral portion. The urethra descends from the apex of the prostate to the bulb of the penis and traverses the sphincter urethrae muscle and the perineal membrane. A small bulbourethral gland is located on each side of this part of the urethra. *(Moore, p 373)*

94.  **(D)** The ejaculatory ducts are slender tubes formed by the union of the duct of the seminal vesicle and the ductus deferens. The ejaculatory ducts open by slitlike apertures into the prostatic urethra, one on each side of the orifice of the prostatic utricle or just inside this vestigial organ. *(Moore, pp 373,376)*

95.  **(B)** The female urethra is a short muscular tube, much shorter than the male urethra. It lies anterior to the vagina and is separated from it superiorly by a vesicovaginal space. Inferiorly it is so intimately associated with the vagina that it appears to be embedded in it. All the other items are correct. *(Moore, pp 374–375)*

96.  **(E)** The prostate is an accessory genital gland. The prostatic secretion is a thin, milky fluid that is discharged into the urethra by contraction of its smooth muscle. The other structures listed comprise the male genital organs. *(Moore, p 375)*

97.  **(C)** The ductus deferens is a thick-walled muscular tube that begins in the tail of the epididymis and ends in the ejaculatory duct. It carries sperm from one to the other. It joins the duct of the seminal vesicle in the groove between the prostate and the bladder to form the ejaculatory duct. *(Moore, p 375)*

98.  **(A)** Ducts of each seminal vesicle joins the ductus deferens to form the ejaculatory duct. Formerly it was thought that the seminal vesicles stored seminal fluid because sperm were observed in them in cadavers. It is now widely accepted that they do not store sperm in living persons. They secrete a thick secretion that mixes with the sperm as they pass along the ejaculatory ducts. *(Moore, p 376)*

99. **(E)** To relieve the urethral obstruction, surgery usually is a successful form of treatment. This disease can begin at the mid-40 age range and results in varying degrees of obstruction of the neck of the bladder. In most males the prostate enlarges progressively as the individual ages. In some males it becomes more fibrous and undergoes atrophy. *(Moore, p 379)*

100. **(C)** "Vertical proflexion" is erroneus terminology. The uterus normally is bent anteriorly, or is anteflexed, between the cervix and the body, and the entire uterus normally is bent or inclined forward, or anteverted. Many women, however, have retroflexed (bent backward) and retroverted uteri that produce no symptoms. *(Moore, p 385)*

101. **(C)**

102. **(A)**

103. **(D)**

104. **(E)**

101–104. The uterus consists of two major parts: (1) the expanded upper two thirds, known as the body; and (2) the cylindrical lower third, known as the cervix. The fundus of the uterus is the rounded upper part of the body superior to the line joining the points of entrance of the two uterine tubes. The region of the body of the uterus on each side where the uterine tubes enter is called the cornu(horn). The cervix of the uterus projects into the uppermost part of the anterior wall of the vagina. As a result of this, the uterus lies almost at a right angle to the axis of the vagina in its normal anteverted position. Because more of the posterior part of the cervix enters the vagina than does the anterior part, the recess, or cul-de-sac, between the vaginal wall and the cervix is deeper posteriorly than anteriorly. The recess anterior to the cervix is called the anterior fornix. The recess posterior to the cervix is the posterior fornix. The recesses on each side are the lateral fornices. These four fornices are parts of a continuous recess surrounding the cervix. *(Moore, pp 383–385)*

105. **(E)**

106. **(B)**

107. **(C)**

108. **(A)**

105–108. The principal supports of the uterus are the pelvic floor and the structures surrounding the uterus. The levator ani and coccygeus muscles and the muscles of the urogenital diaphragm are particularly important structures of support. The peritoneum is reflected anteriorly from the uterus onto the bladder and posteriorly over the posterior fornix of the vagina onto the rectum. Laterally the peritoneum forms the folds, called the broad ligaments. The broad ligaments are folds of the peritoneum, with mesothelium on their anterior and posterior surfaces. They extend from the sides of the uterus to the side walls and floor of the pelvis. They hold the uterus in a relatively normal position. The uterine tube is enclosed in the free edge of each broad ligament. The ligament of the ovary lies posterosuperiorly, and the round ligament of the uterus lies anteroinferiorly within the broad ligament. The broad ligament gives attachment to the ovary through the mesovarium. This short peritoneal fold connects the anterior border of the ovary with the posterior layer of the broad ligament. Anteriorly the body of the uterus is separated from the bladder by the vesicouterine pouch of peritoneum. *(Moore, pp 386–387)*

109. **(B)**

110. **(D)**

111. **(A)**

112. **(C)**

109–112. The uterine tubes are a pair of ducts that extend laterally from the cornu, or horns, of the uterus. They are designed to receive the oocytes discharged from the ovarian follicles and to convey the dividing xygote to the uterine cavity. Each tube opens at its proximal end into the horn of the uterus and at its distal end into the peritoneal cavity near the ovary. The uterine tube has been described as having four parts. (1) The infundibulum is the funnel-shaped distal end of the tube. The margins of the infundibulum are drawn out into numerous fringed folds called fimbria. These fingerlike processes spread over most of the surface of the ovary. During ovulation the fimbria trap the oocyte, and the cilia of its mucosal lining sweep it through the abdominal ostium of the uterine tube. (2) The ampulla of the uterine tube receives the oocyte from the infundibulum; fertilization of the oocyte by a sperm usually occurs here. The ampulla is the widest and longest part of the uterine tube. (3) The isthmus of the uterine tube is the short, narrow, thick-walled part that joins the cornu of the uterus. (4) The uterine part of the tube is the short segment that pierces the wall of the uterus. *(Moore, pp 389–390)*

**113.** **(C)**

**114.** **(A)**

**115.** **(A)**

**116.** **(B)**

**113–116.** The puborectalis muscle, a part of the levator ani, forms a sling at the junction of the rectum and the anal canal, producing the anorectal angle. The rectum is continuous superiorly with the sigmoid colon and begins on the pelvic surface of the third piece of the sacrum. About 12 cm long in both sexes, the rectum follows the curve of the sacrum and coccyx to about 3 cm beyond the tip of the coccyx. Here the rectum ends by turning posteroinferiorly to become the anal canal. In the female the peritoneum is reflected from the rectum to the posterior fornix of the vagina, where it forms the floor of the rectouterine pouch.

The anal canal begins where the rectal ampulla narrows abruptly at the level of the U-shaped sling formed by the puborectalis muscle. The superior half of the anal canal is characterized by a series of five to ten longitudinal ridges or folds of mucosa called anal columns. The inferior ends of the anal columns are united to each other by small semilunar folds of mucosa called anal valves. The inferior comb-shaped limit of the anal valves is known as the pectinate line. Inferior to the pectinate line is the anal pecten. About 2 cm superior to the anus there is an abrupt transition from simple columnar to stratified squamous epithelium. This transitional zone between the anal mucosa and the anal skin, the anocutaneous line, lies at the interval between the subcutaneous part of the external anal sphincter and the inferior border of the internal anal sphincter. This line is of particular interest to the clinician. *(Moore, pp 393–394,400)*

**117.** **(B)**

**118.** **(B)**

**119.** **(A)**

**120.** **(B)**

**117–120.** A rectocele is herniation, or prolapse, of the rectum in females. It occurs when there is weakness of the fibromuscular layer of the posterior wall of the vagina; the vagina tends to bulge through the vaginal orifice with the attached wall of the rectum. In some cases defecation is difficult unless the patient presses on the rectocele with fingers in the vagina.

Items 117, 118, and 120 are all related to the condition of hemorrhoids (piles). Internal hemorrhoids are varicosities of the tributaries of the superior rectal veins and are covered by mucous membrane. Mixed hemorrhoids are varicosities of the superior, communicating, and inferior rectal veins. The anastomosis between the superior and the middle rectal veins forms a clinically important communication between the portal and systemic systems because the superior rectal vein drains into the hepatic portal system, and the middle and inferior rectal veins drain into the systemic system. Any abnormal increase in the pressure in the valveless portal system may cause enlargement of the superior rectal veins contained in the anal columns; this results in internal hemorrhoids. In portal hypertension, as in hepatic cirrhosis, the tiny anastomotic veins in the anal canal and elsewhere become varicose and may rupture. *(Moore, pp 397,399–401)*

**121.** **(A)**

**122.** **(D)**

**123.** **(B)**

**124.** **(C)**

**121–124.** All items except item 122 are related to the autonomic nervous system in the pelvis. Entering somatic nerves include the lumbosacral trunk, the obturator nerve on the ala of the sacrum, and the ventral rami of sacral nerves passing through the pelvic sacral foramina. The sacral plexus gives off the chief somatic nerve of the perineum, the pudendal nerve. Entering autonomic nerves include the two sympathetic trunks and the hypogastric nerves; the ovarian plexus around the ovarian vessels; the superior rectal nerve plexus; and the pelvic splanchnic nerves. The superior hypogastric plexus (presacral nerve) is the direct extension of the aortic plexus below the aortic bifurcation. It consists of a mixture of preganglionic and postganglionic sympathetic fibers, small ganglia, and visceral afferents that mediate pain sensation from the fundus and upper part of the uterus; these fibers follow the route of sympathetic efferents. The inferior hypogastric (pelvic) plexus is formed by lateral extensions of the superior hypogastric plexus, known as the hypogastric nerves, and the pelvic splanchnic nerves. The hypogastric nerves and the superior hypogastric plexus apparently do not contain any parasympathetic fibers. The pelvic splanchnic nerves represént the sacral parasympathetic outflow and are known as nervi erigentes because they are the nerves capable of causing erection of the penis or clitoris. *(Moore, p 403; Hollinshead, pp 748–749,753–754)*

125. **(E)** All of the structures listed are boundaries of the perineum. When the thighs are abducted, the perineum is seen as a diamond-shaped region extending from the symphysis to the tip of the coccyx. Boundaries omitted in this question are the ischial tuberosities and the coccyx. *(Moore, p 294)*

126. **(A)** The urogenital triangle is the anterior portion of the perineum located anterior to the line joining the midpoints of the two ischial tuberosities. The anal triangle is the portion of the perineum posterior to this line. The urogenital triangle contains the urethra, the root of the scrotum, and the penis. The anus is located in the anal triangle. *(Moore, pp 294,299)*

127. **(D)** The ischiorectal fossa is the only item listed found in the anal region. This fossa is located on each side of the anal canal and rectum. The space is filled with soft fat that supports the anal canal but that is readily displaced to allow feces to pass through the terminal part of the digestive tract. The ischiorectal fossa contains the internal pudendal vessels and the pudendal nerve. The other items listed are related to the urogenital, not the anal, region. *(Moore, pp 297–299,301–303)*

128. **(D)** The superficial perineal space is the fascial space between the superficial perineal fascia and the inferior fascia of the urogenital diaphragm (perineal membrane). It contains the root of the penis and the muscles associated with this organ. These muscles are the ischiocavernosus, the bulbospongiosus, and the superficial transverse perineal muscle. The sphincter urethrae and the deep transverse perineal muscle are found in the deep perineal space. The levator ani muscle forms a large part of the floor of the pelvis. *(Moore, pp 307–308,309)*

129. **(A)** The blood supply of the scrotum is via the external pudendal arteries (anterior aspect of the scrotum) and the internal pudendal arteries (posterior aspect of the scrotum). Branches of the testicular and cremasteric arteries also supply the scrotum. The inferior vesical artery is a branch of the anterior division of the internal iliac artery; it passes forward to the base of the bladder and supplies the seminal vesicles, the prostate, and the posteroinferior parts of the bladder but not the scrotum. *(Moore, pp 311,358–359)*

130. **(B)** The anterior part of the scrotum is supplied by the ilioinguinal nerve; its posterior part is supplied by the medial and lateral scrotal branches of the perineal nerve and by the perineal branch of the posterior femoral cutaneous nerve. The lateral femoral cutaneous nerve supplies the skin over the anterior and lateral parts of the thigh. The obturator nerve leaves the pelvis via the obturator foramen to supply the adductor region of the thigh. *(Moore, pp 264,311,356–357)*

131. **(E)** All of the statements about the labia minora are correct. They lie between the labia majora, enclose the vestibule of the vagina, and lie on each side of the vaginal orifice. The labia minora extend posteriorly from the clitoris for about 4 cm, and their medial surfaces are in contact with each other. Posteriorly they may be united by a small fold of skin called the frenulum pudendi or frenulum of the labia minora. *(Moore, pp 325–326)*

132. **(C)** The external urethral orifice is located 2 to 3 cm posterior, not anterior, to the clitoris. The paraurethral glands (Skene's ducts or glands) open on each side of it. Usually it is a median slit that has prominent margins in contact with each other. The hymen is a thin, incomplete fold of mucous membrane surrounding the vaginal, not the urethral, orifice. *(Moore, p 327)*

133. **(D)** The only correct statement is item 4. The greater vestibular glands are two in number. They are located in the superficial, not the deep, perineal space, one on each side just posterior to the bulb of the vestibule. These glands are homologous with the bulbourethral glands, not the prostate, of the male. *(Moore, p 330)*

134. **(C)** The coxae, or hip bones, consist of three parts: the ilium, ischium, and pubis. Fusion of the bones occurs at 15 to 17 years of age, and the bones are firmly joined in the adult. The three parts meet at the acetabulum. The sacrum and coccyx are parts of the vertebral column interposed dorsally between the two hip bones. The ilium joins the sacrum at the sacroiliac joint on each side (synovial joints). The pubic bones are joined together by a symphysis, not a synovial joint; this is a connection by ligaments and a fibrocartilaginous disk. *(Moore, p 334)*

135. **(B)** The pelvis major lies above the superior pelvic aperture and the linea terminalis. It is bounded anteriorly by the abdominal wall, laterally by the iliac fossae, and posteriorly by L5 and S1 vertebrae. The posterior wall of the pelvis minor is formed by the sacrum and coccyx; the symphysis pubis is part of its anterior wall along with the body of the pubis and the pubic rami. *(Moore, p 335)*

**136.** **(A)** A gynecoid pelvis has the characteristics noted in the first three items. Usually a woman with a gynecoid pelvis has a reasonably uneventful delivery. A platypelloid pelvis is broad or flat and is described as resembling a shallow, flat bowl. In patients with this type of pelvis, there may be difficulty with the fetal head engaging in the superior pelvic aperture. *(Moore, pp 341–343)*

**137.** **(C)** The iliolumbar ligament, strong and triangular, connects the tip and lower anterior part of each transverse process of L5 vertebra to the internal lip of each iliac crest posteriorly. The lateral lumbosacral ligament is an attachment of the lower fibers of the lumbosacral ligament to the lateral part of the sacrum. The iliolumbar ligaments are important because of their functions, as listed in items 2 and 4. *(Moore, p 345)*

**138.** **(E)** All of the items listed about the sacrotuberous ligament are correct. Both it and the sacrospinous ligament bind the sacrum to the ischium and resist backward rotation of the inferior end of the sacrum. They permit some movement of the sacrum, which gives resilience to the region when sudden weights are applied to the vertebral column (as when landing on the feet during a fall). *(Moore, p 345)*

**139.** **(A)** The obturator internus muscle is a thick, fan-shaped muscle situated partly within the pelvis minor and partly posterior to the hip joint. It covers most of the side wall of the pelvis minor. The first three items listed are correct. This muscle leaves the pelvis through the lesser, not the greater, sciatic foramen. It makes a right angle turn around the lesser sciatic notch to enter the gluteal region and pass to its insertion. *(Moore, pp 346–349)*

**140.** **(D)** The two levatores ani muscles and the two coccygeus muscles, with their superior and inferior investing fasciae, form the funnel-shaped pelvic diaphragm. The piriformis and obturator are muscles of the lateral pelvic wall. The deep transverse perineal muscle is situated in the urogenital diaphragm. *(Moore, p 349)*

**141.** **(C)** The coccygeus muscle lies against the iliococcygeus part of the levator ani muscle and is continuous with it. The coccygeus forms the posterior and smaller part of the pelvic diaphragm. It probably supports the coccyx and pulls it forward after it has been pressed back during childbirth and in defecation. The levator ani, not the coccygeus, is the portion of the pelvic diaphragm that supports the fetal head during childbirth while the cervix is dilating to permit delivery of the baby. The puborectalis is the part of the levator ani that forms a sling around the rectum. *(Moore, p 351)*

**142.** **(A)** The sacral plexus is formed by the lumbosacral trunk and the ventral rami of the first three and the descending part of the fourth sacral nerves. The lumbosacral trunk is a thick cord formed by the ventral nerve rami of L4 and L5; it joins S1 as this nerve passes to join the sacral plexus. The sciatic nerve is formed by the ventral rami of L4 through S1. The inferior gluteal nerve is formed by L5, S1, and S2 and supplies the gluteus maximus muscle. The pudendal nerve arises from ventral rami of S2,S3,S4. The obturator nerve (L2,L3,L4) is a branch of the lumbar plexus, not the sacral plexus. *(Moore, pp 353–354)*

**143.** **(B)** Branches of the internal iliac artery include both visceral branches and those supplying the body wall. The internal iliac artery often divides into anterior and posterior divisions before giving off its named branches. Included in a number of branches of the anterior division are the umbilical and obturator arteries. Other arteries entering the pelvis are the median sacral and the superior rectal vessels. *(Moore, pp 357–358)*

**144.** **(C)** The uterine artery is an anterior branch of the internal iliac artery. Usually it arises separately from the internal iliac, but it may arise from the umbilical artery. It enters the root of the broad ligament and passes anterior to and above the ureter near the lateral fornix of the vagina. The superior rectal artery, not the uterine, is a direct continuation of the inferior mesenteric artery. *(Moore, pp 359–360,365)*

**145.** **(B)** The internal pudendal artery is a branch of the anterior, not the posterior, division of the internal iliac artery. The internal pudendal artery leaves the pelvis between the piriformis and coccygeus muscles by passing through the lowest part of the greater sciatic foramen. It passes around the posterior aspect of the ischial spine or the sacrospinous ligament to enter the ischiorectal fossa; it travels with the internal pudendal veins and branches of the pudendal nerve through the pudendal canal in the lateral wall of the ischiorectal fossa. Just before it reaches the symphysis pubis, it divides into its terminal branches, the deep and dorsal arteries of the penis or clitoris. *(Moore, pp 363–365)*

**146.** **(E)** All of the items listed are correct. The superior rectal vein drains into the inferior mesenteric vein and forms one of the clinically important communications between the portal and systemic venous systems. Portal obstruction, associated with cirrhosis of the liver, frequently results in development of rectal varicosities. In some people, veins forming the internal rectal plexus become varicose and form internal hemorrhoids. The prostatic venous plexus may drain, via the sacral veins, into the vertebral venous plexus. Thus these large, valveless veins may transport cancer cells to the vertebral column where they metastasize with the vertebrae. *(Moore, pp 365–367)*

**147. (A)** All of the arteries listed, except the middle rectal, are main sources of blood to the ureters. In the pelvis the arteries supplying the ureters approach from the lateral side. Usually these long branches form such an excellent anastomotic chain that some branches may be ligated without interfering with the blood supply to the ureter. *(Moore, pp 246,369)*

**148. (B)** The wall of the bladder chiefly is made up of smooth muscle called the detrusor urinae muscle. It consists of three layers running in many directions. The ureters pass obliquely through the bladder wall in an inferomedial direction. The openings of the urethra and the ureters are located at the base, not the superior surface, of the bladder and form the angles of the trigone. The pelvic splanchnic nerves (parasympathetic fibers), not the sympathetics, are motor to the detrusor muscle and inhibitory to the internal sphincter. *(Moore, pp 371–372)*

**149. (C)** The male urethra, for purposes of description, is divided into three parts: the prostatic part, the membranous part, and the spongy part. The prostatic part begins at the internal urethral orifice of the bladder; the ejaculatory ducts open into this portion. The membranous second part is the shortest portion of the urethra; the spongy part is the longest. The bulbourethral glands lie on each side of the membranous portion of the urethra; their ducts, however, open into the proximal part of the spongy urethra. *(Moore, pp 372–373)*

**150. (D)** The short female urethra is very distensible because it contains much elastic tissue as well as smooth muscle. Urethral stricture usually does not occur in the female as it does in the male as a result of trauma or infection. The commonest congenital abnormality of the urethra in the male, not the female, is hypospadias. In these males there is a defect in the ventral wall of the spongy urethra, so that it is open for a greater or lesser distance. *(Moore, pp 374–375)*

**151. (B)** The ductus deferens is a thick-walled, muscular tube. It begins in the tail of the epididymus and ends in the ejaculatory duct. The ductus deferens is innervated richly by autonomic nerve fibers. The tiny artery to the ductus deferens is applied closely to its surface; it arises from the umbilical artery. *(Moore, p 375)*

**152. (B)** The seminal vesicles consist of long tubes that are coiled to form vesiclelike masses on the base of the bladder. It is now accepted widely that they do not store sperm in living persons. They secrete a thick secretion that mixes with the sperm as they

pass along the ejaculatory ducts. The secretion probably is concerned with activation of the sperm. *(Moore, pp 375–376)*

**153. (B)** The arteries of the prostate are derived from the internal pudendal, inferior vesical, and middle rectal arteries. The veins of the prostate are wide and thin walled and form the prostatic venous plexus around the sides and base of the bladder. This plexus drains into the internal iliac veins. It communicates also with the vesical plexus and the vertebral venous plexus. *(Moore, p 379)*

**154. (A)** Benign prostatic hypertrophy is a common condition in older males. The condition results in varying degrees of obstruction of the neck of the bladder. In most males the prostate progressively undergoes hypertrophy. It is a common cause of urethral obstruction leading to nocturia (need to urinate during the night), dysuria (difficulty and pain on urination), and urgency (sudden desire to void). *(Moore, p 379)*

**155. (E)** All of the items listed are correct. From above downward, the anterior wall of the vagina is in contact with the cervix, the base of the bladder, the terminal part of the ureters, and the urethra. It is intimately connected to the neck of the bladder and to the urethra. The superior limit usually is covered with peritoneum; thus injuries to this part of the vagina may involve the peritoneal cavity. *(Moore, p 383)*

**156. (A)** The wall of the uterus consists of three layers: (1) the outer serosa, or perimetrium; (2) the middle muscular layer, or myometrium; and (3) the inner mucosal layer, or endometrium. The parametrium is a condensation of loose areolar tissue and smooth muscle at the base of the broad ligament and around the inferior end of the cervix. *(Moore, p 390)*

**157. (B)** Arteries of the uterine tube are derived from anastomoses between the uterine and ovarian arteries. These tubal branches pass along the tube between the layers of the mesosalpinx. The ampullary and uterine parts of the tube are the most vascularized portions. *(Moore, p 390)*

**158. (C)** The ovarian arteries arise from the abdominal aorta and descend on the posterior abdominal wall. On reaching the pelvic brim, the ovarian arteries cross over the external iliac vessels internal to the ureter. They run medially in the suspensory ligament of the ovaries to enter the broad ligament inferior to the uterine tubes. At the level of the ovary, the ovarian artery sends branches through the mesovarium to the ovary. This artery continues medially in the broad ligament to supply the uterine tube and to anastomose with the uterine artery. *(Moore, p 392).*

159. **(D)** The only correct item is item 4. The rectum is continuous superiorly with the sigmoid colon and begins on the pelvic surface of the third piece of the sacrum. The puborectalis, not the iliococcygeus, forms a sling at the junction of the rectum and the anal canal, producing the anorectal angle. The rectum lies immediately posterior to the prostate in the male and the vagina in the female. The inferior third of the rectum has no peritoneal covering. *(Moore, pp 393–394)*

160. **(E)** All of the structures listed are related anteriorly to the rectum. The two layers of the rectovesical septum lie in the median plane between the bladder and the rectum. The rectovesical septum represents a potential cleavage plane between the rectum and the prostate. In the female the anterior relation is the vagina. *(Moore, p 396)*

161. **(B)** There are five rectal arteries that anastomose freely with one another. The superior rectal artery, the continuation of the inferior mesenteric, supplies the terminal part of the sigmoid colon and the superior part of the rectum. The two middle rectal arteries, branches of the internal iliacs, supply the middle and inferior parts of the rectum. The inferior rectal arteries are branches of the internal pudendal arteries, not directly of the internal iliacs. *(Moore, p 397)*

162. **(E)** All of these structures or conditions can be palpated through the walls of the rectum; additional structures that also can be palpitated are the seminal vesicles, pelvic surfaces of the sacrum and coccyx, ischial tuberosities, pathological thickening of the ureters, and abnormal contents of the rectouterine pouch (female). Tenderness of an inflamed vermiform appendix can be detected rectally if this organ lies in the pelvis. *(Moore, p 398)*

163. **(C)** The interior of the anal canal is characterized by a series of five to ten longitudinal ridges or folds of mucosa called the anal columns. The terminal branches of the superior rectal vessels are within the anal columns; here the superior rectal veins of the portal system anastomose with the middle and inferior rectal veins of the caval system. The inferior ends of the anal columns are united to each other by small semilunar folds of mucosa called anal valves. Transverse folds are characteristic of the rectum, not the anal canal. *(Moore, pp 397,399–400)*

164. **(D)** The iliolumbar artery is the only one listed that does not supply the anal canal. The superior rectal artery supplies the superior part of the canal. The median sacral artery gives off small branches that supply the posterior wall of the anorectal junction. The two middle rectal arteries help to supply the superior part of the anal canal by anastomosing with the superior rectal arteries. *(Moore, pp 364,401)*

165. **(A)** The lymph vessels from the part of the anal canal superior to the pectineal line drain into the internal iliac lymph nodes and through them to the common iliac and aortic lymph nodes. The lymph vessels inferior to the pectinate line drain into the superficial inguinal lymph nodes. *(Moore, pp 401–402)*

166. **(A)** The nerve supply of the anal canal superior to the pectinate line is the same as for the rectum. The sympathetic nerves pass along the superior rectal vessels and partly form the inferior hypogastric (pelvic) plexus. The parasympathetic nerves, from S2 to S4, run in the pelvic splanchnic nerves to join the inferior hypogastric plexus. The sensory nerves, sensitive only to stretching, are derived from the inferior rectal branches of the pudendal nerve. *(Moore, pp 401–402)*

167. **(C)** Internal hemorrhoids are varicosities of the tributaries of the superior rectal vein. External hemorrhoids are varicosities of the inferior rectal vein. Thrombus formation is commoner in external than in internal hemorrhoids. The anastomosis between the superior and middle rectal veins forms a clinically important communication between the portal and systemic systems. In portal hypertension, as in hepatic cirrhosis, the tiny anastomotic veins in the anal canal and elsewhere become varicose and may rupture. *(Moore, p 402)*

168. **(E)** The two sacral sympathetic trunks converge as they pass along the sacrum and unite in the small median ganglion impar. Sympathetic trunks send gray rami communicantes to each ventral ramus of sacral and coccygeal nerves. The superior hypogastric plexus lies just below the bifurcation of the aorta; branches from it enter the pelvis as the right and left hypogastric nerves. *(Moore, p 403)*

169. **(B)** The right and left hypogastric nerves from the superior hypogastric plexus descend on the lateral walls of the pelvis where they mingle with the pelvic splanchnic nerves to form the right and left inferior hypogastric plexuses. The superior hypogastric, not the inferior hypogastric plexus, is the direct downward prolongation of the intermesenteric plexus. The ganglion impar sends a few branches to the coccygeal body, which lies anterior to the apex of the coccyx. *(Moore, p 403)*

170. **(D)** The iliolumbar ligaments unite each thick transverse process of L5 vertebra to the internal lip of the iliac crest, posteriorly. These ligaments help to stabilize the lumbosacral joint and limit axial rotation of L5 vertebra on the sacrum. Ligamenta flava join laminae of adjacent vertebral arches. The

adjacent edges of the spinous processes are joined by weak interspinous ligaments, and their tips are joined by strong supraspinous ligaments. *(Moore, pp 404,630)*

171. **(D)** The sacroiliac joints are strong joints. The sacrum is held firmly to the ilium by very strong interosseous and dorsal sacroiliac ligaments. Movement of the joints is limited to a slight gliding and rotary movement because they are designed primarily for weight bearing, not movement. They are synovial rather than cartilaginous joints and do not contain a joint disc. *(Moore, p 406)*

172. **(E)** All of the items listed are correct. Each articular surface of the pubic symphysis is covered by a thin layer of hyaline cartilage that is connected to the cartilage of the other side by a thick fibrocartilaginous interpubic disc. The superior pubic ligament connects the pubic bones superiorly; the arcuate pubic ligament connects the inferior borders of the joint. *(Moore, p 407)*

173. **(C)** The amniotic and chorionic sacs (amniochorionic membrane) rupture, permitting the amniotic fluid to escape. Usually this rupture occurs during the first stage of labor or at the end of it. The amniotic and chorionic sacs protrude into the cervical canal during the first stage of labor and help to dilate the cervix. *(Moore, pp 411,416)*

174. **(D)** The umbilical artery, a branch of the internal iliac, usually supplies the upper part of the bladder through its branches, the superior vesical arteries. It also gives off the artery of the ductus deferens that supplies, in addition to the ductus, the ureter, the seminal vesicles, and part of the bladder. The rectal arteries supply the rectum; the uterine artery supplies the uterus and helps to supply the ovary. *(Hollinshead, pp 745–747)*

175. **(E)** All of the listed branches of the internal iliac artery supply pelvis and lower limb. The iliolumbar artery supplies the iliacus muscle, and a lumbar branch replaces the fifth lumbar arteries. The lateral sacral artery disappears into the first sacral foramen and sends a branch into the pelvic sacral foramina. The obturator artery gives branches to the obturator internus muscle; it disappears into the thigh after passing through the obturator canal. The inferior gluteal artery passes through the greater sciatic foramen and supplies the gluteal region. *(Hollinshead, p 747)*

176. **(C)** Somatic nerves entering the pelvis include the lumbosacral trunk, the obturator nerve, and the ventral rami of sacral nerves. The lumbar plexus is joined to the sacral plexus by the lumbosacral trunk that is composed of part of L4 and the whole of L5 ventral ramus. Part of L2 anterior division contrib-

utes to the obturator nerve. The femoral nerve enters the thigh, not the pelvis, behind the inguinal ligament. The hypogastric nerves are autonomic, not somatic, nerves. *(Hollinshead, pp 384,694,748)*

177. **(D)** The sciatic is the only nerve listed that is a branch of the sacral plexus; it is the largest branch of the plexus. The sciatic nerve consists of two nerves: the common peroneal and the tibial. The other nerves named are branches of the lumbar, not the sacral, plexus. *(Hollinshead, pp 694,750–751)*

178. **(C)** The pelvic splanchnic nerves, not considered to be a part of the sacral plexus, are given off by S2–4 ventral rami. They are preganglionic parasympathetics and also convey visceral afferents from pelvic viscera to sacral segments of the spinal cord. They contribute to the inferior hypogastric plexus. *(Hollinshead, p 751)*

179. **(E)** All of these statements about the inferior hypogastric plexus are correct. It is the chief autonomic plexus of the pelvis. It is formed by lateral extensions of the superior hypogastric plexus known as the hypogastric nerves, and the pelvic splanchnic nerves. *(Hollinshead, pp 752–753)*

180. **(C)** The superior hypogastric plexus is the direct extension of the aortic plexus. It consists of a mixture of preganglionic and postganglionic sympathetic fibers, small ganglia, and visceral afferents that mediate sensation from the fundus and upper part of the uterus that follow the route of sympathetic efferents. The visceral efferents are from T10–L2 segments (sympathetics), not from L3 and L4 as stated. The superior hypogastric plexus ends by bifurcating into right and left hypogastric nerves. The hypogastric nerves and the superior hypogastric plexus apparently do not contain any parasympathetic fibers. *(Hollinshead, pp 753–754)*

181. **(D)** The pelvic splanchnic nerves represent the sacral parasympathetic outflow. The visceral efferents in these nerves have their cell bodies in the intermediolateral column of S2–4 segments of the spinal cord. These preganglionic fibers do not pass through the ganglia of the sacral sympathetic trunk; they synapse either in ganglia located in the inferior hypogastric plexus or in the walls of the viscera that they innervate. It is stated, generally, that their visceral afferents mediate pain from all pelvic organs; the uterus is an exception to this general rule. *(Hollinshead, p 754)*

182. **(A)** The rectum begins at the rectosigmoid junction as the continuation of the sigmoid colon. The rectum is not straight; it presents an anteroposterior sacral flexure and also three lateral curves or bends. In the rectum there is disappearance of the

haustra, sacculations that characterize most of the colon. (Hollinshead, p 260)

**183. (E)** All of the items are correct. The rectum receives its nerves from two plexuses. The superior rectal plexus may supply rectal blood vessels but is not important to the physiology of the rectum. The motor fibers to the rectum appear to be entirely parasympathetic and are conveyed in the middle rectal plexus. The afferent supply to the rectum concerned with pain and pressure of feces or gas in the rectum belongs to the parasympathetic system. (Hollinshead, p 763)

**184. (A)** The arteries of the prostate are derived from the inferior vesical, middle rectal, and internal pudendal arteries. The inferior gluteal is not identified as an arterial supply to the prostate. The inferior vesical artery reaches the bladder and the prostate along the lateral ligament of the bladder and supplies both. (Hollinshead, pp 772,745)

# CHAPTER 7

# The Upper Limb
## Questions

DIRECTIONS (Questions 1 through 50): Each of the numbered items or incomplete statements in this section is followed by answers or by completions of the statement. Select the ONE lettered answer or completion that is BEST in each case.

1. The arm can be defined correctly as the

   (A) upper extremity
   (B) brachium
   (C) antebrachium
   (D) upper limb
   (E) shoulder girdle

2. In relation to the vascular system of the upper limb, all of the following are correct EXCEPT the

   (A) named arteries are all deep ones
   (B) arteries are regularly accompanied by one or two deep veins
   (C) subclavian artery is the arterial stem to the limb
   (D) basilic is a deep vein
   (E) cephalic vein runs up the radial side of the limb

3. Correct description of the lymphatics of the upper limb includes which of the following?

   (A) superficial lymphatics begin with brachial vessels
   (B) deep lymphatics are more numerous than are superficial lymphatics
   (C) superficial lymphatics usually remain on the palmar side of the limb
   (D) lymphatics usually do not accompany deep blood vessels
   (E) lymphatics end in axillary lymph nodes

4. In the development of nerve–muscle relations, which of the following statements is correct?

   (A) the nerve supply of the latissimus dorsi originates at lumbar levels
   (B) nerves retain original connection to their muscle mass

   (C) most muscles of the upper limb receive their nerve supply from a single spinal nerve
   (D) nerves do not parallel muscles in their development
   (E) fibers of upper limb muscles are innervated by dorsal branches of spinal nerves

5. Which of the following is a correct statement regarding distribution of nerves within muscle?

   (A) no fixed pattern of distribution has been noted
   (B) distribution is not influenced by the shape of the muscle
   (C) in skeletal muscle each individual muscle fiber receives a nerve fiber
   (D) nerve fibers to skeletal muscle are known as adrenergic fibers
   (E) all nerve fibers entering a voluntary muscle are motor to muscle fibers

6. All the following items are true in relation to development of the upper limb EXCEPT the

   (A) limb early projects at approximately a right angle to the body
   (B) anatomical position is with the forearm supinated and palm facing forward
   (C) limb later in growth adducts to almost parallel the trunk
   (D) little finger develops on the preaxial border
   (E) developing musculature of the limb divides into ventral and dorsal parts

7. Correct patterns of cutaneous nerve supply of the upper limb include which of the following?

   (A) the third and fourth cervical nerves supply a limited area of skin over the pectoral region and shoulder
   (B) the third thoracic nerve usually sends a branch to the skin of the medial and upper parts of the arm
   (C) the cervical plexus supplies most of the skin of the arm
   (D) antebrachial cutaneous nerves supply the region of the humerus
   (E) branches of the brachial plexus supply only the arm and forearm

8. Characteristics of muscles of the upper limb include which of the following?

   (A) anterior muscles are extensors
   (B) the anterior group of forearm muscles arises chiefly on the radial side of the arm
   (C) the anterior group of forearm muscles extends the wrist and fingers
   (D) the anterior group of forearm muscles pronates the forearm
   (E) the posterior group of forearm muscles flexes the wrist and fingers

9. Which of the following is an accurate description of nerve supply to upper limb muscles?

   (A) the radial nerve supplies the supinator muscle of the forearm
   (B) the musculocutaneous nerve supplies the extensor muscles of the arm
   (C) the median nerve supplies flexor muscles of the arm
   (D) the axillary nerve supplies pectoral muscles
   (E) the ulnar nerve supplies the coracobrachialis muscle

10. Arterial blood supply to the upper limb is correctly described by which of the following?

   (A) the axillary artery is a branch of the arch of the aorta
   (B) the brachial artery is a direct branch of the subclavian artery
   (C) the radial artery is a direct branch of the axillary artery
   (D) the humeral circumflex arteries are branches of the brachial artery
   (E) the ulnar artery is a direct division of the brachial artery

11. The pectoral girdle consists of the

   (A) humerus and scapula
   (B) sternum and humerus
   (C) sternum and scapula
   (D) clavicle and scapula
   (E) clavicle and sternum

12. The clavicle has attachment to the scapula at which of the following locations?

   (A) through the scapuloclavicular ligament
   (B) at the sternoclavicular joint
   (C) through the coracoacromial ligament
   (D) through the coracohumeral ligament
   (E) through the conoid and trapezoid ligaments

13. All the following items characterize structure of the scapula EXCEPT

   (A) a thickened medial border adjacent to the coracoid
   (B) a subscapular fossa on its costal surface
   (C) a spine continuing into the acromion
   (D) three angles and three borders
   (E) the glenoid cavity at its lateral angle

14. The humerus presents which of the following structural characteristics?

   (A) an olecranon fossa on its anterior surface
   (B) a radial fossa on its medial lower end
   (C) a capitulum on its lower lateral condyle
   (D) a deltoid tuberosity adjoining its head
   (E) a greater tubercle at its midregion

15. Knowledge of the lymphatic drainage of the breast is of particular practical importance mainly because

   (A) the lymphatics may become dense and cause breast enlargement
   (B) the lymphatics frequently form benign breast tumors
   (C) the lymphatics often are invaded by carcinoma of the breast
   (D) obstruction of lymphatics may cause serious edema of the breast
   (E) lymphatic blockage may cause milk to become toxic

16. The pectoralis major muscle inserts on the

   (A) lesser tubercle of the humerus
   (B) lateral third of the clavicle
   (C) coracoid process of the scapula
   (D) crest of the greater tubercle of the humerus
   (E) head of the humerus

17. The chief action of the pectoralis major is

   (A) abduction and lateral rotation of the humerus
   (B) elevation and upward rotation of the scapula
   (C) elevation of the sternum
   (D) adduction and medial rotation of the humerus
   (E) elevation of the clavicle

18. The pectoralis minor muscle usually has its insertion on the

(A) coracoid process of the scapula
(B) anterior manubrium
(C) medial third of the clavicle
(D) acromion of the scapula
(E) greater tubercle of the humerus

19. Innervation of the triceps brachii involves which of the following nerves?

(A) long thoracic
(B) lower subscapular
(C) musculocutaneous
(D) axillary
(E) radial

20. Which of the following statements correctly describes the axillary artery?

(A) it originates from the arch of the aorta
(B) it gives origin to the thoracoacromion artery
(C) it has the lateral thoracic artery as its first branch
(D) it divides into radial and ulnar arteries
(E) it provides no blood supply to the humerus

21. The central structure of the axilla is stated to be the

(A) subscapular artery
(B) brachial plexus
(C) axillary artery
(D) lateral thoracic artery
(E) axillary nerve

22. Which of the following statements correctly describes the lateral cord of the brachial plexus?

(A) it represents lateral divisions of the plexus
(B) typically it has five branches
(C) it contains nerve fibers from C8 and T1
(D) it gives rise to the ulnar nerve
(E) it gives rise to the musculocutaneous nerve

23. The posterior cord of the brachial plexus is described correctly by all the following statements EXCEPT

(A) it is formed by union of all the posterior divisions of the plexus
(B) it gives off the upper subscapular nerve
(C) it gives off the median nerve
(D) it gives off the axillary nerve
(E) it has the radial nerve as a terminal branch

24. The long thoracic nerve innervates which of the following muscles?

(A) triceps brachii
(B) serratus anterior
(C) pectoralis minor
(D) latissimus dorsi
(E) deltoid

25. Injury to the upper part of the brachial plexus usually results

(A) from violent separation of head and shoulder
(B) in spastic paralysis of the upper limb
(C) in Klumpke's paralysis
(D) in paralysis of the hand
(E) in involvement of nerves C8 and T1

26. Attachments of the trapezius muscle may correctly be described as

(A) an origin on spinous processes of all lumbar vertebrae
(B) an origin from the spine of the scapula
(C) an origin from the vertebral border of the scapula
(D) an origin from the occipital bone
(E) an insertion on the thoracic spinous processes of all thoracic vertebrae

27. Which of the following nerves provides motor innervation for the latissimus dorsi muscle?

(A) accessory
(B) dorsal scapular
(C) thoracodorsal
(D) transverse cervical
(E) axillary

28. The axillary nerve innervates which of the following muscles?

(A) supraspinatus
(B) infraspinatus
(C) deltoid
(D) teres major
(E) trapezius

29. Primary lateral rotators of the arm are described as including the

(A) infraspinatus and teres minor
(B) subscapularis and teres major
(C) supraspinatus and anterior deltoid
(D) triceps brachii and latissimus dorsi
(E) serratus anterior and rhomboidus major

30. An injury to the spinal accessory nerve primarily would affect which of the following movements?

(A) adduction of the arm at the glenohumeral joint
(B) lateral rotation of the arm
(C) depression of the scapula
(D) protraction of the scapula
(E) upward rotation of the scapula

31. The deltoid is assisted in its primary action of abduction of the arm by the

(A) teres major
(B) latissimus dorsi
(C) levator scapulae
(D) supraspinatus
(E) serratus anterior

32. An important artery to the shoulder that arises in the neck is the

    (A) thoracoacromial
    (B) subscapular
    (C) scapular circumflex
    (D) posterior humeral circumflex
    (E) thyrocervical trunk

33. All of the following muscles comprise the musculo-tendinous cuff of the shoulder joint EXCEPT the

    (A) subscapularis
    (B) supraspinatus
    (C) infraspinatus
    (D) deltoid
    (E) teres minor

34. When the arm is abducted, the strength of the shoulder joint is largely dependent on the

    (A) depth of the glenoid fossa
    (B) heavy joint capsule
    (C) transverse humeral ligament
    (D) musculotendinous cuff
    (E) glenohumeral ligaments

35. The biceps muscle is the only member of the anterior group of arm muscles that

    (A) originates on the supraglenoid tubercle of the scapula
    (B) originates on the coracoid process of the scapula
    (C) inserts on the ulnar tuberosity
    (D) is innervated by the musculocutaneous nerve
    (E) partly arises from the body of the humerus

36. Correct description of innervation of the upper limb includes which of the following?

    (A) musculocutaneous nerve, a derivative of the medial cord of the brachial plexus
    (B) median nerve, innervating no muscles in the arm
    (C) ulnar nerve, a derivative of the lateral cord of the brachial plexus
    (D) radial nerve, innervating the coracobrachialis muscle
    (E) lateral antebrachial cutaneous nerve, terminating above the elbow

37. The anatomical snuffbox is described correctly by which of the following?

    (A) being located medial to the extensor pollicis longus
    (B) having the abductor pollicis longus as part of its posterior boundary
    (C) having the extensor pollicis longus as its anterior boundary

    (D) being located between the tendons of the extensor pollicis brevis and the extensor indicis
    (E) being a hollow at the base of the first metacarpal, distal to the end of the radius

38. A structure found at the distal end of the ulna is the

    (A) olecranon
    (B) trochlear notch
    (C) styloid process
    (D) ulnar tuberosity
    (E) coronoid process

39. All of the following muscles attach to the radius EXCEPT the

    (A) biceps
    (B) supinator
    (C) brachioradialis
    (D) brachialis
    (E) pronator quadratus

40. Which of the following bones is found in the proximal row of carpals?

    (A) lunate
    (B) trapezium
    (C) trapezoid
    (D) capitate
    (E) hamate

41. Which of the following statements is true concerning the radial nerve?

    (A) it carries only motor fibers
    (B) it supplies the flexor carpi ulnaris muscle
    (C) it is accompanied by the profunda brachii artery
    (D) it innervates the biceps muscle
    (E) it is a terminal branch of the medial cord of the brachial plexus

42. Which of the following structures crosses superficial to the flexor retinaculum at the wrist?

    (A) median nerve
    (B) flexor pollicis longus
    (C) flexor carpi radialis
    (D) palmaris longus
    (E) flexor digitorum profundus

43. Which one of the following muscles originates on the humerus, the radius, and the ulna?

    (A) flexor digitorum superficialis
    (B) flexor carpi radialis
    (C) palmaris longus
    (D) flexor pollicis longus
    (E) flexor digitorum profundus

44. In the usual pattern of digital synovial tendon sheaths, infection can travel into the common flexor tendon sheath at the wrist from the

   (A) thumb
   (B) index finger
   (C) middle finger
   (D) ring finger
   (E) little finger

45. The radial artery is described correctly as

   (A) arising as the medial terminal branch of the brachial artery
   (B) giving rise to the common interosseous artery
   (C) giving rise to the superficial palmar artery
   (D) providing the major components of the superficial palmar arch
   (E) giving rise to all four common digital arteries

46. Which of the following statements correctly describes the median nerve?

   (A) it gives rise to the posterior interosseous nerve
   (B) it passes superficial to the flexor retinaculum
   (C) it innervates all flexor muscles on the anterior forearm
   (D) it gives a muscular branch to the thenar muscles
   (E) it innervates all the lumbrical muscles

47. In a typical innervation pattern of the hand, which of the following muscles is supplied by branches of the median nerve?

   (A) abductor pollicis brevis
   (B) abductor pollicis longus
   (C) adductor pollicis
   (D) palmaris brevis
   (E) second and third interossei

48. Which of the following statements correctly describes the arteries of the palm?

   (A) the ulnar artery is the source of the superficial palmar arch
   (B) palmar metacarpal arteries arise from the superficial palmar arch
   (C) the princeps pollicis artery arises from the ulnar artery
   (D) deep palmar arch branches do not connect with common digital arteries
   (E) proper digital arteries run only on the radial side of the fingers

49. A muscle that is both a strong flexor at the elbow and a supinator of the forearm is the

   (A) brachialis
   (B) supinator
   (C) biceps brachii
   (D) brachioradialis
   (E) flexor carpi radialis

50. Anomalous innervation of the hand muscles has been found in all the following forms EXCEPT

   (A) the opponens pollicis being supplied by both median and ulnar nerves
   (B) the short flexor of the thumb being supplied by only the ulnar nerve
   (C) the median nerve sending a branch into the first dorsal interosseous
   (D) deep penetration of the ulnar nerve into the thenar mass
   (E) the median nerve sending a direct branch across the palm to the hypothenar muscles

DIRECTIONS (Questions 50 through 56): Each group of items in this section consists of lettered headings followed by a set of numbered words or phrases. For each numbered word or phrase, select the ONE lettered heading that is most closely associated with it. Each lettered heading may be selected once, more than once, or not at all.

Questions 51 through 54

   (A) radial nerve
   (B) median nerve
   (C) ulnar nerve
   (D) musculocutaneous nerve
   (E) axillary nerve

51. nerve that supplies the deltoid

52. nerve that supplies the coracobrachialis

53. nerve that supplies the interossei

54. nerve that supplies the abductor pollicis brevis

Questions 55 and 56

   (A) palmar metacarpal artery
   (B) princeps pollicis artery
   (C) ulnar artery
   (D) radial artery
   (E) common digital artery

55. arterial origin of the superficial palmar arch

56. arterial origin of the deep palmar arch

**DIRECTIONS (Questions 57 through 67):** Each group of items in this section consists of lettered headings followed by a set of numbered words or phrases. For each numbered word or phrase, select

A if the item is associated with (A) only,
B if the item is associated with (B) only,
C if the item is associated with both (A) and (B),
D if the item is associated with neither (A) nor (B).

**Questions 57 through 60**

    (A) adduction of the scapula
    (B) upward rotation of the scapula
    (C) both
    (D) neither

57.  serratus anterior

58.  trapezius

59.  rhomboids

60.  latissimus dorsi

**Questions 61 through 64**

    (A) posterior wall of the axilla
    (B) medial rotation of the humerus
    (C) both
    (D) neither

61.  pectoralis major

62.  serratus anterior

63.  subscapularis

64.  teres major

**Questions 65 through 67**

    (A) quadrangular space
    (B) triangular space
    (C) both
    (D) neither

65.  teres major muscle

66.  axillary nerve

67.  scapular circumflex artery

**DIRECTIONS (Questions 68 through 88):** For each of the items in this section, ONE or MORE of the numbered options is correct. Choose the answer

A if only 1, 2, and 3 are correct,
B if only 1 and 3 are correct,
C if only 2 and 4 are correct,
D if only 4 is correct,
E if all are correct.

68.  The brachial artery has which of the following characteristics?

    (1) has palpable pulsations along the medial bicipital furrow
    (2) divides into terminal branches just below the crease of the elbow
    (3) gives off the profunda brachii artery
    (4) has the ulnar artery as a terminal branch

69.  The radial nerve is described correctly by which of the following?

    (1) in the axilla, it lies in front of the axillary artery
    (2) it descends along the origin of the lateral head of the triceps
    (3) it terminates in a motor branch, the superficial radial nerve
    (4) it lies on the capsule of the elbow joint

70.  The anterior interosseous nerve supplies the

    (1) pronator quadratus
    (2) abductor pollicis brevis
    (3) flexor pollicis longus
    (4) flexor carpi ulnaris

71.  The medial epicondyle of the humerus serves as the site of origin for the

    (1) pronator teres
    (2) flexor pollicis longus
    (3) palmaris longus
    (4) flexor digitorum profundus

72.  In a complete section of the ulnar nerve at the wrist, disability of the hand results from paralysis of the

    (1) opponens pollicis
    (2) first and second lumbricals
    (3) abductor pollicis brevis
    (4) third dorsal interosseous

73.  In the usual pattern of motor innervation, two nerves supply the

    (1) flexor pollicis brevis
    (2) lumbricals
    (3) flexor digitorum profundus
    (4) extensor pollicis longus

74. In elevation of the arm at the glenohumeral joint, primary muscle action includes which of the following?

    (1) the deltoid is a prime mover
    (2) the supraspinatus acts progressively throughout the entire movement
    (3) the infraspinatus assists with the movement
    (4) the subscapularis assists with the movement

75. Muscle action in flexion at the elbow is described correctly by which of the following?

    (1) the chief flexor is the brachialis
    (2) the biceps muscle acts best when the forearm is in pronation
    (3) the brachioradialis is a useful adjunct flexor
    (4) the pronator teres has a more advantageous location for flexion than other flexors of the elbow

76. Which of the following statements is true of Colles' fracture?

    (1) it occurs more frequently in men than in women
    (2) it involves the distal end of the radius
    (3) it causes displacement of the lunate
    (4) it usually shows a "dinner fork" deformity

77. Which of the following statements is true of wrist joint structure?

    (1) the radius articulates with the scaphoid, lunate, and triquetrum
    (2) the articular disc at the distal ulna articulates with the proximal row of carpals
    (3) articular nerves are derived from the anterior interosseous nerve
    (4) no articular structure is present between the radius and ulna

78. Correct description of the elbow joint includes which of the following?

    (1) the "carrying angle" is about 150 degrees
    (2) extension is limited by collateral ligaments
    (3) two articulations complete the joint
    (4) extension is limited by impingement of the ulnar olecranon on the olecranon fossa

79. The elbow joint is described correctly as

    (1) an articulation of the trochlea of the humerus with the head of the radius
    (2) an articulation of the capitulum of the humerus with the ulna
    (3) having close relation to the ulnar nerve at the lateral epicondyle of the humerus
    (4) being subject to injuries of the subcutaneous olecranon bursa

80. The order of superficial flexor muscles across the forearm beginning on the radial side and moving toward the ulnar side is

    (1) first, the pronator teres
    (2) second, the flexor carpi radialis
    (3) third, the palmaris longus
    (4) fourth, the flexor carpi ulnaris

81. When the median nerve is severed in the elbow region, there is loss of

    (1) flexion of the proximal interphalangeal joints of all the digits
    (2) flexion of the distal interphalangeal joints of the index and middle fingers
    (3) flexion of the metacarpophalangeal joints of the index and middle fingers
    (4) flexion of the distal interphalangeal joints of the ring and little fingers

82. Symptoms of carpal tunnel syndrome may include

    (1) loss of adduction of the thumb
    (2) absence of tactile sensation on the palmar surface of the index finger
    (3) loss of flexion at the distal interphalangeal joint of the fourth and fifth digits
    (4) paresthesia on the palmar surface of the middle finger

83. The ulnar artery has which of the following characteristics?

    (1) it is the larger terminal branch of the brachial artery
    (2) it gives off the anterior ulnar recurrent branch
    (3) it gives off the common interosseous artery
    (4) it has palpable pulsations anterior to the head of the ulna

84. Dupuytren's contracture symptoms involve

    (1) initially, the ring and little fingers
    (2) impairment of flexion of the thumb
    (3) the index finger rarely
    (4) those similar to the carpal tunnel syndrome

85. Which of the following statements is true in relation to the scapula?

    (1) its medial border crosses ribs 4 through 9
    (2) its superior angle lies at about the level of the spinous process of T12
    (3) the acromion can be palpated deep to the lateral part of the clavicle
    (4) the inferior angle lies at about the level of the spinous process of T7

| SUMMARY OF DIRECTIONS | | | | |
|---|---|---|---|---|
| A | B | C | D | E |
| 1, 2, 3 only | 1, 3 only | 2, 4 only | 4 only | All are correct |

86. Incorrect use of crutches may cause

(1) injury to the posterior cord of the brachial plexus
(2) paralysis of extensors of the wrist
(3) paralysis of the anconeus
(4) Dupuytren's contracture

87. Which of the following correctly describes the latissimus dorsi muscle?

(1) it arises from the lower six thoracic vertebral spinous processes
(2) it is an important lateral rotator of the humerus
(3) it inserts into the floor of the intertubercular sulcus
(4) it is supplied by the long thoracic nerve

88. Characteristics of the deltoid muscle include

(1) an origin on the medial third of the clavicle
(2) an innervation by the thoracodorsal nerve
(3) an action of initiation of the movement of abduction of the humerus from the adducted position
(4) actions of both medial and lateral rotation of the humerus

# Answers and Explanations

1. **(B)** The chief named parts of the upper limb are the acromion, or tip of the shoulder; the axilla, or armpit; the brachium, or arm; the cubitus, or elbow; the antebrachium, or forearm; the manus, or hand; the carpus, or wrist; and the digits, the fingers, and the thumb. In anatomical descriptions, the arm refers to the region of the humerus and related structures. *(Hollinshead, p 148)*

2. **(D)** The superficial veins of the upper limb begin as networks on the digits and drain mostly onto the dorsum of the hand. They unite to form two chief channels: the cephalic vein from the radial side and the basilic vein from the ulnar side. The named arteries of the limb are all deep ones, and they are accompanied regularly by one or two deep veins. The subclavian artery is the arterial stem to the limb. *(Hollinshead, p 161)*

3. **(E)** Both superficial and deep lymphatics of the upper limb end in axillary lymph nodes. Superficial lymphatics begin as a dense capillary plexus in the skin of the digits. Deep lymphatics are much less, not more, numerous than superficial lymphatics. Lymphatic vessels from the palmer surface tend to run onto the dorsum; thus infections of the palm may be evidenced by swelling on the dorsum of the hand. The deep lymphatics follow the deep arteries and veins. *(Hollinshead, pp 161–164)*

4. **(B)** Muscle masses migrate, but nerves retain their original connection to their muscle mass. The latissimus dorsi, for example, spreads as far caudally as the pelvis but actually is derived from mesenchyme originating at the lower cervical region, and its nerve contains fibers from lower cervical nerves. Most muscles of the limbs receive their nerve supply from two or more spinal nerves. Since the relationship between nerves and muscle is so close and so nearly constant, nerves tend to parallel muscle in their development. Muscle fibers of the limb muscles are supplied by ventral (motor) branches of spinal nerves. *(Hollinshead, pp 102–103)*

5. **(C)** In skeletal muscle every individual muscle fiber receives a nerve fiber. The distribution of a nerve within a muscle typically follows a fixed pattern: the nerve branches and rebranches in the connective tissue of the muscle, and the branches follow such courses as to bring them into contact with all the fibers within the muscle. The pattern of nerve distribution depends in part on the shape of the muscle: in a long muscle major nerve branches run longitudinally, and in a short muscle they run transversely. Nerve fibers to skeletal muscle release acetylcholine at the neuromuscular junction, an essential part of the mechanism for muscle contraction. These nerve fibers therefore are known as cholinergic, not adrenergic, fibers. Not all nerve fibers entering a voluntary muscle are motor fibers destined for muscle fibers: sympathetic fibers innervate blood vessels, and afferent fibers conduct sensory impulses toward the spinal cord. *(Hollinshead, pp 105–107)*

6. **(D)** In the original position of the developing limb, the palm faces forward, with the thumb on the cranial or preaxial border and the little finger on the caudal or postaxial border. The limb first projects out at approximately a right angle but later adducts to become almost parallel to the trunk. The anatomical position, therefore, is described as the palm facing forward, with the ulnar side medial and the radial side lateral. The developing musculature of the limb divides into ventral and dorsal parts. This division is accompanied by a corresponding division of the nerves growing into the muscle. *(Hollinshead, p 147)*

7. **(A)** The third and fourth cervical nerves supply a limited area of skin over the pectoral region and shoulder. The second, not the third, thoracic nerve usually sends a branch to the skin of the medial and upper parts of the arm. Most of the skin of the upper limb is supplied by branches of the brachial plexus, not the cervical plexus. Antebrachial cutaneous nerves supply skin of the forearm, not the arm (humeral region). *(Hollinshead, p 151)*

8. **(D)** The anterior muscles of the arm originally are ventral ones and are flexors. The anterior muscles of

the forearm arise chiefly on the ulnar side of the arm or forearm and flex the wrist and fingers; they also pronate the forearm. The posterior group of forearm muscles arises chiefly on the radial side of the arm and on the back of the forearm and extends the wrist and digits; it also supinates the forearm. *(Hollinshead, pp 156–157)*

9.  **(A)** With minor exceptions, all the major nerves of the upper limb are derived from the brachial plexus. The brachial plexus is divisible into anterior and posterior parts. Anterior nerves are the musculocutaneous, the median, and the ulnar; posterior nerves are the axillary and the radial. The radial nerve supplies the supinator muscle. The musculocutaneous nerve supplies anterior (flexor) muscles of the arm. The median and ulnar nerves run through the arm without supplying any muscles there. The axillary nerve supplies two posterior muscles of the shoulder (deltoid and teres minor). The ulnar nerve, as previously stated, is derived from anterior portions of the brachial plexus. *(Hollinshead, pp 156–157)*

10. **(E)** The subclavian artery derives its blood directly or indirectly from the arch of the aorta. As it enters the axilla, its name changes to the axillary artery, continued into the arm as the brachial artery. The brachial artery ends in front of the elbow by dividing into the radial and ulnar arteries. The anterior and posterior humeral circumflex arteries are branches of the axillary artery. *(Hollinshead, p 161)*

11. **(D)** The anteriorly situated clavicle and the posteriorly situated scapula constitute the shoulder girdle. Neither the sternum nor the humerus is considered part of the shoulder girdle. *(Hollinshead, p 165)*

12. **(E)** The clavicle has attachment to the coracoid process of the scapula through the coracoclavicular ligament, composed of the conoid and trapezoid ligaments. There is no scapuloclavicular ligament. The clavicle articulates with the sternum only at the sternoclavicular joint. The coracoacromial ligament is situated between the coracoid process and acromion of the scapula. The coracohumeral ligament is not attached to the clavicle. *(Hollinshead, pp 167–168)*

13. **(A)** The scapula is a rather thin, triangular, flat bone strengthened by the scapular spine and a thickened lateral, not a medial, border. The coracoid process is located superiorly near the lateral angle. The subscapular fossa is located on the costal surface of the bone and the spine on the dorsal surface continuing laterally into the acromion. The scapula has three borders (medial, lateral, and superior) and three angles (inferior, lateral, and superior) where these borders meet. The lateral angle is expanded to form the glenoid cavity, which articulates with the head of the humerus. *(Hollinshead, p 169)*

14. **(C)** The humerus has at its expanded lower end the condyle, consisting of the capitulum on the lateral anterior inferior surface and the trochlea, the pullylike medial surface. The olecranon fossa is located on the lower posterior surface for reception of the olecranon of the ulna. The radial fossa for reception of the head of the radius is located just above the capitulum. The deltoid tuberosity is located on the anterolateral surface of the humerus about midway its length, not adjoining the head, and the greater tubercle adjoins the head. *(Hollinshead, pp 172–173)*

15. **(C)** Knowledge of the lymphatic drainage of the breast is of practical importance because of the frequent invasion of lymphatics by carcinoma. The chief drainage is into axillary lymph nodes, which normally receive 75% of the total drainage from the breast. They are therefore particularly likely to be involved when there is mammary carcinoma. *(Hollinshead, p 176)*

16. **(D)** The pectoralis major muscle inserts on the crest of the greater tubercle of the humerus (lateral lip of the intertubercular groove). It originates from the medial third of the clavicle, from the lateral part of the entire length of the anterior surface of the manubrium and body of the sternum, and from the cartilages of about the first six ribs. It has no attachment to the head of the humerus. *(Hollinshead, pp 177–178)*

17. **(D)** The chief action of the pectoralis major is adduction and medial rotation of the humerus. It assists the anterior deltoid in flexion of the arm. The lower fibers can help to extend the limb until it is by the side. It is not in a position to abduct the humerus, does not attach to the scapula, and cannot elevate the clavicle. *(Hollinshead, p 180)*

18. **(A)** The pectoralis minor usually takes origin from three ribs: the second or third to the fifth or sixth. It inserts into the medial side of the coracoid process of the scapula. It does not attach to the greater tubercle of the humerus. *(Hollinshead, p 179)*

19. **(E)** The radial nerve supplies the triceps muscle. The long thoracic innervates the serratus anterior muscle; the lower subscapular innervates the subscapularis and the teres major; the musculocutaneous nerve innervates the anterior muscles of the arm; and the axillary nerve innervates the deltoid and teres minor muscles. *(Hollinshead, p 181)*

20. **(B)** The axillary artery gives off the two chief arteries of the pectoral region: the thoracoacromial and the lateral thoracic. The axillary artery is the continuation of the subclavian as the vessel crosses the first rib. The axillary artery becomes the brachial artery as it leaves the axilla at the lower border of the teres major muscle. The axillary artery usually is

described as giving off six branches, the first being the supreme thoracic artery and the last the anterior and posterior humeral circumflex arteries to the upper end of the humerus. *(Hollinshead, pp 188–189)*

21. **(C)** The axilla houses the great vessels and nerves of the limb, which are closely grouped together and enclosed in a layer of fascia, the axillary sheath. Most of the nerve trunks appearing in the axilla are the lower part of the brachial plexus and its major branches and rather closely surround the axillary artery. This artery is therefore thought of as the central structure of the axilla. The cords of the brachial plexus are named lateral, medial, and posterior according to their position relative to the axillary artery. *(Hollinshead, pp 183–185)*

22. **(E)** The lateral cord of the brachial plexus typically has only three branches, the musculocutaneous nerve being one of these. The lateral and medial cords represent anterior divisions of the brachial plexus, not lateral divisions. The lateral cord contains fibers from C5, C6, and C7 and perhaps from C4 if these fibers join the plexus. The ulnar nerve continues into the arm from the medial cord of the plexus. *(Hollinshead, p 185)*

23. **(C)** The posterior cord of the brachial plexus does not give off the median nerve; this nerve is formed from branches of the lateral and medial cords. The posterior cord is formed by union of all posterior divisions of the brachial plexus; the upper subscapular and axillary nerves are among its branches and the radial nerve is a terminal branch. *(Hollinshead, pp 186–187)*

24. **(B)** The long thoracic nerve is located on the medial wall of the axilla on the surface of the serratus anterior muscle, which it supplies. This nerve arises from the brachial plexus in the neck before the trunks are formed. None of the other muscles listed are supplied by the long thoracic nerve. *(Hollinshead, p 186)*

25. **(A)** The brachial plexus often is injured by violent separation of the head and shoulder, as may occur from a fall from a motorcycle. These injuries commonly affect the upper part of the plexus. This is also a common type of birth injury resulting from a difficult delivery. Such lesions involving the upper part of the plexus (especially C5 and C6) are called Erb's or Erb-Duchenne paralysis and especially affect the shoulder and arm. Injuries to the lower part of the plexus (C8 and T1) frequently are called Klumpke's paralysis and especially affect the distal part of the limb. This injury does not result in spastic paralysis, as it is a peripheral, not a central nervous system, injury. *(Hollinshead, p 187)*

26. **(D)** The trapezius muscle originates from the superior nuchal line on the occipital bone and from the external occipital protuberance. It originates from the ligamentum nuchae of the cervical vertebral spinous processes, from the spinous process of the 7th cervical vertebra and from all the thoracic vertebral spinous processes and their connecting supraspinous ligaments. This muscle inserts on approximately the distal third of the clavicle and on the acromion and spine of the scapula. *(Hollinshead, p 194)*

27. **(C)** The thoracodorsal nerve supplies the latissimus dorsi muscle. It leaves the axilla on the costal surface of the muscle and runs downward on this surface with the thoracodorsal artery. None of the other nerves listed supply this muscle. *(Hollinshead, p 195)*

28. **(C)** The axillary nerve runs deep to the deltoid, circling the surgical neck of the humerus and sending branches into the deltoid. The supraspinatus and infraspinatus are innervated by the suprascapular nerve and the trapezius by the accessory nerve. *(Hollinshead, pp 198–201)*

29. **(A)** The infraspinatus and teres minor both are lateral rotators of the arm. The supraspinatus primarily is an abductor and is active in supporting the arm against a downward pull. The subscapularis and teres major both are medial rotators, as are the anterior deltoid and latissimus dorsi. The serratus anterior and rhomboidus major have no attachment on the humerus and therefore cannot rotate it. *(Hollinshead, p 201)*

30. **(E)** The spinal accessory nerve innervates the trapezius. This muscle as a whole retracts the scapula, but its upper and lower fibers, acting together, upwardly rotate the scapula (turn the glenoid cavity upward). The other actions listed would not be major results of an injury to the spinal accessory nerve. *(Hollinshead, p 195)*

31. **(D)** The supraspinatus assists the deltoid in abduction of the arm. This muscle arises from the supraspinatous fossa of the scapula and passes laterally over the top of the shoulder to attach to the uppermost of the three facets of the greater tubercle of the humerus. None of the other muscles listed are in a position to accomplish abduction of the arm. *(Hollinshead, p 199)*

32. **(E)** The transverse cervical artery and the suprascapular artery, typically both branches of the thyrocervical trunk, arise in the neck. The other arteries listed arise from the axillary artery in the axilla. *(Hollinshead, pp 203–204)*

33. **(D)** All of the muscles listed except the deltoid are intimately related to the shoulder joint. Since most of them are rotators, the musculotendinous cuff also is called the rotator cuff. The deltoid muscle com-

pletely covers the shoulder joint and the muscles of the rotator cuff. *(Hollinshead, p 204)*

34. **(D)** The coracohumeral ligament apparently prevents downward displacement of the humeral head when the arm is by the side. With the arm abducted, the ligament is lax, and the strength of the joint depends entirely on the musculotendinous cuff. The glenoid fossa is shallow and does not offer a stable articulating surface for the relatively large head of the humerus. The other ligaments listed do not appear to be of importance. *(Hollinshead, p 205)*

35. **(A)** The biceps brachii muscle along with the coracobrachialis and the brachialis muscles are located on the anterior arm. The biceps is the only one of this group that originates on the supraglenoid tubercle of the scapula. The biceps has a short head originating on the coracoid process, the site of origin of the coracobrachialis. The brachialis is the only one of the group originating on the anterior surface of the humerus and inserting on the ulnar tuberosity. All of these muscles are innervated by the musculocutaneous nerve. *(Hollinshead, p 209)*

36. **(B)** The median nerve runs down the arm on the axillary and brachial arteries. It gives off no branches to structures of the arm. The musculocutaneous nerve is a derivative of the lateral, not the medial cord, of the brachial plexus. The ulnar nerve is a derivative of the medial, not the lateral, cord of the plexus. The radial nerve does not innervate the coracobrachialis muscle, and the lateral antebrachial cutaneous nerve, a continuation of the musculocutaneous nerve, continues into the forearm. *(Hollinshead, pp 212–213)*

37. **(E)** When the extended thumb is separated widely from the index finger, two ridges are raised by tendons that enclose a hollow between them distal to the end of the radius. This hollow is named the anatomical snuff box because of the use to which it was once put. The abductor pollicis longus and extensor pollicis brevis tendons form the anterior boundary of the anatomical snuff box; the extensor pollicis longus tendon forms its posterior border. It is located laterally, not medially, to the extensor pollicis longus and has no relation to the extensor indicis tendon. *(Hollinshead, pp 220,241–243)*

38. **(C)** Of the structures listed, the styloid process is the only one located at the distal end of the ulna. The olecranon, trochlear notch, ulnar tuberosity, and coronoid process are all at or near the proximal end of the ulna. *(Hollinshead, p 223)*

39. **(D)** The brachialis muscle inserts on the ulna, not the radius. The biceps, supinator, brachioradialis, and pronator quadratus all have insertions on the radius. *(Hollinshead, p 223)*

40. **(A)** The lunate is the only bone listed that is found in the proximal row of carpals. All of the others listed are found in the distal row of carpals. *(Hollinshead, p 226)*

41. **(C)** The radial nerve is accompanied in the arm by the first major branch of the brachial artery, the profunda brachii. The radial nerve ends by dividing into superficial and deep branches. The superficial branch gives cutaneous nerves to the posterior forearm and much of the dorsum of the hand. The deep branch of the radial nerve gives motor fibers to the extensor muscles of the arm and forearm. This nerve is a terminal branch of the posterior cord of the brachial plexus. It does not innervate the biceps or the flexor carpi ulnaris muscles. *(Hollinshead, pp 214,232,245)*

42. **(D)** All of the structures listed, except the palmaris longus tendon, cross the wrist deep to the flexor retinaculum. The palmaris longus tendon crosses superficial to the retinaculum and ends by becoming continuous with (inserting into) the superficial fibers of the palmar aponeurosis. *(Hollinshead, pp 249,251)*

43. **(A)** The flexor digitorum superficialis arises by two heads: the humeroulnar head arises from the medial epicondyle of the humerus and from the medial border of the base of the coronoid process of the ulna; the radial head arises from the upper part of the anterior border of the radius. The flexor carpi radialis arises from the medial epicondyle of the humerus, as does the palmaris longus. The flexor pollicis longus arises from the anterior surface of the radius and adjacent interosseous membrane; it does not originate on the humerus. Neither does the flexor digitorum profundus reach the humerus, originating only below the elbow on the ulna. *(Hollinshead, pp 250–251,260–261)*

44. **(E)** The fibrous digital sheaths end proximally over the metacarpal heads, as do the synovial sheaths, except for the little finger. Therefore if this usual pattern is present, an infection within the sheath of the little finger easily can travel into the common flexor tendon sheath at the wrist. Infections within the other tendon sheaths must be retained within them or must break through their walls. *(Hollinshead, pp 252,264)*. The tendon of the flexor pollicis longus usually is not involved in the common flexor synovial tendon sheath, although it also may join the sheath. *(Basmajian, p 382)*

45. **(C)** The radial artery usually gives off a superficial palmar branch that enters the thenar muscles; this artery frequently joins the superficial palmar arch. The radial artery arises as the lateral terminal branch of the brachial artery. All of the other items involve the ulnar artery, which gives rise to the common interosseous and provides the major components

of the superficial palmar arch from which arise common digital arteries. *(Hollinshead, pp 255,257)*

46. **(D)** The muscular branch of the median nerve turns laterally into the thenar muscles, where it usually is distributed to about two and one half muscles. The radial nerve gives rise to the posterior interosseous nerve. The median nerve passes deep to the flexor retinaculum; it innervates flexor muscles on the forearm except for the flexor carpi ulnaris and the ulnar side of the flexor digitorum profundus. The median nerve supplies only the first and second lumbrical muscles, not all of them. *(Hollinshead, pp 258–259,261)*

47. **(A)** The median nerve innervates all the short muscles of the thumb except the adductor and the deep head of the short flexor, which typically are innervated by the deep branch of the ulnar nerve. The palmaris brevis also is supplied by the ulnar nerve. Typically all of the interossei are supplied by the ulnar nerve; perhaps in 10% of cases the first dorsal interosseous is supplied by the median nerve. The abductor pollicis longus is supplied by the radial nerve. *(Hollinshead, pp 266–267,271,244)*

48. **(A)** The superficial branch of the ulnar artery forms the superficial palmar arch. The radial artery divides into the deep palmar arch and the princeps pollicis artery. The palmar metacarpal arteries arise from the deep palmar arch and join the common digital arteries from the superficial arch to form the proper digital arteries. Proper digital arteries run on each side of the fingers. *(Hollinshead, pp 251, 268–270)*

49. **(C)** The biceps brachii is both a flexor at the elbow and a strong supinator when the arm is flexed or more power is demanded. The brachialis is a flexor of the elbow and primarily is responsible for maintaining a flexed position. The flexor carpi radialis only assists in flexion at the elbow. The brachioradialis contracts if the movement is a quick one or if resistance is offered to flexion while the forearm is pronated. The supinator participates in all supination and is equally effective whether the forearm is flexed or extended. *(Hollinshead, pp 273–274)*

50. **(E)** Anomalous innervation of hypothenar muscles apparently is never a result of the median nerve sending a direct branch across the palm to the hypothenar muscles. Recent electromyographic clinical evidence is said to show all the other anomalous innervations listed. *(Hollinshead, p 280)*

51. **(E)**

52. **(D)**

53. **(C)**

54. **(B)**

51–54. The axillary and radial nerves are terminal branches of the posterior division of the brachial plexus. The axillary nerve supplies the deltoid and the skin covering the deltoid. The coracobrachialis, the biceps, and the brachialis all are supplied by the musculocutaneous nerve, the continuation of the lateral cord of the brachial plexus. All the interossei muscles are innervated by the ulnar nerve, the continuation of the medial cord of the brachial plexus. The recurrent branch of the median nerve supplies the abductor pollicis brevis. *(Basmajian, pp 341,357,377–379,393)*

55. **(C)**

56. **(D)**

55–56. The origin of the superficial palmar arch is the ulnar artery. The ulnar artery is the larger terminal branch of the brachial artery. As the ulnar artery enters the hand, it gives off a small deep branch into the hypothenar muscles and then continues as the superficial palmar arch. The origin of the deep palmar arch is the radial artery. The radial artery leaves the anterior aspect of the forearm by turning laterally around the wrist, deep to the extensor tendons of the thumb, thereby attaining the dorsum of the hand. This artery enters the palm from the dorsum, emerging between the two heads of the first dorsal interosseous muscle, and divides as it does so into the princeps pollicis artery and the deep palmar arch. The superficial palmar arch usually is completed on the radial side by one or more communications from the radial artery. The palmar metacarpal arteries are branches of the deep palmar arch that contribute to the common digital arteries from the superficial palmar arch to form proper digital arteries. *(Hollinshead, pp 254–256,266,268–269)*

57. **(B)**

58. **(C)**

59. **(A)**

60. **(D)**

57–60. The movements of the scapula are described as being elevation, depression, protraction, retraction, upward rotation, and downward rotation. In upward rotation the inferior angle swings laterally and forward so that the glenoid cavity is turned up. In downward rotation it swings medially and backward so that the glenoid is turned down. The serratus anterior muscle advances the inferior angle of the scapula laterally to rotate the scapula upward. The trapezius both upwardly rotates the scapula and retracts it. The rhomboids, major and minor, both adduct the scapula and turn it downward. The latissimus dorsi adducts, medially rotates, and extends the arm, not the scapula. *(Hollinshead, pp 197,195)*

**61. (B)**

**62. (D)**

**63. (C)**

**64. (C)**

**61–64.** The pectoralis major and two muscles that lie behind it form the anterior wall of the axilla; this muscle is inserted into the crest of the greater tubercle of the humerus (lateral lip of the intertubercular groove) putting it into a position to medially rotate the humerus. The serratus anterior is part of the medial wall of the axilla; it does not attach to the humerus. Both the subscapularis and teres major help to form the posterior wall of the axilla; their tendons insert near each other: the subscapularis on the lesser tubercle of the humerus and the teres major on the medial lip of the intertubercular groove. Both of these muscles medially rotate the humerus. *(Basmajian, pp 328–332,340–343)*

**65. (C)**

**66. (A)**

**67. (B)**

**65–67.** The lower lateral border of the subscapularis parallels the lateral border of the scapula and is also paralleled by the lower border of the teres minor posteriorly; however, the upper border of the teres major muscle is more nearly horizontal. So between this border, on the one hand, and the lower border of the subscapularis and teres minor, on the other hand, there is a triangular gap, crossed by the long head of the triceps. The part of the gap between the long head of the triceps and the surgical neck of the humerus is called the quadrangular space. Through this space the axillary nerve and the posterior humeral circumflex vessels leave the axilla. The medial part of the gap is called the triangular space, and the scapular circumflex vessels turn around the lower border of the subscapularis in this space. The teres major commonly is a lower border of the whole triangular space and therefore of the part called the quadrangular space. *(Hollinshead, pp 200–201)*

**68. (E)** The brachial artery can be palpated along the medial bicipital furrow throughout the length of the arm to the point where it disappears behind the bicipital aponeurosis. Just below the crease of the elbow, the brachial artery divides into its terminal branches, the ulnar and the radial arteries. The profunda brachii artery arises from the brachial artery just below the teres major. *(Basmajian, pp 358,361)*

**69. (D)** The radial nerve in the axilla lies behind, not in front of, the axillary artery. It descends along the origin of the medial, not the lateral, head of the triceps. The superficial radial nerve is a terminal sensory, not motor, branch of the radial nerve. Distal to where it vertically descends between the brachioradialis and the brachialis, it lies on the capsule of the elbow joint and on the supinator muscle. *(Basmajian, p 360)*

**70. (B)** The anterior interosseous nerve arises from the median nerve just below the elbow and clings to the interosseous membrane. It supplies the three deep muscles of the forearm: flexor pollicis longus, half of the flexor digitorum profundus, and the pronator quadratus. The abductor pollicis brevis is innervated by a branch of the median nerve that enters the thenar eminence. The flexor carpi ulnaris is supplied by the ulnar nerve. *(Basmajian, pp 365,369,377–379)*

**71. (B)** The pronator teres and the palmaris longus originate on the medial epicondyle of the humerus. The flexor pollicis longus arises below the elbow on the anterior surface of the radius and adjacent part of the interosseous membrane. The flexor digitorum profundus arises from the anterior and medial surfaces of the ulna, the adjacent part of the interosseous membrane, and the medial surface of the olecranon. *(Basmajian, pp 363–365)*

**72. (D)** When the ulnar nerve is cut at the wrist, innervation of the third and fourth lumbrical muscles (not the first and second innervated by the medial nerve) and all the dorsal and palmer interossei is lost resulting in paralysis of these muscles. The result of this loss of muscle power, along with the overaction of remaining normal muscles, results in a "claw hand." Claw hand deformity is most evident on the ulnar side. See references for detailed description of muscle action contributing to the deformity. *(Basmajian, pp 379,392; Hollinshead, p 280)*

**73. (A)** The flexor pollicis brevis, half of the flexor digitorum profundus, and the lumbrical muscles are supplied by both median and ulnar nerves. The two lumbrical muscles on the radial side are supplied by the median nerve; the two on the ulnar side are supplied by the ulnar nerve. The extensor pollicis longus is innervated by only the radial nerve. *(Basmajian, pp 379,381,386)*

**74.** **(E)** The supraspinatus and the deltoid act together and progressively in elevation of the arm. The supraspinatus alone is unable to initiate abduction. These elevator muscles are assisted by the three short depressors: subscapularis, infraspinatus, and teres minor. The two groups act as a force couple to elevate the arm, the one elevating and the other depressing. *(Basmajian, p 402)*

**75.** **(B)** The chief flexor of the elbow joint is the brachialis muscle. The biceps acts best as a flexor when the forearm is supinated, not pronated. When the brachialis and biceps are paralyzed, the brachioradialis is a useful flexor. The pronator teres is not as advantageously situated for flexion as is the brachioradialis and is only a weak flexor. *(Basmajian, pp 404–405)*

**76.** **(C)** Colles' fracture, involving the distal end of the radius, occurs more frequently in women than in men. The distal fragment usually is tilted backward and slightly to the lateral side. This dorsal displacement produces a characteristic hump, described as the "dinner fork" deformity. Usually Colles' fracture does not involve displacement of the lunate bone. *(Moore, pp 835,852)*

**77.** **(A)** The distal end of the radius and the articular disc of the distal radioulnar joint articulate with the proximal row of carpal bones (scaphoid, lunate, and triquetrum). The articular nerves are derived from the anterior interosseous nerve, a branch of the median nerve. The head of the ulna articulates with the ulnar notch in the distal end of the radius to form the distal radioulnar joint. The articular disc binds the lower end of the ulna and radius together and is the main uniting structure of the radioulnar joint. *(Moore, pp 832–833)*

**78.** **(C)** When the forearm is fully extended and supinated, the arm and forearm are not in a straight line. Normally the forearm is directed somewhat laterally, forming a "carrying angle" of about 163 degrees (not 150° as listed). Elbow extension is limited both by the collateral ligaments of the joint and by impingement of the olecranon process of the ulna on the olecranon fossa of the humerus. The elbow joint consists of three (not two as listed) different articulations: the humeroulnar articulation, the humeroradial articulation, and the proximal radioulnar joint. *(Moore, pp 821–823)*

**79.** **(D)** At the elbow joint the spool-shaped trochlea and the spheroid capitulum of the humerus articulate with the trochlear notch of the ulna and the proximal surface of the head of the radius, respectively. The ulnar nerve runs posteriorly to the medial epicondyle (not the lateral) and is closely applied to the ulnar collateral ligament. The subcutaneous olecranon bursa is exposed to injury from falls on the elbow and to infection from skin abrasions. *(Moore, pp 821–823,827)*

**80.** **(E)** The order of superficial muscles on the flexor (anterior) forearm is correct as listed in all items. This can be illustrated by placing fingers of one's left hand on the right forearm anterior surface just below the elbow. The left thumb is placed posterior to the right elbow around the medial epicondyle of the humerus. The forefinger then lies on the position of the right pronator teres; the middle finger lies over the flexor carpi radialis; the ring finger lies over the palmaris longus; and the little finger lies over the flexor carpi ulnaris. *(Moore, pp 760,755–758)*

**81.** **(A)** When the median nerve is severed in the elbow region, there is loss of flexion of the proximal interphalangeal joints of all the digits; this is because of paralysis of the flexor digitorum superficialis muscle, which flexes at these joints. There is also loss of the median nerve innervation to the flexor digitorum profundus. The ability to flex at the metacarpophalangeal joints of the index and middle fingers is affected because the digital branches of the median nerve supply the first and second lumbrical muscles. Flexion of the distal interphalangeal joints of the ring and little fingers is not affected because the part of the flexor digitorum profundus producing these movements is supplied by the ulnar nerve. *(Moore, pp 758–759,765–767)*

**82.** **(C)** The carpal bones articulate with one another and are bound together with ligaments to form a compact mass with a posterior convexity and an anterior concavity. The carpal sulcus thus formed on the anterior side is converted into an osseofibrous carpal tunnel (canal) by the flexor retinaculum. This tunnel is filled completely by tendons and the median nerve. Any lesion that significantly reduces the size of the carpal tunnel may cause compression of the median nerve, resulting in absence of tactile sensation on the palmar surface of the index finger and paresthesia on the palmar surface of the middle finger; these effects are seen on the palmar skin of the thumb and lateral half of the ring finger. Motor effects are seen in those thumb muscles innervated by the median nerve. Adduction of the thumb is not affected because the pollicis adductor muscle usually is supplied by the ulnar nerve, as is the ulnar portion of the flexor digitorum profundus that flexes at the distal interphalangeal joints of these digits. *(Moore, pp 752,767–769)*

**83.** **(E)** The brachial artery divides into two terminal branches, the radial artery and the large ulnar artery. The ulnar artery gives off the anterior ulnar recurrent branch just below the elbow; it also gives off the common interosseous artery that branches into anterior and posterior interosseous arteries. The pulsation of the ulnar artery can be felt where it passes anterior to the head of the ulna. *(Moore, pp 771–772)*

84. **(B)** Dupuytren's contracture is a progressive fibrosis of the palmar aponeurosis resulting in formation of abnormal bands of fibrous tissue that extend from the aponeurosis to the base of the phalanges. These bands may pull the fingers into such marked flexion at the metacarpophalangeal joints that they cannot be straightened. Usually the ring and little fingers are affected initially. The index finger rarely is involved, nor is the thumb. The symptoms of this disorder are quite different from the carpal tunnel syndrome. *(Moore, p 792)*

85. **(C)** The medial border of the scapula crosses ribs 2 to 7 (not ribs 4 to 9). The superior angle of the scapula lies at about the spinous process of vertebra T2 and the inferior angle at about the level of the spinous process of vertebra T7. The coracoid process (not the acromion) can be palpated anteriorly deep to the lateral clavicle; the acromion is palpated laterally and on the dorsal aspect of the scapula. *(Moore, pp 679–680)*

86. **(A)** Incorrect use of crutches may cause injury to the posterior cord of the brachial plexus. This results in paralysis of the wrist extensors and of the anconeus, both innervated by the radial nerve, a branch of the posterior cord. There is no particular relation to Dupuytren's contracture. *(Moore, p 699)*

87. **(B)** The latissimus dorsi muscle arises from the spinous processes of the lower six thoracic vertebrae and inserts into the floor of the intertubercular sulcus. It is a strong medial (not a lateral) rotator of the humerus. It is supplied by the thoracodorsal nerve (not the long thoracic nerve). *(Moore, p 715)*

88. **(D)** The only correct answer is item 4. The deltoid is both a medial and lateral rotator of the humerus. It originates on the lateral, not the medial third, of the clavicle and is supplied by the axillary, not the thoracodorsal, nerve. It cannot initiate the movement of abduction of the humerus from the adducted position without the assisted action of the supraspinatus muscle. *(Moore, pp 719–721)*

# CHAPTER 8

# The Lower Limb
## Questions

**DIRECTIONS (Questions 1 through 50): Each of the numbered items or incomplete statements in this section is followed by answers or by completions of the statement. Select the ONE lettered answer or completion that is BEST in each case.**

1. All of the following statements on development of the lower limb are correct EXCEPT

    (A) the limb bud at an early stage projects at almost a right angle to the body
    (B) the limb bud undergoes adduction and lateral rotation
    (C) the original dorsal surface of the limb comes to lie almost anteriorly
    (D) the big toe originally is directed cranialward
    (E) the lower limb bud lies at the level of a greater number of spinal nerves than does the upper limb

2. Parts and regions of the lower limb are defined correctly by which of the following?

    (A) malleolus, the knee
    (B) clunis, the gluteals
    (C) tarsus, the calf
    (D) sura, the ankle
    (E) hallux, digit 5

3. The skeleton of the lower limb can be described correctly by all of the following EXCEPT

    (A) the pelvic girdle is attached firmly to the vertebral column
    (B) the bones that compose the pelvic girdle fuse together in development
    (C) the two coxal bones are firmly articulated with each other at the midline
    (D) the fibula does not enter into the knee joint
    (E) the hip joint is identical in all respects to the shoulder joint

4. Which of the following statements is true in relation to muscles of the lower limb?

    (A) many muscles attach the pelvic girdle to the vertebral column

    (B) muscles of the buttock are all from original dorsal musculature
    (C) in the thigh, relations are distorted by early rotation of the limb
    (D) muscles of the calf represent original dorsal muscles
    (E) original ventral muscles become innervated by dorsal branches of the lumbosacral plexus

5. Nerves supplying the lower limb can be described correctly by all of the following statements EXCEPT

    (A) innervation is almost entirely through the lumbosacral plexus
    (B) cutaneous nerves to the front of the thigh are from the lumbar plexus
    (C) posterior muscles in the thigh are supplied by the sacral plexus
    (D) the obturator nerve is derived from posterior divisions of the sacral plexus
    (E) the femoral nerve is derived from posterior divisions of the lumbar plexus

6. The lumbar plexus can be described correctly by which of the following statements?

    (A) the second, third, and fourth lumbar nerves contribute to its branches
    (B) it gives rise to the sciatic nerve
    (C) most of its branches enter the buttock
    (D) it gives rise to the tibial nerve
    (E) it sends innervation to the muscles of the calf

7. In relation to vessels of the lower limb, which of the following statements is true?

    (A) the popliteal artery is a branch of the external iliac artery
    (B) the anterior tibial artery is a direct branch of the deep femoral artery
    (C) the superficial veins rarely become varicose
    (D) their arterial stem is the femoral artery
    (E) communicating veins normally conduct blood from the deep veins to the superficial veins

8. Which of the following statements is correct about veins and lymphatics of the lower limb?

   (A) large superficial veins originate on the sole of the foot
   (B) the great saphenous vein is seen on the lateral side of the knee
   (C) the small saphenous vein commonly ends in the popliteal vein
   (D) there are many deep lymph nodes
   (E) the general direction of superficial lymphatics is toward the foot

9. All of the following statements about nerves and vessels of the thigh are correct EXCEPT

   (A) the femoral artery terminates in the anterior thigh
   (B) the obturator nerve is distributed almost entirely to the thigh
   (C) arteries entering the buttock are distributed almost entirely to the buttock
   (D) the sciatic nerve supplies all the posterior muscles of the leg
   (E) through its branches the sciatic nerve supplies all muscles of the foot

10. Which of the following statements is true in relation to pelvic girdle bones?

   (A) their ossification is complete by 2 years of age
   (B) the pubis forms the lateral part of the acetabulum
   (C) the ischium forms the medial inferior part of the acetabulum
   (D) the ilium forms the upper part of the acetabulum
   (E) the acetabular notch is found on the superior wall of the acetabulum

11. Structures adding strength to the pelvic girdle consist of all the following EXCEPT the

   (A) transverse acetabular ligament
   (B) sacroiliac ligaments
   (C) sacrotuberous ligament
   (D) sacrospinous ligament
   (E) iliolumbar ligament

12. Blood supply to the upper end of the femur is derived mostly from the

   (A) artery of the ligament of the head of the femur
   (B) inferior gluteal artery
   (C) medial femoral circumflex artery
   (D) superior gluteal artery
   (E) superficial iliac circumflex artery

13. The gluteus maximus can be described correctly by which of the following statements?

   (A) it passes through the greater sciatic foramen
   (B) it inserts into the head of the femur
   (C) it is innervated by the superior gluteal nerve
   (D) it has a chief action of abduction at the hip joint
   (E) it arises partly from the sacrotuberous ligament

14. Structures lying deep to the gluteus maximus can be described correctly by which of the following statements?

   (A) the obturator internus originates on the external ilium
   (B) the quadratus femoris is the key to the region
   (C) the piriformis largely fills the greater sciatic foramen
   (D) the gluteus medius arises from the sacrum
   (E) short hip rotator muscles are innervated by the second and third lumbar nerves

15. The chief action of the gluteus medius is

   (A) extension at the hip joint
   (B) flexion at the hip joint
   (C) medial rotation of the femur
   (D) abduction at the hip joint
   (E) adduction at the hip joint

16. Distribution of which of the following arteries primarily is confined to the buttock?

   (A) internal pudendal
   (B) inferior gluteal
   (C) femoral
   (D) profunda femoris
   (E) posterior femoral circumflex

17. The gluteus medius and gluteus minimus muscles have similar characteristics EXCEPT for which of the following?

   (A) both arise from the ilium
   (B) both insert on the greater trochanter of the femur
   (C) both are supplied by the superior gluteal nerve
   (D) both are particularly important in walking
   (E) both are strong flexors and medial rotators

18. Which of the following statements is true in regard to the sciatic nerve?

   (A) it consists of two nerves
   (B) it passes anterior to the obturator internus
   (C) it gives off several fibers to the buttock
   (D) it runs superficial to the thigh muscles that arise from the ischial tuberosity
   (E) it generally runs through the piriformis muscle

19. All of the following statements correctly describe the fascia lata of the thigh EXCEPT

(A) it consists of dense deep fascia
(B) it is reinforced by a strong medial part, the iliotibial tract
(C) it splits to cover both sides of the tensor fasciae latae muscle
(D) at the knee it blends with expansions from muscle tendons
(E) just below the inguinal ligament, it presents the saphenous hiatus

20. Which of the following statements is true in relation to the femoral triangle?

(A) its medial border is the sartorius
(B) the femoral nerve is its most medial structure
(C) its upper border is the inguinal ligament
(D) the femoral canal is its most lateral structure
(E) the iliopsoas muscle is its anterior wall

21. Which of these statements is true in regard to the sartorius muscle?

(A) it inserts on the lateral upper tibial surface
(B) it forms the floor of the femoral triangle
(C) it lies lateral to the adductor canal
(D) it arises from the anterior superior iliac spine
(E) it is supplied by the superior gluteal nerve

22. All of the following statements correctly describe the adductor canal EXCEPT

(A) it lies deep to the sartorius
(B) it is bounded anteriorly by the adductor longus
(C) it begins above at the apex of the femoral triangle
(D) the femoral artery and vein run through the canal
(E) the canal ends below at the tendinous hiatus

23. The only one of the quadriceps femoris muscles originating above the hip joint is the

(A) rectus femoris
(B) vastus lateralis
(C) vastus medialis
(D) vastus intermedius
(E) rectus lateralis

24. Which of the following statements is correct in relation to extension at the knee joint with the hip extended?

(A) the sartorius begins the movement
(B) the rectus femoris is the most efficient actor
(C) the vastus medialis does not assist in the movement
(D) the vastus lateralis does not participate
(E) the articularis genus muscle is a primary actor

25. Correct description of the femoral nerve includes which of the following?

(A) it supplies no cutaneous branches
(B) usually it contains fibers from L5, S1, and S2
(C) in the femoral triangle it lies medial to the femoral vessels
(D) it supplies the adductor magnus muscle
(E) it supplies the rectus femoris muscle

26. Which of the following statements is correct about the compartments of the femoral sheath?

(A) the intermediate compartment contains the femoral artery
(B) the lateral compartment contains the femoral vein
(C) the medial compartment is the femoral canal
(D) the femoral canal contains the genitofemoral nerve
(E) the upper end of the intermediate compartment forms the femoral ring

27. All of the following statements about the femoral blood vessels are correct EXCEPT

(A) both femoral circumflex veins typically enter the femoral vein
(B) the medial femoral circumflex artery is a branch of the profunda femoris artery
(C) the descending genicular artery is the last named femoral branch
(D) the lateral femoral circumflex artery usually is a direct branch of the femoral artery
(E) the profunda femoris artery is a large branch of the femoral artery

28. All of the following muscles arise from the ischial tuberosity EXCEPT the

(A) short head of the biceps femoris
(B) semitendinosus
(C) long head of the biceps femoris
(D) semimembranosus
(E) posterior adductor magnus

29. All of the following muscles cross the knee joint EXCEPT the

(A) gracilis
(B) adductor magnus
(C) sartorius
(D) biceps femoris
(E) semitendinosus

30. The thigh muscle that receives innervation from both tibial and common peroneal nerves is the

(A) semimembranosus
(B) biceps femoris
(C) adductor magnus
(D) semitendinosus
(E) gracilis

31. Which of the following muscles can extend the thigh at the hip without at the same time producing flexion at the knee?

    (A) biceps femoris, long head
    (B) semitendinosus
    (C) biceps femoris, short head
    (D) adductor magnus
    (E) semimembranosus

32. Which of the following is the most anterior of the structures in the popliteal fossa?

    (A) popliteal artery
    (B) popliteal vein
    (C) common peroneal nerve
    (D) sciatic nerve
    (E) tibial nerve

33. Which of the following arteries forms the chief blood supply of the posterior muscles of the thigh?

    (A) inferior gluteal artery
    (B) transverse branch of the medial femoral circumflex artery
    (C) perforating branches of the profunda femoris artery
    (D) popliteal vessels
    (E) genicular arteries

34. All the following statements about the hip joint are correct EXCEPT

    (A) the acetabulum has a labrum attached to its margin
    (B) the head of the femur fits deeply into the acetabulum
    (C) the iliofemoral ligament strongly resists hyperextension at the hip
    (D) the joint is most stable when the femur is extended
    (E) the ligament of the head of the femur strongly resists adduction of the thigh

35. Dislocation of the hip most usually occurs as a result of which of these conditions?

    (A) a severe blow upon the knee while the hip is flexed
    (B) the body weight landing heavily on one foot as the body falls
    (C) a severe blow on the buttock while the hip is flexed
    (D) severe forced hyperextension of the hip while the knee is extended
    (E) an abnormal stretch upon the adductor muscles as the limb is forced into abduction

36. Which of the following statements correctly describes the tibia?

    (A) the fibular notch is on its upper lateral side
    (B) The intercondylar eminence is located on its anterior border
    (C) lateral and medial condyles are located at its upper end
    (D) The tuberosity forms the medial malleolus at the ankle
    (E) the soleal line is located between the malleoli

37. Which of the following statements is correct regarding the tarsal bones?

    (A) the cuboid lies medially in the foot
    (B) the calcaneus is the smallest tarsal bone
    (C) the navicular lies just in front of the cuneiforms
    (D) the cuneiforms lie lateral to the cuboid
    (E) the talus lies above the calcaneus

38. All of the following statements regarding the arches of the foot are correct EXCEPT

    (A) the medial part of the longitudinal arch is higher than the lateral part
    (B) the lateral longitudinal arch passes forward from the calcaneus through the cuboid to the heads of the two lateral metatarsals
    (C) the transverse arch lies at the level of the proximal row of tarsals
    (D) the two arches interlock to form a functional single arch
    (E) the arch is loaded from the top through the talus

39. Which is the primary movement that occurs at the talocrural joint?

    (A) inversion and eversion
    (B) flexion and extension
    (C) eversion and abduction
    (D) inversion and adduction
    (E) equinovarus

40. Which of the following fixed deformities is NOT characteristic of the typical congenital clubfoot?

    (A) dorsiflexion
    (B) plantarflexion
    (C) inversion
    (D) adduction
    (E) equinovarus

41. Cutaneous innervation of the leg includes all the following nerves EXCEPT the

    (A) saphenous
    (B) superficial peroneal
    (C) lateral sural cutaneous
    (D) obturator
    (E) posterior femoral cutaneous

42. Which of the following relationships is correct about structures passing behind the medial malleolus at the ankle?

   (A) the tendon of the posterior tibial muscle is posterolateral
   (B) the tendon of the flexor digitorum longus muscle is most anteromedial
   (C) the tibial nerve is most posterior
   (D) the posterior tibial vessels lie next to the tibialis posterior muscle
   (E) the flexor hallucis longus is the most inferior

43. Which of the following muscles is both a dorsiflexor and an inverter of the foot?

   (A) peroneus brevis
   (B) peroneus longus
   (C) extensor hallucis longus
   (D) peroneus tertius
   (E) extensor digitorum longus

44. Muscles that originate predominately on the fibula consist of all the following EXCEPT the

   (A) tibialis anterior
   (B) extensor hallucis longus
   (C) peroneus brevis
   (D) extensor digitorum longus
   (E) peroneus tertius

45. The common peroneal nerve is most easily injured at which of the following locations?

   (A) where it leaves the sciatic nerve in the popliteal fossa
   (B) as it branches into the short head of the biceps femoris
   (C) just behind the head of the fibula
   (D) where it emerges below the piriformis muscle
   (E) at the place it divides into the superficial and deep peroneal nerves

46. Muscles innervated by the deep peroneal nerve include all of the following EXCEPT the

   (A) tibialis anterior
   (B) extensor digitorum longus
   (C) extensor hallucis longus
   (D) peroneus brevis
   (E) peroneus tertius

47. Which of the following arteries is the continuation of the anterior tibial artery in the foot?

   (A) recurrent
   (B) dorsalis pedis
   (C) medial plantar
   (D) posterior tibial
   (E) peroneal

48. Which of the following muscles is NOT a plantar flexor at the ankle?

   (A) gastrocnemius
   (B) soleus
   (C) peroneus longus
   (D) peroneus tertius
   (E) posterior tibial

49. Which of the following arteries is a terminal branch of the popliteal artery?

   (A) anterior tibial
   (B) dorsalis pedis
   (C) peroneal
   (D) medial superior genicular
   (E) sural

50. A branch of the femoral artery that enters into collateral circulation around the knee is the

   (A) lateral femoral circumflex
   (B) lateral superior genicular
   (C) descending genicular
   (D) anterior tibial
   (E) anterior tibial recurrent

**DIRECTIONS (Questions 51 through 60): Each group of items in this section consists of lettered headings followed by a set of numbered words or phrases. For each numbered word or phrase, select the ONE lettered heading that is most closely associated with it. Each lettered heading may be selected once, more than once, or not at all.**

**Questions 51 through 54**

   (A) fibular collateral ligament
   (B) tibial collateral ligament
   (C) oblique popliteal ligament
   (D) anterior cruciate ligament
   (E) posterior cruciate ligament

51. Stands well away from the capsule of the knee joint

52. Closely applied to the medial capsule of the knee joint

53. Runs across the posterior aspect of the knee

54. Helps the posterior knee structures to resist hyperextension

**Questions 55 through 58**

    (A) quadratus plantae
    (B) flexor digitorum longus
    (C) tibialis anterior
    (D) flexor digitorum brevis
    (E) peroneus longus

**55.** tibial nerve

**56.** medial plantar nerve

**57.** lateral plantar nerve

**58.** deep peroneal nerve

**Questions 59 through 60**

    (A) posterior tibial artery
    (B) lateral plantar artery
    (C) dorsalis pedis artery
    (D) medial plantar artery
    (E) plantar metatarsal arteries

**59.** Forms the plantar arterial arch

**60.** Completes the plantar arterial arch medially

**DIRECTIONS (Questions 61 through 66): Each group of items in this section consists of lettered headings followed by a set of numbered words or phrases. For each numbered word or phrase, select**

    **A if the item is associated with (A) <u>only</u>,**
    **B if the item is associated with (B) <u>only</u>,**
    **C if the item is associated with <u>both</u> (A) <u>and</u> (B),**
    **D if the item is associated with <u>neither</u> (A) <u>nor</u> (B).**

**Questions 61 through 64**

    (A) inverter of the foot
    (B) adductor of the foot
    (C) both
    (D) neither

**61.** tibialis anterior

**62.** tibialis posterior

**63.** triceps surae

**64.** peroneus longus

**Questions 65 through 66**

    (A) deltoid ligament
    (B) spring ligament (calcaneonavicular ligament)
    (C) both
    (D) neither

**65.** supports the head of the talus

**66.** resists eversion of the foot

**DIRECTIONS (Questions 67 through 88): For each of the items in this section, <u>ONE</u> or <u>MORE</u> of the numbered options is correct. Choose the answer**

    **A if only <u>1, 2, and 3</u> are correct,**
    **B if only <u>1 and 3</u> are correct,**
    **C if only <u>2 and 4</u> are correct,**
    **D if only <u>4</u> is correct,**
    **E if <u>all</u> are correct.**

**67.** In the standing position, support of the arch of the foot depends primarily upon the

    (1) tibialis anterior
    (2) plantar ligaments
    (3) peroneus longus
    (4) plantar aponeurosis

**68.** Correct statements regarding structures of the knee joint include which of the following?

    (1) the lateral condyle of the tibia is shorter from front to back than the medial condyle
    (2) the lateral meniscus is shorter than the medial meniscus
    (3) the medial meniscus is shaped like a capital C
    (4) the biceps femoris tendon crosses the fibular collateral ligament

**69.** Bony components of the knee joint include which of the following?

    (1) femur
    (2) tibia
    (3) patella
    (4) fibula

**70.** Correct statements about the iliofemoral ligament include which of the following?

    (1) it is shaped like an inverted Y
    (2) it resists flexion at the hip
    (3) it is attached above to the anterior inferior iliac spine
    (4) its attachment to the femur creates the intertrochanteric crest

**71.** The first layer of plantar muscles includes the

    (1) flexor digitorum longus
    (2) quadratus plantae
    (3) flexor hallucis brevis
    (4) flexor digitorum brevis

**72.** The fibula is characterized by

    (1) giving origin to the flexor hallucis longus
    (2) holding the talus in its socket

(3) moving down to deepen the ankle joint at the strike phase of running

(4) helping the tibia to carry weight to the ground at the ankle

73. Muscles that pass under the inferior part of the extensor retinaculum at the ankle include the

(1) peroneus tertius
(2) extensor digitorum longus
(3) extensor hallucis longus
(4) posterior tibial

74. Muscles supplied by the superficial peroneal nerve include the

(1) peroneus longus
(2) extensor digitorum brevis
(3) peroneus brevis
(4) anterior tibial

75. Which of the following statements correctly describes the popliteal artery?

(1) it gives rise to the profunda artery
(2) it originates from the femoral artery
(3) it is the chief arterial supply of the adductor muscles
(4) the posterior tibial artery is one of its major branches

76. The peroneal artery can correctly be described as

(1) arising from the anterior tibial artery
(2) descending in front of the fibula
(3) forming the plantar arch
(4) ending as the lateral calcaneal artery

77. Which of the following statements correctly describes the triceps surae?

(1) it includes the gastrocnemius muscle
(2) it includes the plantaris muscle
(3) it inserts into the tendo calcaneus
(4) it includes the posterior tibial muscle

78. The profunda femoris artery is an important blood supply to the thigh that

(1) branches from the femoral artery
(2) gives off the lateral femoral circumflex artery
(3) gives off the medial femoral circumflex artery
(4) usually gives off four perforating branches

79. The adductor muscles of the thigh all share which of the following characteristics?

(1) they extend downward as far as the tibia
(2) they arise from the anterior hip bone and the obturator membrane
(3) they are innervated by the femoral nerve
(4) they adduct and medially rotate the femur

80. With the knee flexed, the leg is rotated laterally by the

(1) semitendinosus
(2) semimembranosus
(3) gracilis
(4) biceps femoris

81. The popliteal fossa contains which of the following structures?

(1) tibial nerve
(2) popliteal artery
(3) popliteal vein
(4) common peroneal nerve

82. Muscles that are amalgamations of two muscles include the

(1) semitendinosus
(2) biceps femoris
(3) semimembranosus
(4) adductor magnus

83. Which of the following muscles receives innervation from the superior gluteal nerve?

(1) gluteus medius
(2) gluteus minimus
(3) tensor fasciae latae
(4) quadriceps femoris

84. The adductor muscles are innervated solely by the obturator nerve EXCEPT for which of the following?

(1) pectineus
(2) adductor longus
(3) adductor magnus
(4) gracilis

85. Features of the femoral triangle can be identified correctly by which of the following surface landmarks?

(1) the lateral edge of the sartorius is its lateral boundary
(2) the inguinal ligament is the base of the triangle
(3) the lateral border of the adductor longus is its medial boundary
(4) the pulse of the femoral artery is felt between the femoral nerve and the femoral vein

86. The femoral vein can be described correctly as

(1) ending posterior to the inguinal ligament
(2) becoming the external iliac vein
(3) receiving the profunda femoris vein
(4) passing through the saphenous opening in the fascia lata

**SUMMARY OF DIRECTIONS**

| A | B | C | D | E |
|---|---|---|---|---|
| 1, 2, 3 only | 1, 3 only | 2, 4 only | 4 only | All are correct |

87. Characteristics of a femoral hernia include which of the following?

    (1) weakness in the abdominal wall at the femoral ring
    (2) protrusion of the intestine through the femoral ring into the femoral triangle
    (3) possible strangulation resulting in impairment of blood supply to the herniated bowel
    (4) a palpable mass inferolateral to the pubic tubercle

88. Which of the following statements is true in relation to the common peroneal nerve?

    (1) it is the most often injured nerve in the lower limb
    (2) it is susceptible to pressure exerted at the head of the fibula
    (3) its severance results in paralysis of all the dorsiflexor muscles of the foot
    (4) its severance results in paralysis of all the invertor muscles of the foot

# Answers and Explanations

1. **(B)** All of the items are correct except B. Instead of lateral rotation the limb bud undergoes medial rotation, adduction, and extension. The limb bud at an early stage of development projects at about a right angle to the body, with the big toe directed cranialward and the little toe directed caudally. At this stage the borders and surfaces of the lower limb correspond to those of the upper limb. With the rotation mentioned above of almost 80 degrees, the original dorsal surface comes to lie almost anteriorly and laterally and the originally ventral surface posteriorly and medially. The lower limb bud has a wider base than does the upper one and consequently lies at the level of a greater number of spinal nerves than does the upper limb. *(Hollinshead, pp 333–334)*

2. **(B)** The gluteal region is the natis, clunis, or buttock. The malleoli are bony projections of the ankle. Sura is the calf of the leg; tarsus is the region of the ankle; and hallux is the big toe. *(Hollinshead, p 334)*

3. **(E)** All the items are correct except E. The hip joint is not at all identical to the shoulder joint, as it is far stronger structurally than the shoulder. The upper end of the femur fits into a deep, cuplike cavity, the acetabulum, while the glenoid cavity is shallow and does not hold the head of the humerus tightly. The femur, corresponding to the humerus, is the largest bone of the body. The pelvic girdle is directly and firmly attached to the vertebral column, and the two coxal bones firmly articulate with each other at the midline. *(Hollinshead, p 336)*

4. **(C)** Relations of the muscles of the thigh are distorted by rotation of the limb during its development. Muscles comparable to those of the upper limb attaching the girdle to the vertebral column are sparse because of the firm articulation of the pelvic girdle to the sacrum. The larger and more posterior muscles of the buttock are dorsal muscles, but some of the deeper smaller muscles of the buttock, judging by their innervation, are representative of the ventral muscular group. Muscles of the calf and plantar surface represent original ventral muscles. Whether in thigh, leg, or foot, the originally ventral muscles are innervated through the ventral branches of the lumbosacral plexus while the originally dorsal muscles are innervated by its dorsal branches. *(Hollinshead, pp 337–338)*

5. **(D)** The obturator nerve is derived from anterior divisions of the lumbar plexus, and the femoral nerve is derived from posterior divisions of the lumbar plexus. With exception of some skin of the buttock supplied by segmental nerves, the cutaneous innervation of the lower limb is entirely through branches of the lumbosacral plexus. Cutaneous nerves to the front of the thigh are from the lumbar plexus through the femoral nerve. Posterior muscles in the thigh and all the muscles below the knee are supplied by the sacral plexus. *(Hollinshead, pp 339–340)*

6. **(A)** The second, third, and fourth lumbar nerves typically contribute to the two major branches of the lumbar plexus. The sacral plexus, not the lumbar plexus, gives rise to the sciatic nerve. Most of the branches of the sacral, not the lumbar plexus, enter the buttock. The sacral plexus gives rise to the tibial nerve, a component of the sciatic nerve, that is distributed to muscles of the calf and the plantar aspect of the foot. *(Hollinshead, pp 339–340)*

7. **(D)** The arterial stem of the lower limb is the femoral artery, the continuation of the external iliac artery at the inguinal ligament. The popliteal artery is the continuation of the femoral artery as it passes into the posterior aspect of the thigh behind the knee. The anterior and posterior tibial arteries are branches of the popliteal artery. Superficial veins are subject to becoming varicose. Communicating veins normally conduct blood from superficial into deep veins. *(Hollinshead, pp 343–345)*

8. **(C)** The small saphenous vein begins along the lateral margin of the foot, passes upward along the posterior aspect of the calf, and commonly ends in the popliteal vein. The superficial veins (greater and lesser saphenous veins) originate on the dorsum, not the sole, of the foot. The great saphenous vein runs up the medial side of the limb, not the lateral side; it can be

seen on the medial side of the knee. The deep lymphatic vessels follow blood vessels, and there are few deep lymph nodes. The superficial lymphatics are numerous and run up all surfaces of the leg. *(Hollinshead, p 345)*

9.  **(A)** The femoral artery continues into the leg and foot, but the anterior nerves, chiefly femoral and obturator, are distributed almost entirely to the thigh. Arteries entering the buttock are distributed almost entirely to this part, but the sciatic nerve continues through the buttock and the thigh to supply all the muscles of the leg and foot and most of the skin of these parts *(Hollinshead, pp 349–350)*

10. **(D)** Ossification of the coxal bone occurs from three primary centers, one each for the ilium, ischium, and pubis. These centers fuse over a considerable range of time, but usually ossification of the coxal bone is complete by age 20 to 21 years. The ilium forms the upper part of the acetabulum. The pubis forms the anterior part of the acetabulum and the anteromedial part of the hip bone. The ischium forms the posterior inferior part of the acetabulum and the lower posterior part of the hip bone. The acetabular notch is in the inferior, not the superior, wall of the acetabulum. *(Hollinshead, pp 350,353)*

11. **(A)** The continuation of the labrum of the acetabulum across the acetabular notch is called the transverse acetabular ligament. Between it and the edge of the notch is loose connective tissue through which one or more acetabular arteries enter the joint. The transverse acetabular ligament does not add to stability of the joint. The sacroiliac joint is reinforced by heavy ligaments. The sacrotuberous ligament is a particularly strong bracing ligament of the sacroiliac joint; it forms the medial border of both the greater and lesser sciatic foramina. The sacrospinous ligament runs between the sacrum and coccyx and the ischial spine; it separates the greater sciatic foramen from the lesser sciatic foramen. The sacrotuberous and sacrospinous ligaments are so placed as to resist rotation of the sacrum between the coxal bones. The iliolumbar ligament attaches to the anterior or pelvic surface of the ilium and sacrum. *(Hollinshead, pp 353–356,395–396)*

12. **(C)** The arteries to the upper end of the femur are derived mostly from the medial and lateral femoral circumflex arteries, the two largest branches of the profunda femoris artery. The artery of the ligament of the head of the femur, derived from the obturator or the medial femoral circumflex artery, enters the head through the ligament of the head and supplies a variable amount of bone adjacent to the fovea. Otherwise the head and neck are supplied by branches of the two circumflex arteries. *(Hollinshead, pp 359–360,388)*

13. **(E)** The gluteus maximus muscle arises from the outer surface of the ilium behind the posterior gluteal line, from the dorsal surface of the sacrum and coccyx, and from the adjacent sacrotuberous ligament. It is inserted into both the iliotibial tract and the gluteal tuberosity of the femur. It receives its nerve supply from the inferior gluteal nerve, and its primary action is extension at the hip joint. It does not pass through the greater sciatic foramen. *(Hollinshead, pp 362–365)*

14. **(C)** The piriformis muscle occupies a key position in the structures deep to the gluteus maximus. The piriformis arises from the sacrum, passes through and largely fills the greater sciatic foramen. The obturator internus arises on the internal surface of the coxal bone, from the obturator membrane and the bone around that; it converges to a tendon that passes through the lesser sciatic foramen to insert into the medial surface of the greater trochanter. The piriformis, not the quadratus femoris, is the key to the region. The gluteus medius arises from the wing of the ilium, not from the sacrum. The short rotator muscles of the hip receive innervation from the fourth or fifth lumbar nerves and first or second sacral nerves, not the second and third lumbar nerves as stated. *(Hollinshead, pp 362–365)*

15. **(D)** The chief action of the gluteus medius is abduction of the free limb. More importantly, together with the gluteus minimus, it keeps the contralateral side of the pelvis from sagging markedly when the weight of the body is put upon one limb; therefore these muscles are exceedingly important in walking. Only the gluteus minimus is active in flexion and medial rotation of the femur. Apparently the gluteus medius assists with extension and lateral rotation of movements. It is not a chief actor in the movements of adduction, medial rotation, or flexion at the hip joint. *(Hollinshead, p 365)*

16. **(B)** The distribution of the superior and inferior gluteal vessels primarily is confined to the buttock. The internal pudendal artery primarily passes through the buttock to another distribution. The femoral artery and its branch, the profunda femoris, are not vessels of the gluteal region. There is no posterior femoral circumflex artery. *(Hollinshead, pp 366,387)*

17. **(E)** Neither of the muscles, gluteus medius or gluteus minimus, is a strong flexor at the hip joint, although the gluteus minimus may act in the movement. The gluteus minimus also has been shown by electromyography to be active in medial rotation of the thigh while the gluteus medius apparently is active only in lateral rotation and extension of the thigh. These muscles are similar in all the other items listed. *(Hollinshead, pp 363–365)*

18. **(A)** The sciatic nerve consists of two nerves, the tibial and common peroneal, that are bound together closely by connective tissue. The sciatic nerve normally emerges below the piriformis muscle, although it may come through the muscle or the nerve may be divided by the muscle. It always runs across the posterior surfaces of the obturator internus, gemelli, and quadratus femoris muscles. It gives off no fibers to muscles of the buttock. As it runs down the thigh, it disappears deep to the hamstring muscles that arise from the ischial tuberosity. *(Hollinshead, p 368)*

19. **(B)** The fascia lata of the thigh consists of dense, deep fascia reinforced by a particularly strong lateral band, the iliotibial tract. The fascia lata is rather weak over the medial adductor muscles but is stronger both anteriorly and posteriorly. Laterally it splits to go on both sides of the tensor fasciae latae muscle. At the knee, in general, the fascia lata loses its identity, for it blends with expansions from the tendons at the knee to help form the patellar retinacula and then becomes continuous with the fascia of the leg. Just below the inguinal ligament, the fascia lata presents a defect, the saphenous hiatus, through which passes the great saphenous vein to enter the femoral vein. *(Hollinshead, pp 372–373)*

20. **(C)** The upper border of the femoral triangle is the inguinal ligament. The medial border of the triangle is the medial (sometimes said to be the lateral) border of the adductor longus muscle. The lateral border of the triangle is the sartorius. Within the triangle the femoral nerve is most lateral. The femoral artery and vein, enclosed within the femoral sheath of connective tissue that also contains the essentially empty femoral canal, lie in that order medial to the femoral nerve. The iliopsoas muscle is part of the posterior wall of the triangle. *(Hollinshead, pp 373–374)*

21. **(D)** The sartorius has its origin on the anterior superior iliac spine and inserts on the upper medial, not lateral, surface of the body of the tibia close to the insertions of the gracilis and the semitendinosus muscles. The sartorius forms the lateral border of the femoral triangle and covers the adductor canal. This muscle is supplied by branches of the femoral, not the superior gluteal, nerve. *(Hollinshead, pp 374–375)*

22. **(B)** The adductor canal is not bounded anteriorly but posteriorly by the adductor longus and magnus and anterolaterally by the vastus medialis. The adductor canal is, in cross-section, a triangular space between the quadriceps muscle and the adductor group, lying deep to the sartorius. The femoral artery and vein run through the canal. The canal ends at the tendinous (adductor) hiatus, a gap in the adductor magnus muscle through which pass the femoral artery

and vein to run downward behind the knee as the popliteal vessels. *(Hollinshead, pp 375–376)*

23. **(A)** The rectus femoris is the only portion of the quadriceps group that arises from above the hip; this muscle originates on the anterior inferior iliac spine and from the ilium just above the acetabulum. The three vasti muscles have their origins on parts of the shaft of the femur. The combined tendon of the rectus and the vasti muscles forms the quadriceps tendon. This tendon attaches to the upper border of the patella; the patella then is attached to the tibial tuberosity by a heavy tendon known as the patellar ligament. There is no such muscle as the rectus lateralis. *(Hollinshead, pp 376–377)*

24. **(B)** The quadriceps is the only muscle that can extend the knee; the sartorius can only flex it. Of the quadriceps group, the rectus femoris is most efficient as a knee extensor when the hip is extended because it attaches above the hip joint. It does not, however, assist in the last 10 to 15 degrees or more of knee extension, leaving this to the vasti. The vastus medialis, vastus lateralis, and vastus intermedius all participate in extension at the knee joint. The articularis genus inserts into the suprapatellar bursa of the knee joint and pulls this up as the knee is extended to prevent it getting caught in the joint. *(Hollinshead, p 381)*

25. **(E)** The femoral nerve usually sends two branches to the rectus femoris muscle. The nerve contains both muscular and cutaneous branches that are given off in no particular order. Its cutaneous branches to the thigh are the anterior femoral cutaneous nerves. The other cutaneous branch of the femoral nerve, the saphenous nerve, runs through most of the length of the adductor canal and is distributed to the leg and the foot. The femoral nerve usually contains fibers from the second, third, and fourth lumbar nerves; it lies lateral to the femoral vessels in the femoral triangle. The obturator nerve supplies the anterior portion of the adductor magnus and the tibial nerve its posterior portion. *(Hollinshead, pp 384–385)*

26. **(C)** In the femoral triangle, the femoral vein and artery are together, surrounded by a continuation of the fascial layer lining the abdominal cavity. This fascial investment of the femoral vessels is the femoral sheath. Within it are three compartments separated by septa that pass between its anterior and posterior walls. The lateral compartment contains the femoral artery and the femoral branch of the genitofemoral nerve. The intermediate compartment contains the femoral vein, and the medial compartment is the femoral canal, which contains only a slight amount of loose connective tissue and a few lymphatics and lymph nodes. The upper end of the femoral canal is the femoral ring. A femoral hernia

descends through the femoral ring into the femoral canal. *(Hollinshead, p 386)*

27. **(D)** The profunda femoris artery normally gives off both the lateral femoral circumflex artery and the medial femoral circumflex artery. The femoral artery gives off the profunda artery. Both femoral circumflex veins typically enter the femoral vein instead of the deep femoral vein. The descending genicular artery is the last named branch of the femoral artery. *(Hollinshead, pp 386–389)*

28. **(A)** The short head of the biceps femoris is the muscle listed that does not arise from the ischial tuberosity. Its long head arises from the ischial tuberosity; the short head arises from the lateral lip of the linea aspera of the femur. *(Hollinshead, p 389)*

29. **(B)** The posterior part of the adductor magnus, arising from the ischial tuberosity, extends vertically downward and reaches the lowest attachment of the muscle, the adductor tubercle on the femur. It therefore does not cross the posterior knee but is an extensor of the hip. All of the other muscles listed have attachments on the tibia below the knee. *(Hollinshead, pp 374,381–382,389)*

30. **(B)** The biceps femoris muscle receives innervation from both tibial and common peroneal nerves of the sciatic, the long head from the tibial nerve and the short head from the common peroneal nerve. The semitendinosus and semimembranosus muscles receive innervation from only the tibial nerve. The posterior part of the adductor magnus receives innervation from the tibial nerve; its anterior part is supplied by the obturator nerve. The gracilis receives fibers from neither the tibial or common peroneal nerve; it is innervated by the anterior branch of the obturator nerve. *(Hollinshead, pp 381,383,390)*

31. **(D)** The posterior part of the adductor magnus can extend the thigh along with the other muscles arising from the ischial tuberosity. It can extend the thigh without at the same time producing flexion at the knee because it inserts above the knee and therefore cannot act over the knee. The long head of the biceps, the semitendinosus, and the semimembranosus all extend the hip and flex the knee. The short head of the biceps only can flex the knee because it arises on the femur below the hip joint. *(Hollinshead, p 392)*

32. **(A)** The popliteal fossa is the area posterior to the knee. Its lower borders are formed by the two heads of the gastrocnemius muscle that arise from the medial and lateral condyles of the femur and converge to a union in the upper part of the calf. In the upper part of the fossa, the sciatic nerve lies posterolateral to the popliteal vessels. The popliteal vein is next anteriorly and the popliteal artery is the most ante-

rior, lying directly on the popliteal surface of the femur. The common peroneal nerve diverges laterally to pass around the lateral side of the leg. The tibial nerve descends almost straight down through the fossa. *Hollinshead, p 392)*

33. **(C)** The perforating branches of the profunda femoris vessels, usually four in number, form the chief blood supply of the thigh. The inferior gluteal artery gives twigs into the upper end of the muscles attaching to the ischial tuberosity. The transverse branch of the medial femoral circumflex artery appears between the quadratus femoris and the upper border of the adductor magnus muscle, and the popliteal vessels (from the femorals) reach the popliteal fossa by passing through the tendinous hiatus. The branches of the profunda artery, however, provide the major blood supply to the region. *(Hollinshead, p 395)*

34. **(E)** The ligament of the head of the femur runs from the acetabular fossa and transverse ligament to the fovea of the head. Theoretically it should resist adduction at the hip, but apparently it does not become taut enough to function as a checking ligament in the adult. It has been said to check posterior-superior displacement of the head of the femur in the fetus. This ligament conducts the artery of the ligament of the head from the obturator artery into the head of the femur; here it supplies a variable amount of blood to the area adjacent to the fovea. The other items listed are all correct. *(Hollinshead, pp 360,395–397)*

35. **(A)** Traumatic dislocation of the hip is not a frequent occurrence because of the deep acetabulum. When it takes place, usually it is the result of a severe blow upon the knee while the hip is flexed. The head of the femur is thus dislocated posteriorly, with a tearing of the posterior part of the capsule. Anterior dislocation is much rarer than posterior dislocation; in this case the head of the femur passes around the medial edge of the iliofemoral ligament and lodges against the body of the pubis or the obturator foramen. *(Hollinshead, p 399)*

36. **(C)** The upper end of the tibia consists of lateral and medial condyles; they are separated by the intercondylar eminence. The fibular notch is found on the lateral side of the lower end of the tibia for accommodation of the fibula. The tibial tuberosity receives the insertion of the quadriceps muscle. The soleal line, on the upper third of the posterior surface of the bone, marks the tibial origin of the soleus muscle. *(Hollinshead, pp 402–405)*

37. **(E)** The talus lies above the calcaneus and receives body weight from the tibia. The cuboid lies laterally in the foot, in front of the calcaneus. The calcaneus is the largest, not the smallest, tarsal bone. The navi-

cular lies just posterior to the cuneiforms and in front of the talus. The cuneiforms lie medial to the cuboid and in front of the navicular. *(Hollinshead, pp 405–406)*

**38.** **(C)** The transverse arch lies at the level of the distal row of tarsals rather than at the proximal row. All of the other statements regarding the arches of the foot are correct. Although it is convenient to describe two arches, longitudinal and transverse, they interlock so that they form a functioning single arch of complex form. Weight is distributed in this arch according to the engineering principle that the distribution of stress throughout an arch is strictly proportional to the relative heights of various parts of the arch. This is true only when the arch is loaded from the top, which is, in the normal foot, the talus. *(Hollinshead, pp 411–412)*

**39.** **(B)** Because the trochlea tali is grasped so firmly between the medial and lateral malleoli, little movement is possible at the talocrural joint except flexion and extension. A certain amount of inversion and eversion is possible at the subtalar joint because the calcaneus, underlying the talus, can rock from side to side. Movement occurs also between other tarsal bones, but the greatest amount of movement occurs at the transverse tarsal joint: dorsiflexion, plantar flexion, and two combined movements, one of eversion and abduction and one of inversion and adduction. *(Hollinshead, p 412)*

**40.** **(A)** The term clubfoot is used now to describe a condition of plantar flexion, inversion, and adduction (talipes equinovarus). This is the commonest type of deformity seen in congenital clubfoot. The cause is not understood. A dorsiflexion deformity is not seen in this type of deformity. *(Hollinshead, p 412)*

**41.** **(D)** The obturator nerve does not extend down into the leg area and therefore does not provide cutaneous innervation to that region. It does give rise to an articular branch to the knee joint that descends along the femoral artery; it also gives rise to the cutaneous branch of the obturator distributed to skin on the medial side of the thigh. The saphenous nerve, from the femoral, the superficial peroneal nerve, the lateral and medial sural cutaneous nerves, and the posterior femoral cutaneous nerve, all supply areas of skin of the leg. *(Hollinshead, pp 385–386,413–414)*

**42.** **(E)** The most posterior and inferior of the structures passing behind the medial malleolus under the flexor retinaculum is the flexor hallucis longus muscle. The flexor retinaculum runs between the medial malleolus and the calcaneus. It sends three septa to the tibia and thus contains four compartments. In the most anteromedial compartment is the tendon of the tibialis posterior, and in the next is the tendon of the flexor digitorum longus. The third compartment

holds the tibial nerve and the posterior tibial vessels. The most posterior and inferior compartment is occupied by the flexor hallucis longus. *(Hollinshead, pp 416–417)*

**43.** **(C)** The extensor hallucis longus primarily is an extensor of the big toe but also it is in a position to dorsiflex, adduct, and invert the foot. The extensor digitorum longus and the peroneus tertius also dorsiflex the foot but are located more laterally, so that instead of inverting, they evert and abduct the foot. The peroneus longus and brevis both evert the foot. They are also plantar flexors instead of dorsiflexors. *(Hollinshead, pp 416,420,423)*

**44.** **(A)** The tibialis anterior does not arise from the fibula but from the lateral surface of the tibia, from the interosseous membrane, from an upper part of its covering fascia, and from an intermuscular septum between it and the extensor digitorum longus. The extensor digitorum longus has some origin from the tibia, but it arises mostly from the fibula and the anterior intermuscular septum. The other muscles listed arise from the fibula. *(Hollinshead, pp 420–421)*

**45.** **(C)** Of the locations listed, the common peroneal nerve is most likely to be injured as it becomes subcutaneous just behind the head of the fibula. Just before or after it has penetrated the posterior intermuscular septum, it divides into two branches, the deep and superficial peroneal nerves. Extended pressure on the lateral side of the knee, as in keeping the legs crossed for a long time, can result in injury to the common peroneal nerve. *(Hollinshead, p 420)*

**46.** **(D)** All of the muscles listed, except the peroneus brevis, are innervated by the deep peroneal nerve. They are all dorsiflexors of the foot. The peroneus brevis is innervated by the superficial peroneal nerve and is an abductor and weak plantar flexor of the foot. *(Hollinshead, pp 420,423)*

**47.** **(B)** The dorsalis pedis artery typically is the continuation of the anterior tibial artery after it passes under the inferior extensor retinaculum onto the dorsum of the foot. The dorsalis pedis usually appears just medial to the deep peroneal nerve on the dorsum of the foot. It runs distally toward the interspace between the first and second toes. *(Hollinshead, p 424)*

**48.** **(D)** The peroneus tertius arises from the fibula and the anterior intermuscular septum. This origin is continuous with that of the extensor digitorum longus, and the two tendons pass together across the ankle. The tendon of the peroneus tertius diverges laterally to insert into the dorsal surface of the base of the fifth metatarsal. This muscle therefore is placed on the dorsal surface of the foot and becomes a dorsiflexor, not a plantar flexor. The other muscles list-ed are plantar flexors. *(Hollinshead, pp 421,423,429)*

49. **(A)** The popliteal artery, the continuation of the femoral artery when it enters the popliteal fossa, descends through the middle of the popliteal fossa anterior to the popliteal vein; it accompanies the tibial nerve deep to the gastrocnemius and soleus muscles. It ends by dividing into the anterior and posterior tibial arteries; this division usually occurs on the posterior surface of the popliteus muscle. The anterior tibial artery then turns forward below the muscle and above the interosseous membrane to continue down the anterior aspect of the leg. *(Hollinshead, pp 432–433)*

50. **(C)** The femoral artery runs downward in the adductor canal. Close to the lower end of the canal, the artery gives off its last named branch, the descending genicular artery; this artery gives off branches on the medial side of the knee that anastomose with branches of arteries on the lateral side of the knee to form a pattern of collateral circulation around this joint. The popliteal artery gives off the medial and lateral superior genicular arteries that run medially and laterally around the femur to help form collateral circulation of the knee joint. Medial and lateral inferior genicular arteries that encircle the leg also help form the anastomoses around the knee. *(Hollinshead, pp 389,432)*

51. **(A)**

52. **(B)**

53. **(C)**

54. **(C)**

51–54. Laterally and medially the knee joint is strengthened by the fibular and tibial collateral ligaments. The fibular collateral ligament is a rounded cord stretching between the lateral femoral epicondyle and the head of the fibula. It stands well away from the capsule of the joint. The tibial collateral ligament is a broad band attached above to the medial femoral epicondyle and below to the medial tibial condyle. It is applied closely to the joint capsule. The oblique popliteal ligament runs upward and laterally across the posterior aspect of the knee joint. It is a thickening of the posterior part of the joint capsule that helps to resist hyperextension at the knee. The cruciate ligaments had no matching items. *(Hollinshead, pp 434–435)*

55. **(B)**

56. **(D)**

57. **(A)**

58. **(C)**

55–58. The flexor digitorum longus is supplied by the tibial nerve. The flexor digitorum brevis is supplied by the medial plantar nerve. The quadratus plantae is supplied by the lateral plantar nerve, and the tibialis anterior is supplied by the deep peroneal nerve. *(Hollinshead, pp 421,431,443)*

59. **(B)**

60. **(C)**

59–60. The lateral plantar artery and the medial plantar artery are branches of the posterior tibial artery. The lateral plantar artery is larger than the medial plantar and passes obliquely forward and laterally across the foot. It ends by crossing medially across the foot on the proximal ends of the interossei muscles as the plantar arterial arch. The deep plantar branches of the dorsalis pedis artery emerges between the heads of the first dorsal interosseous muscle and completes the plantar arch medially. *(Hollinshead, pp 448–449)*

61. **(C)**

62. **(C)**

63. **(A)**

64. **(D)**

61–64. The best invertors and adductors of the foot are the tibialis posterior and tibialis anterior muscles. The triceps surae, in plantar flexing the foot, inverts it apparently because of the relationship of the calcaneus to the joints of the ankle and foot rather than because of any particular direction of pull of the triceps. The peroneus longus neither inverts nor adducts the foot; it everts and abducts it. *(Hollinshead, p 457)*

65. **(B)**

66. **(A)**

65–66. The heavy deltoid ligament is located on the medial side of the foot. It fans out from an attachment on the medial malleolus into four named ligaments that attach to the navicular, calcaneus, and posterior talus bones. This ligament particularly resists eversion of the foot; ankle sprains frequently result in its tearing. The strong calcaneonavicular (spring) ligament resists the tendency for the head of the talus to be driven downward between the calcaneus and the navicular. The "spring" ligament is continuous medially with the deltoid ligament and supports the head of the talus. The portion of the head of the talus that rests upon the "spring" ligament is at the summit of the medial arch of the foot. *(Hollinshead, pp 453,456; Basmajian, p 308)*

67. **(C)** In the standing position, it appears that the plantar ligaments and aponeurosis bear the greatest stress. The chief function of the evertor and invertor muscles is to preserve a relative constancy in the ratio of weight distribution among the heads of the metatarsals. It has been demonstrated electromyographically that the tibialis anterior, peroneus longus, and intrinsic muscles of the foot play no important role in the normal static support of the long arches of the foot; they usually are completely inactive. *(Basmajian, p 312)*

68. **(E)** All of the items listed are correct. The lateral condyle of the tibia is shorter from front to back than the medial condyle, and the lateral meniscus is shorter than the medial meniscus. The medial meniscus is C shaped, and the lateral meniscus is shaped like a small o. The biceps femoris tendon crosses the fibular collateral ligament. *(Basmajian, p 300)*

69. **(A)** The bones taking part in the knee joint are the femur, the tibia, and the patella. The fibula is associated only indirectly with the knee. Primitively there were three joint cavities, now merged into one. *(Basmajian, p 297)*

70. **(B)** The iliofemoral ligament is a broad, strong band shaped like an inverted Y. Above, it is attached to the acetabular margin. The lower attachment of this ligament on the anterior femur creates the broad, rough intertrochanteric line, not the intertrochanteric crest found on the posterior aspect of the femur between the greater and lesser trochanters. In erect standing, the line of gravity passes behind the hip joints; therefore the trunk tends to fall backward or rather to rotate backward. The iliofemoral ligament checks this backward rotation. *(Basmajian, p 294)*

71. **(D)** The plantar muscles are arranged in four layers beginning superficially. The first layer muscles arise from the calcaneus and include the abductor hallucis, the abductor digiti quinti, and the flexor digitorum brevis. The second layer of muscles consists of the flexor digitorum longus tendon, the quadratus plantae, the lumbricals, and the flexor hallucis longus tendon. The flexor hallucis brevis is a part of the third layer of muscles together with the flexor digiti quinti and the transverse and oblique adductor hallucis muscles. The fourth layer consists of seven interossei muscles and two long tendons on the skeletal plane, the tibialis posterior and the peroneus longus. *(Basmajian, pp 287–289)*

72. **(A)** The fibula transmits no weight to the ground, but it holds the talus in its socket and does this in a resilient manner. A main function of this bone is to give origin to muscles. It has been shown that, contrary to expectation, the fibula moves downward, not up, to deepen the ankle socket at the strike phase of running. *(Basmajian, p 275)*

73. **(A)** The extensor retinaculum has two parts, superior and inferior. The inferior part is placed in front of the ankle and has the appearance of a Y-shaped band. The stem of the Y is attached to the anterior part of the upper surface of the calcaneus; the fibers of the Y form loops on the dorsum of the foot. These loops or slings are especially for the peroneus tertius, extensor digitorum longus, and the extensor hallucis longus to prevent those tendons from bow stringing forward or medially. The posterior tibial tendon passes closely behind the medial malleolus deep to the flexor retinaculum, not deep to the extensor retinaculum. *(Basmajian, pp 273–274,279–280)*

74. **(B)** The muscles supplied by the superficial peroneal nerve include the peroneus longus and the peroneus brevis. The deep peroneal nerve supplies the extensor digitorum brevis and the anterior tibial muscles. *(Basmajian, p 272)*

75. **(C)** The popliteal artery is an extension of the femoral artery as it passes behind the knee. The popliteal artery divides into its two terminal branches, the anterior and posterior tibial arteries. The femoral artery gives rise to the large profunda femoris artery in the femoral triangle several centimeters below the inguinal ligament. The obturator artery assists the profunda artery in supplying the adductor muscles. *(Basmajian, pp 250,252,264,267)*

76. **(D)** The peroneal artery arises from the posterior tibial artery. It descends behind the fibula, the distal tibiofibular joint, and the ankle joint. The lateral plantar artery, the continuation of the posterior tibial artery, runs laterally on the sole of the foot and forms the plantar arch. The peroneal artery ends on the lateral surface of the calcaneus as the lateral calcaneal artery. *(Basmajian, pp 280,290–291)*

77. **(B)** The gastrocnemius and soleus muscles are referred to as the triceps surae. These two muscles, along with the plantaris muscle, insert into the tendo calcaneus (tendon of Achilles). The posterior tibial muscle is not a part of the triceps surae. *(Basmajian, pp 277–278)*

78. **(E)** The profunda femoris artery usually branches from the lateral side of the femoral artery about 4 cm below the inguinal ligament. It gives off the lateral and medial femoral circumflex arteries. Usually it gives off four perforating arteries that encircle the

shaft of the femur tightly, perforating any muscle they encounter. (Basmajian, pp 267–268)

79. (C) The muscles of the adductor region of the thigh arise collectively from the anterior aspect of the hip bone and the obturator membrane. They share their chief actions, adduction and medial rotation of the femur at the hip joint, and also their nerve, the obturator. The gracilis is the only member of this group to cross the knee joint and insert on the tibia. The pectineus muscle is supplied by the femoral nerve, and, as part of the adductor group, by the obturator nerve. (Basmajian, pp 251,254,264,266)

80. (D) With the knee flexed, the only muscle listed that can laterally rotate the leg is the biceps femoris. This muscle, being attached to the head of the fibula, rotates the leg laterally; the semitendinosus and semimembranosus, being attached to the medial side of the tibia, rotate the leg medially. When the knee is flexed, the gracilis medially rotates the leg, and it can flex the hip if the knee is extended. (Basmajian, p 261; Hollinshead, p 384)

81. (E) The popliteal fossa contains all of the structures listed. The order from the surface to the floor of the fossa is nerve, vein, artery. The customary rule (nerve, artery, vein) does not hold here. (Basmajian, p 263)

82. (C) Muscles that are amalgamations of two muscles are the biceps femoris and the adductor magnus. The long head of the biceps femoris developmentally belongs to the front of the limb, the short head to the back. The adductor magnus also is hybrid. The part arising from the pubic arch is supplied by the obturator nerve, while the part arising from the ischial tuberosity is supplied by the tibial division of the sciatic nerve. (Basmajian, pp 261–263)

83. (A) The superior gluteal nerve, accompanied by the superior gluteal vessels, passes through the greater sciatic foramen. This nerve runs between the gluteus medius and the gluteus minimus and supplies, in addition to these two muscles, the tensor fasciae latae. The quadriceps muscles are innervated by the femoral nerve. (Basmajian, pp 254,259)

84. (B) The obturator nerve does innervate all the adductor muscles, but the pectineus and adductor magnus have additional nerve supplies. The pectineus receives its major innervation from the femoral

nerve and also perhaps from the obturator nerve. The posterior portion of the adductor magnus muscle receives a nerve supply from the tibial nerve. (Moore, p 449)

85. (C) When a person stands with the thigh somewhat flexed, abducted, and laterally rotated, the femoral triangle can be observed as a depression in the proximal third of the thigh. Its base, the inguinal ligament, often can be observed and palpated. Its lateral boundary, the medial (not the lateral) edge of the sartorius, is obvious, but its medial boundary, the medial border of the adductor longus, is not so easy to identify. The position of the femoral artery in the triangle can be identified by palpating the femoral pulse. (Moore, pp 453–454)

86. (A) The femoral vein ends posterior to the inguinal ligament, where it becomes the external iliac vein. While within the femoral triangle, the femoral vein receives, as well as other tributaries, the profunda femoris and great saphenous veins. The great saphenous vein, not the femoral vein, passes through the saphenous opening in the fascia lata to join the femoral vein. (Moore, pp 457,435)

87. (E) All of the items listed are characteristic of femoral hernia. The femoral ring is a weak point in the abdominal wall that normally admits the size of the tip of the little finger. A femoral hernia is protrusion of abdominal viscera, often the small intestine, through the femoral ring into the femoral canal. Strangulation of a femoral hernia that interferes with the blood supply to the herniated bowel may occur owing to the sharpness and rigidity of the boundaries of the femoral ring. A femoral hernia presents as a mass inferolateral to the pubic tubercle and medial to the femoral vein. (Moore, pp 459–462,593–594,599,601)

88. (A) All of the items are correct about the peroneal nerve except item 4. This nerve is said to be the most commonly injured nerve in the lower limb, mainly because it is exposed where it winds superficially around the neck of the fibula. Severance of this nerve results in paralysis of all the dorsiflexors of the foot. Some of the dorsiflexor muscles also invert the foot (anterior tibial and extensor hallucis longus), but other invertor muscles (tibialis posterior) are not involved because of innervation by the tibial nerve. (Moore, pp 491,527,591,597,603)

# CHAPTER 9

# Diagram Labeling Exercises
## Questions

**Anterior Aspect of the Skull**
(Questions 1 through 12)

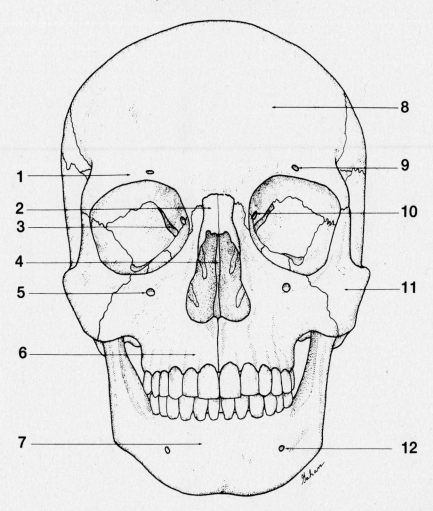

**Directions:** With reference to the diagram, match the numbered structures with their corresponding lettered items.

(A) supraorbital foramen
(B) zygomatic bone
(C) superciliary arch
(D) infraorbital foramen
(E) mental foramen
(F) superior orbital fissure

(G) optic canal
(H) perpendicular plate of the ethmoid
(I) nasal bone
(J) maxilla
(K) mandible
(L) frontal bone

## Branches of the Maxillary Artery
### (Questions 13 through 24)

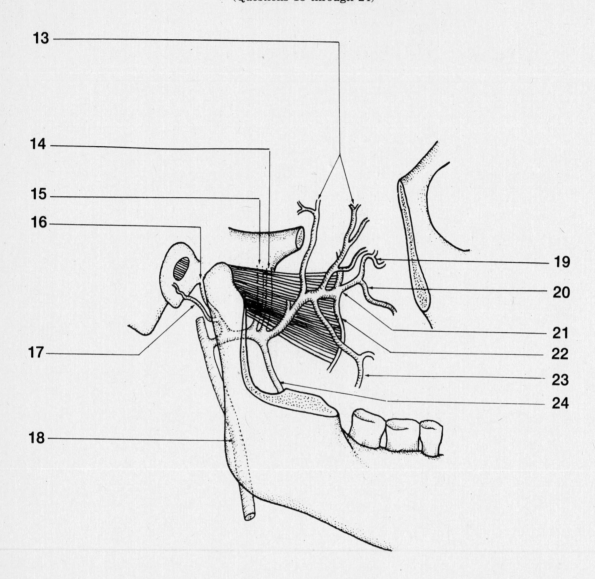

**Directions:** With reference to the diagram, match the numbered structures with their corresponding lettered items.

(A)  external carotid artery
(B)  deep auricular artery
(C)  anterior tympanic artery
(D)  inferior alveolar artery
(E)  buccal artery
(F)  middle meningeal artery

(G)  accessory meningeal artery
(H)  greater palatine artery
(I)  posterior superior alveolar artery
(J)  infraorbital artery
(K)  sphenopalatine artery
(L)  deep temporal arteries

## Triangles of the Neck
(Questions 25 through 34)

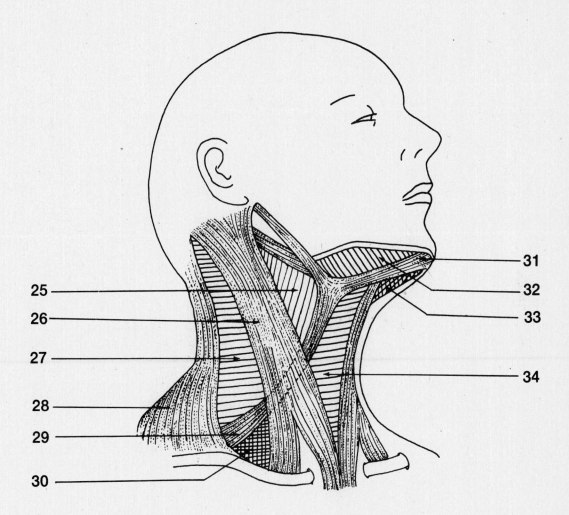

25
26
27
28
29
30

31
32
33

34

**Directions:** With reference to the diagram, match the numbered structures with their corresponding lettered items.

(A)  carotid triangle
(B)  posterior cervical triangle
(C)  omoclavicular triangle
(D)  muscular triangle
(E)  submandibular triangle

(F)  sternocleidomastoid muscle
(G)  trapezius muscle
(H)  omohyoid muscle
(I)  anterior belly of the diagastric muscle
(J)  submental triangle

## Muscles of the Face, Deep Layer
### (Questions 35 through 45)

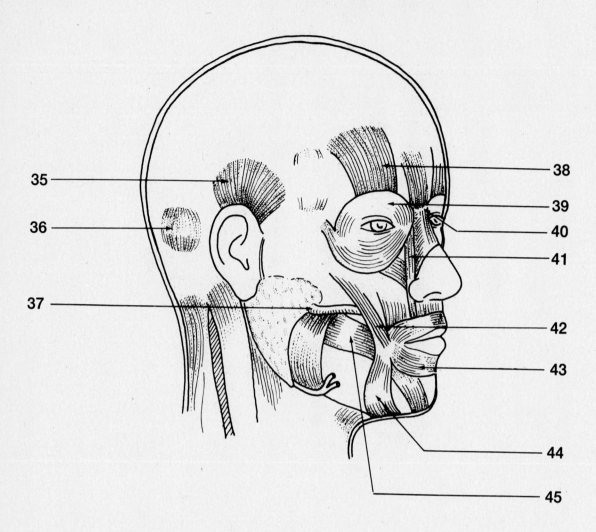

**Directions:** With reference to the diagram, match the numbered structures with their corresponding lettered items.

(A)  frontalis muscle
(B)  orbicularis oculi muscle
(C)  orbicularis oris muscle
(D)  zygomaticus major muscle
(E)  buccinator muscle
(F)  depressor anguli oris muscle

(G)  parotid duct
(H)  superior auricular muscle
(I)  procerus muscle
(J)  levator labii superioris alaeque nasi muscle
(K)  occipitalis muscle

## Muscles of the Head and Deep Face
(Questions 46 through 55)

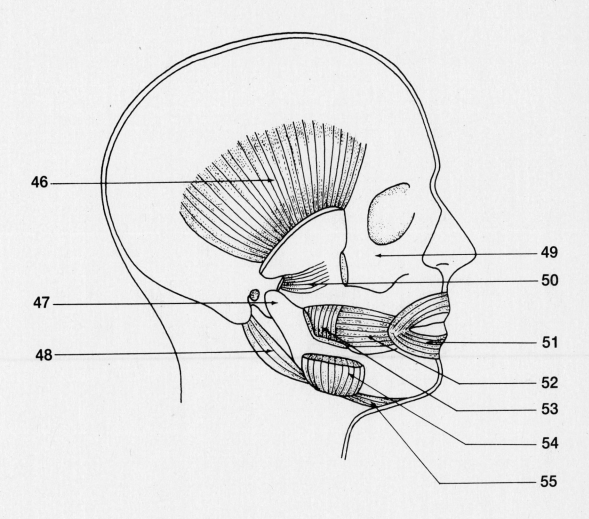

**Directions:** With reference to the diagram, match the numbered structures with their corresponding lettered items.

(A)   temporalis muscle
(B)   condyle of the mandible
(C)   lateral pterygoid muscle
(D)   medial pterygoid muscle
(E)   masseter muscle

(F)   buccinator muscle
(G)   orbicularis oris muscle
(H)   posterior belly of the digastric muscle
(I)   anterior belly of the digastric muscle
(J)   zygomatic bone

## Lateral Aspect of the Skull
### (Questions 56 through 75)

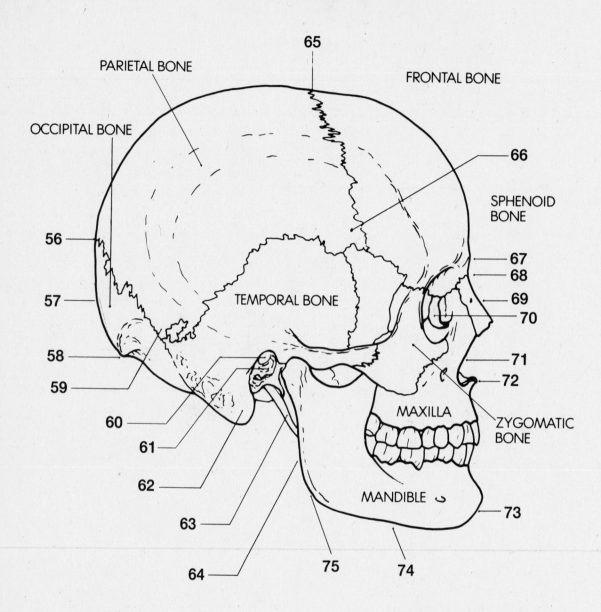

**Directions:** With reference to the diagram, match the numbered structures with their corresponding lettered items.

(A) glabella
(B) angle of mandible
(C) inion
(D) nasal bone
(E) bregma
(F) asterion
(G) base of mandible
(H) posterior pole
(I) lacrimal bone
(J) anterior nasal aperture
(K) nasion

(L) external auditory meatus
(M) mastoid process
(N) pterion
(O) posterior border of ramus
(P) tympanic part
(Q) styloid process
(R) mental protuberance
(S) anterior nasal spine
(T) lambda

## The Mandible, External and Internal
### (Questions 76 through 93)

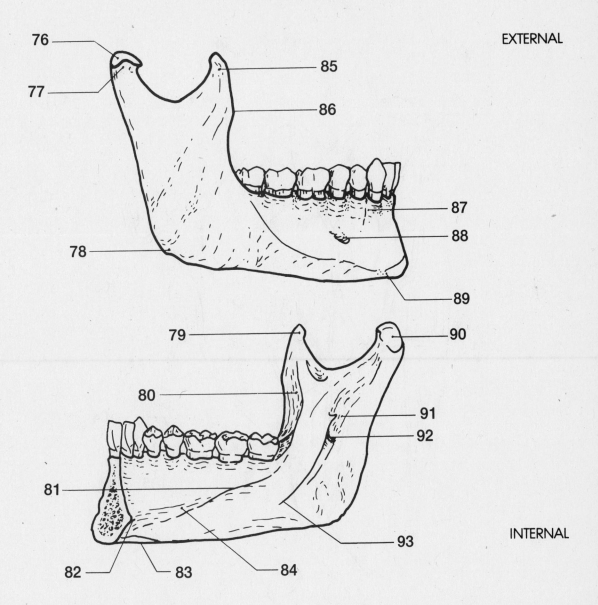

**Directions:** With reference to the diagram, match the numbered structures with their corresponding lettered items.

(A)  angle
(B)  lingula
(C)  neck
(D)  submandibular fossae
(E)  mental tubercle
(F)  alveolar part
(G)  coronoid process
(H)  genial tubercle
(I)  buttress

(J)  head (condyle)
(K)  mandibular foramen
(L)  anterior border
(M)  mylohyoid line
(N)  digastric
(O)  sublingual
(P)  mental foramen
(Q)  head

**Ophthalmic Artery**
(Questions 94 through 106)

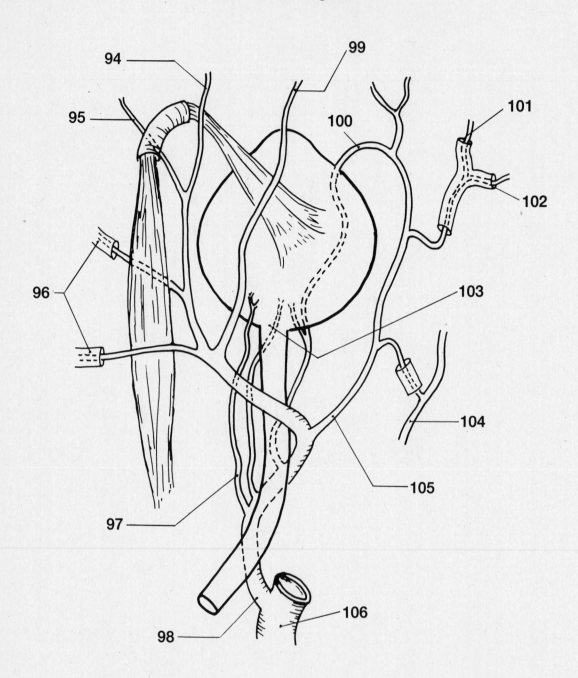

**Directions:** With reference to the diagram, match the numbered structures with their corresponding lettered items.

(A)  middle meningeal
(B)  posterior ciliary
(C)  interior carotid
(D)  zygomatic temporal
(E)  central
(F)  supraorbital
(G)  anterior ciliary

(H)  supratrochlear (frontal)
(I)  posterior and anterior ethmoidal
(J)  lacrimal
(K)  dorsal nasal
(L)  zygomatic facial
(M)  ophthalmic

## Sagittal Section Through the Male Pelvis
### (Questions 107 through 116)

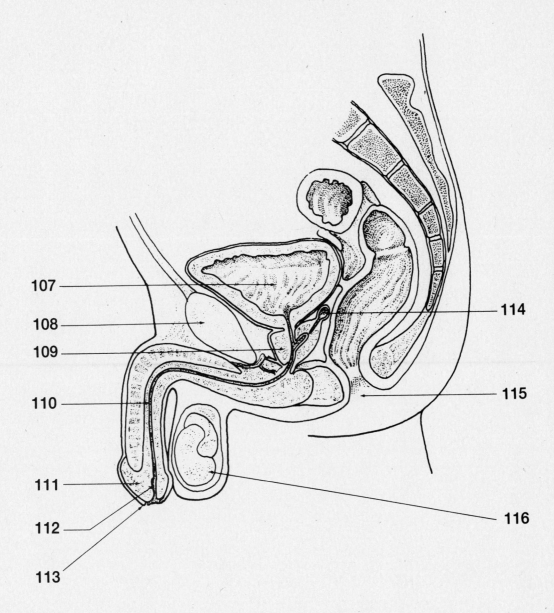

107

108

109

110

111

112

113

114

115

116

**Directions:** With reference to the diagram, match the numbered structures with their corresponding lettered items.

(A) bladder
(B) prostate
(C) urethra
(D) glans penis
(E) scrotum

(F) anus
(G) pubic symphysis
(H) ejaculatory duct
(I) fossa navicularis
(J) external urethral meatus

## Midsagittal Section Through the Female Pelvis
(Questions 117 through 126)

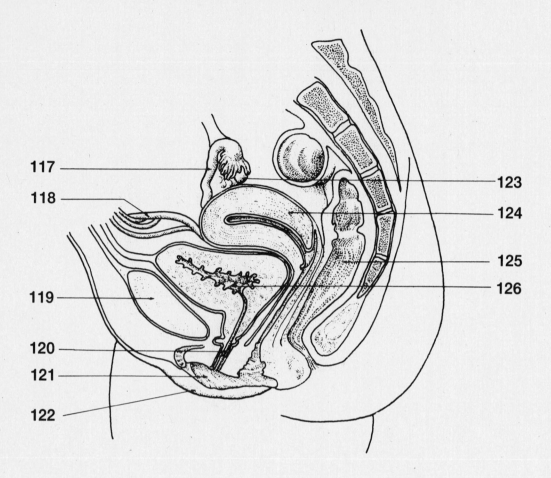

117

118

119

120

121

122

123

124

125

126

**Directions:** With reference to the diagram, match the numbered structures with their corresponding lettered items.

(A)    uterus
(B)    rectum
(C)    bladder
(D)    pubic symphysis
(E)    urethra

(F)    labium minus
(G)    labium majus
(H)    round ligament
(I)    uterine tube
(J)    ovary

## Boundaries and Subdivisions of the Perineum
(Questions 127 through 138)

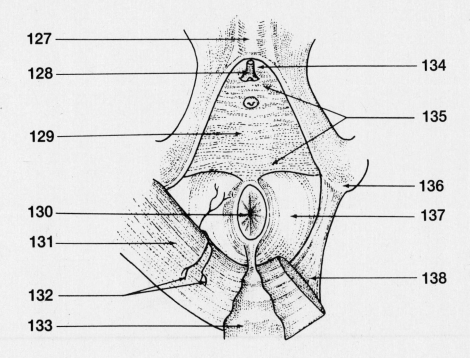

127
128
129
130
131
132
133
134
135
136
137
138

**Directions:** With reference to the diagram, match the numbered structures with their corresponding lettered items.

(A)  anal triangle
(B)  urogenital triangle
(C)  pubic symphysis
(D)  ischial tuberosity
(E)  sacrotuberous ligament
(F)  arcuate pubic ligament

(G)  gluteus maximus
(H)  anus
(I)  coccyx
(J)  deep dorsal vein of the penis
(K)  inferior rectal nerve
(L)  urogenital diaphragm

## Gluteal Region: Ligaments and Bony Parts
### (Questions 139 through 153)

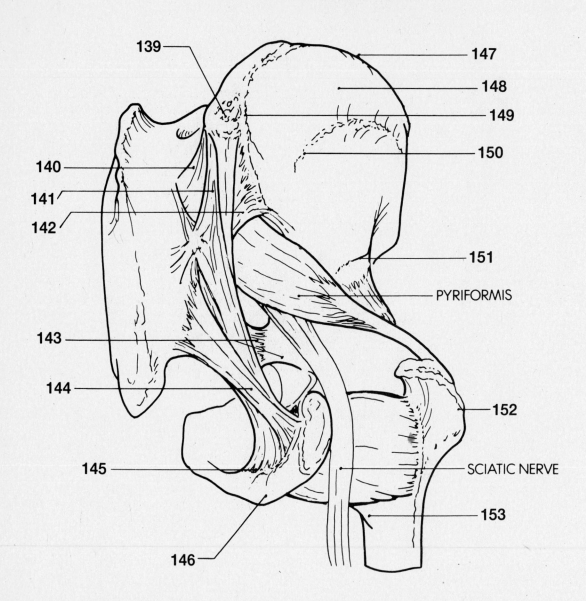

139
147
148
149
150
140
141
142
151
PYRIFORMIS
143
144
152
145
SCIATIC NERVE
153
146

**Directions:** With reference to the diagram, match the numbered structures with their corresponding lettered items.

(A)  greater trochanter
(B)  ischial tuberosity
(C)  dorsum ilii
(D)  tip of coccyx
(E)  lesser trochanter
(F)  posterior sacroiliac (short) ligament
(G)  iliac crest
(H)  inferior gluteal line

(I)  falciform edge
(J)  posterior superior iliac spine
(K)  posterior sacroiliac (long) ligament
(L)  posterior gluteal line
(M)  middle gluteal line
(N)  sacrospinous ligament and ischial spine
(O)  posterior inferior iliac spine

**Right Hip Bone**
(Questions 154 through 167)

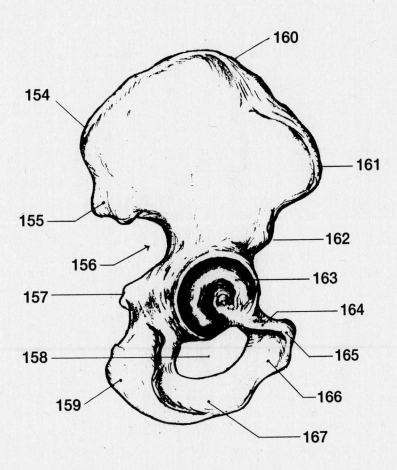

**Directions:** With reference to the diagram, match the numbered structures with their corresponding lettered items.

(A)  posterior superior spine
(B)  ischial tuberosity
(C)  posterior inferior spine
(D)  obturator foramen
(E)  sciatic notch
(F)  spine of ischium
(G)  iliac crest

(H)  ramus of ischium
(I)  anterior superior spine
(J)  inferior ramus of pubis
(K)  anterior inferior spine
(L)  pubic crest
(M)  acetabulum
(N)  superior ramus of pubis

## Anteromedial Aspect of the Arm
(Questions 168 through 177)

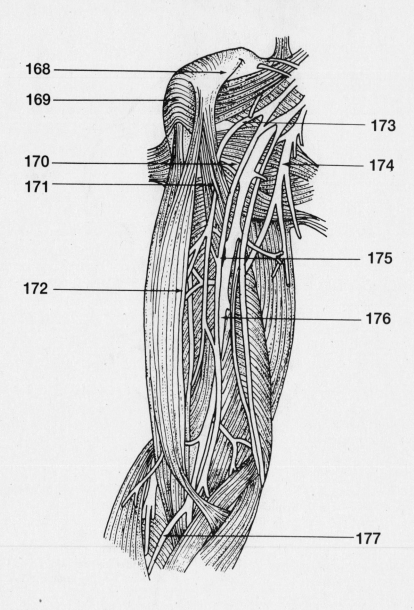

**Directions:** With reference to the diagram, match the numbered structures with their corresponding lettered items.

(A)  biceps brachii
(B)  coracoid process
(C)  coracobrachialis muscle
(D)  musculocutaneous nerve
(E)  median nerve

(F)  radial nerve
(G)  radial artery
(H)  brachial artery
(I)  axillary nerve
(J)  subscapularis muscle

**Posterior Aspect of the Thigh**
(Questions 178 through 187)

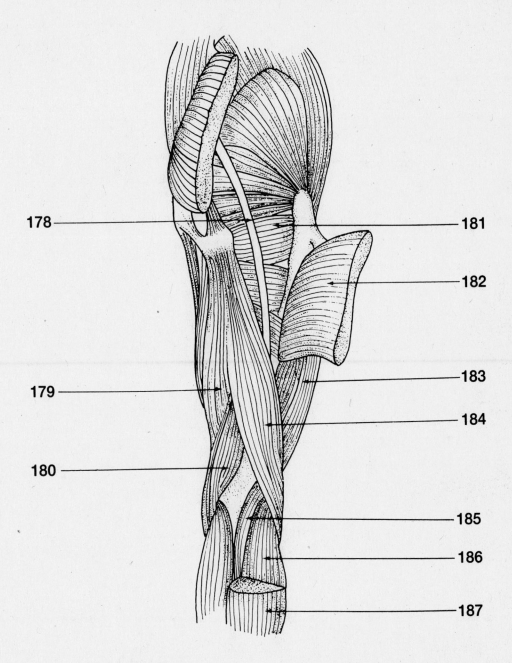

178 —
179 —
180 —
181
182
183
184
185
186
187

**Directions:** With reference to the diagram, match the numbered structures with their corresponding lettered items.

(A)  gluteus maximus
(B)  sciatic nerve
(C)  biceps femoris
(D)  semitendinosus
(E)  semimembranosus

(F)  quadratus femoris
(G)  plantaris
(H)  soleus
(I)  gastrocnemius
(J)  iliotibial tract

## Posterior Aspect, Muscles of the Leg
### (Questions 188 through 205)

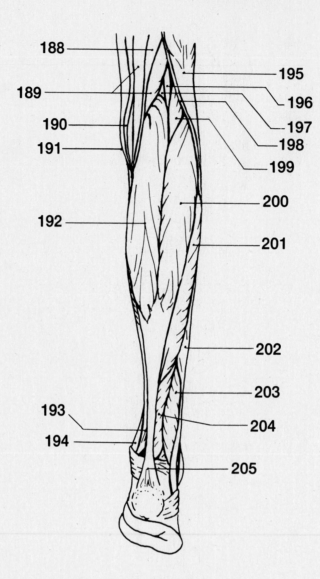

188
189
190
191
192
193
194

195
196
197
198
199
200
201
202
203
204
205

**Directions:** With reference to the diagram, match the numbered structures with their corresponding lettered items.

(A)  plantaris
(B)  popliteal artery
(C)  gastrocnemius (medial head)
(D)  gastrocnemius (lateral head)
(E)  peroneus longus
(F)  biceps femoris
(G)  soleus
(H)  medial popliteal nerve
(I)  semitendinosus

(J)  peroneus brevis
(K)  flexor hallucis longus
(L)  tibialis posterior
(M)  flexor digitorum longus
(N)  gracilis
(O)  sartorius
(P)  lateral popliteal nerve
(Q)  tendocalcaneus (tendon of Achilles)
(R)  semimembranosus

## Anatomic Position of Esophagus and the Stomach
(Questions 206 through 217)

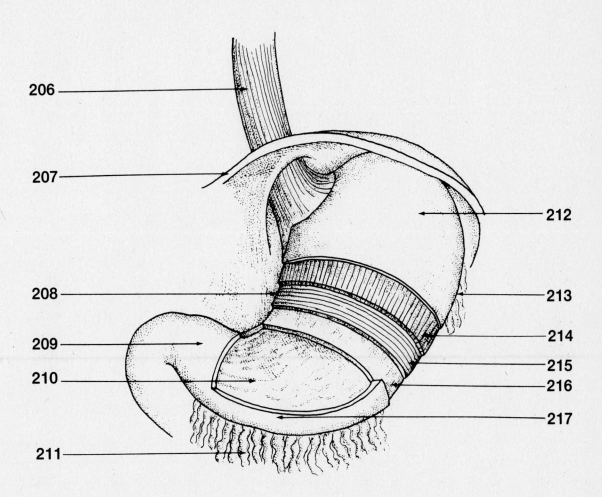

**Directions:** With reference to the diagram, match the numbered structures with their corresponding lettered items.

(A)   esophagus

(B)   diaphragm

(C)   fundus

(D)   lesser curvature

(E)   greater curvature

(F)   greater omentum

(G)   pylorus

(H)   serosa

(I)   longitudinal muscle layer

(J)   circular muscle layer

(K)   submucosa

(L)   mucosa

## The Liver
(Questions 218 through 226)

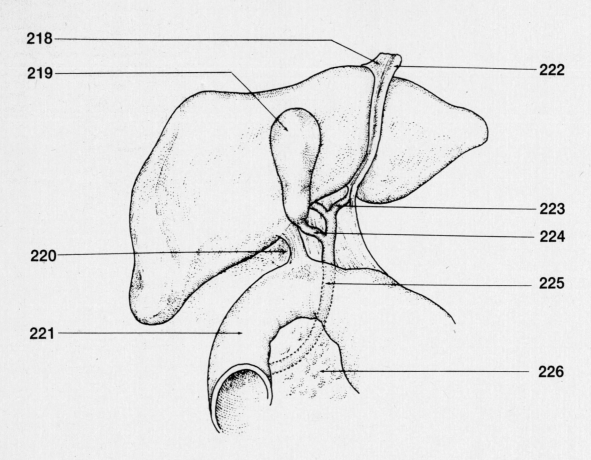

**Directions:** With reference to the diagram, match the numbered structures with their corresponding lettered items.

(A)  falciform ligament          (F)  duodenum
(B)  round ligament              (G)  pancreas
(C)  left hepatic duct           (H)  gallbladder
(D)  cystic duct                 (I)  epiploic foramen
(E)  common bile duct

## A Relationship of the Pancreas to the Duodenum
(Questions 227 through 240)

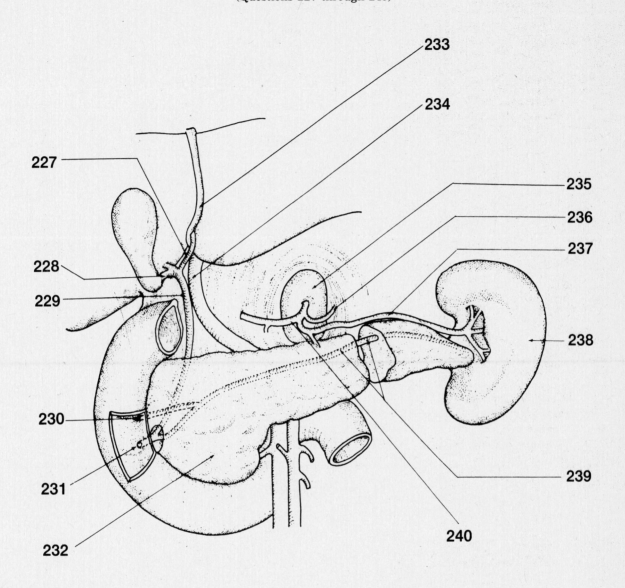

**Directions:** With reference to the diagram, match the numbered structures with their corresponding lettered items.

(A)  hepatic duct
(B)  common bile duct
(C)  cystic duct
(D)  portal vein
(E)  celiac trunk
(F)  hepatic artery
(G)  left gastric artery

(H)  splenic artery
(I)  spleen
(J)  pancreas
(K)  pancreatic duct
(L)  accessory pancreatic duct
(M)  duodenum papilla
(N)  ligamentum teres

## Position and Structure of Large Intestine
(Questions 241 through 252)

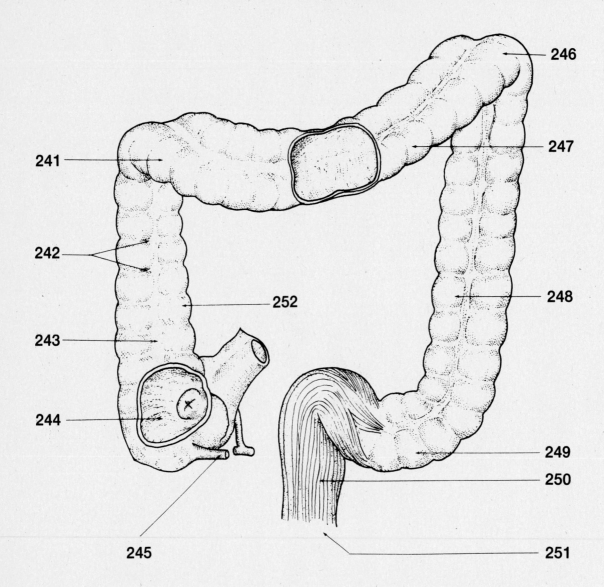

**Directions:** With reference to the diagram, match the numbered structures with their corresponding lettered items.

(A)  epiploic appendages
(B)  taenia coli
(C)  cecum
(D)  vermiform appendix
(E)  right colic flexure
(F)  left colic flexure

(G)  ascending colon
(H)  transverse colon
(I)  descending colon
(J)  sigmoid colon
(K)  rectum
(L)  anus

## Posterior Abdominal Wall
(Questions 253 through 263)

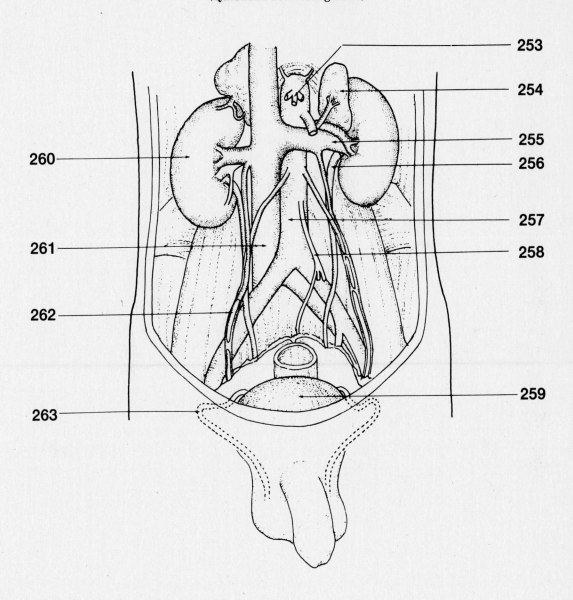

**Directions:** With reference to the diagram, match the numbered structures with their corresponding lettered items.

(A) kidney
(B) inferior vena cava
(C) right testicular artery
(D) vas deferens
(E) celiac trunk
(F) suprarenal

(G) left renal vein
(H) left ureter
(I) aorta
(J) inferior mesenteric artery
(K) bladder

**Celiac Artery Distribution**
(Questions 264 through 274)

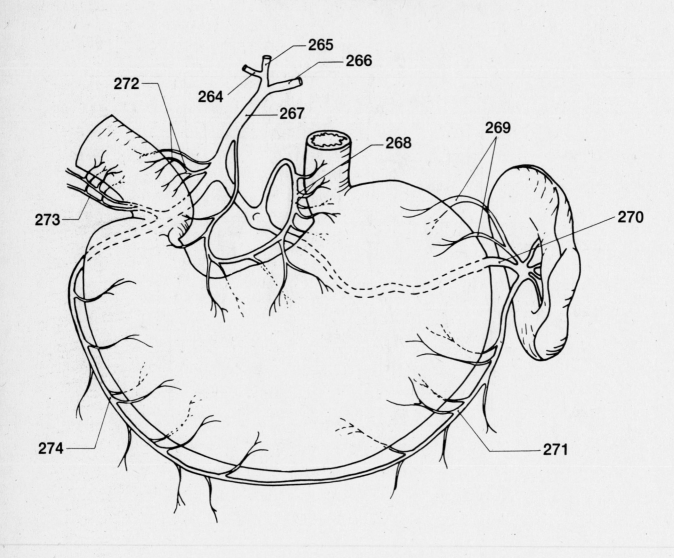

**Directions:** With reference to the diagram, match the numbered structures with their corresponding lettered items.

(A)   short gastric
(B)   left gastroepiploic
(C)   esophageal branch
(D)   superior pancreaticoduodenal
(E)   splenic
(F)   right hepatic

(G)   right gastroepiploic
(H)   hepatic
(I)   left hepatic
(J)   cystic
(K)   superior duodenal

## Abdominal Musculature
### (Questions 275 through 282)

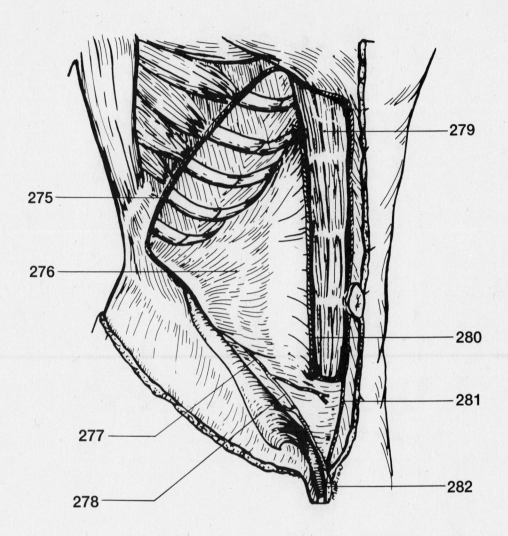

**Directions:** With reference to the diagram, match the numbered structures with their corresponding lettered items.

(A)  external abdominal oblique (cut)
(B)  spermatic cord
(C)  aponeurosis of external oblique (cut)
(D)  rectus abdominis

(E)  internal abdominal oblique
(F)  ilioinguinal nerve
(G)  anterior layer of rectus sheath (cut)
(H)  iliohypogastric nerve

## Anterior View of the Lungs and Pericardium
(Questions 283 through 292)

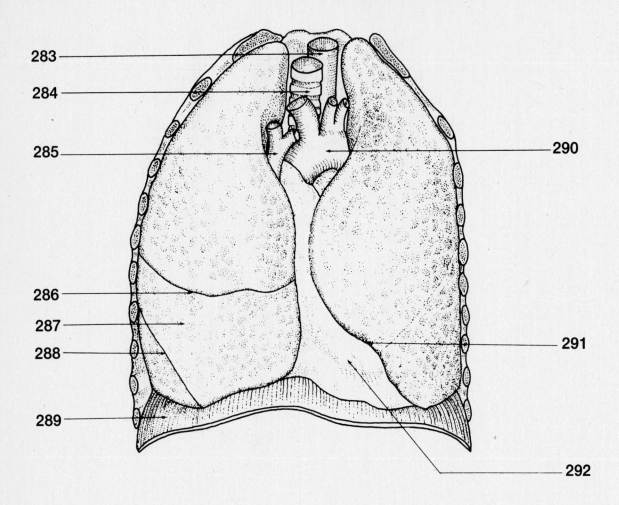

**Directions:** With reference to the diagram, match the numbered structures with their corresponding lettered items.

(A)   esophagus
(B)   trachea
(C)   aortic arch
(D)   horizontal fissure
(E)   oblique fissure

(F)   diaphragm
(G)   pericardium
(H)   superior vena cava
(I)   middle lobe of right lung
(J)   cardiac notch

**Diaphragm View from Below**
(Questions 293 through 300)

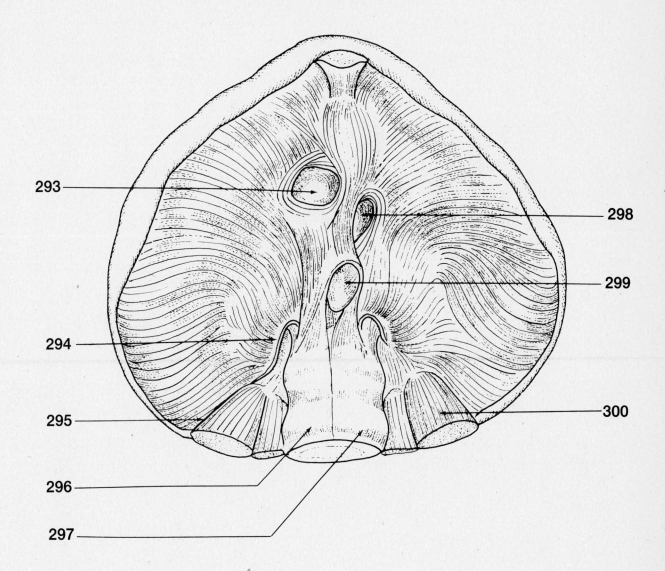

293

294

295

296

297

298

299

300

**Directions:** With reference to the diagram, match the numbered structures with their corresponding lettered items.

(A)   vena caval foramen
(B)   esophageal hiatus
(C)   aortic hiatus
(D)   medial arcuate ligament

(E)   lateral arcuate ligament
(F)   right crus
(G)   left crus
(H)   quadratus lumborum

## Ventral View of the Heart
(Questions 301 through 313)

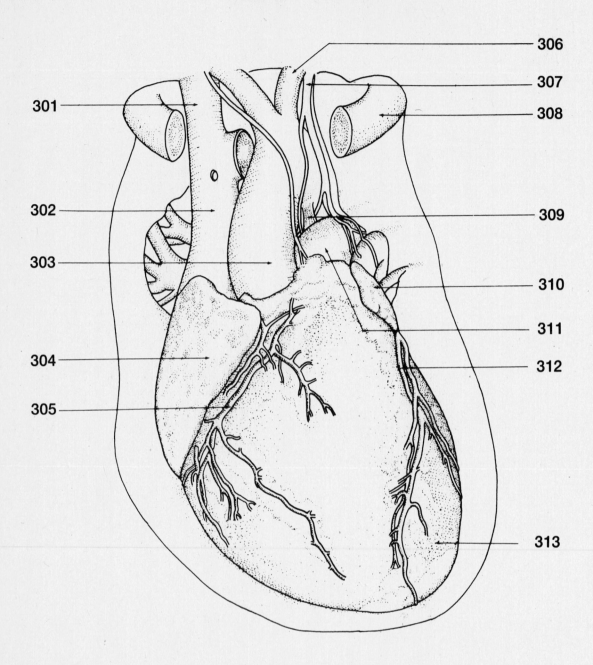

301

302

303

304

305

306

307

308

309

310

311

312

313

**Directions:** With reference to the diagram, match the numbered structures with their corresponding lettered items.

(A)   left vagus nerve
(B)   ligamentum arteriosum
(C)   right coronary artery
(D)   anterior descending branch of left
        coronary artery
(E)   right atrium
(F)   pulmonary trunk

(G)   ascending aorta
(H)   left atrium
(I)    superior vena cava
(J)    left common carotid artery
(K)   left ventricle
(L)   first rib
(M)   right brachiocephalic vein

**Dorsal View of the Heart**
(Questions 314 through 320)

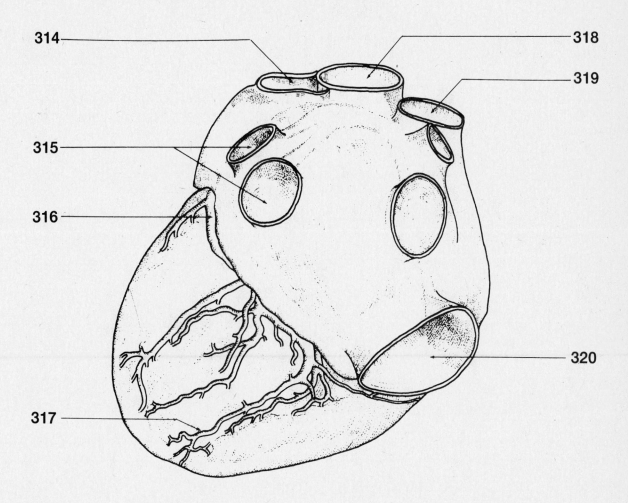

**Directions:** With reference to the diagram, match the numbered structures with their corresponding lettered items.

(A) pulmonary artery
(B) aorta
(C) superior vena cava
(D) left pulmonary veins

(E) middle cardiac vein
(F) inferior vena cava
(G) great cardiac vein

## Semidiagrammatic Representation of the More Anterior Structures of the Posterior Mediastinum
(Questions 321 through 332)

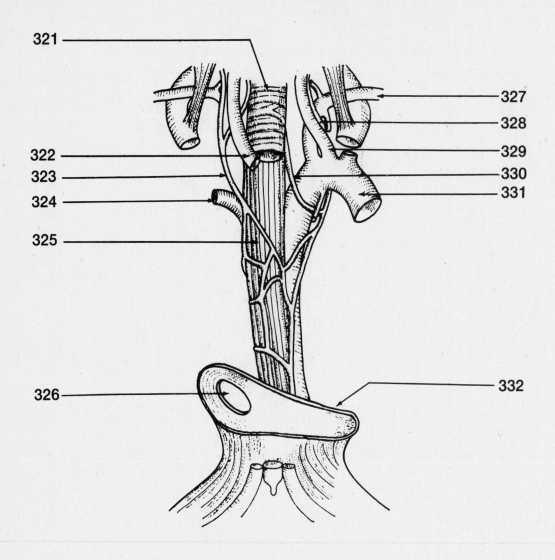

**Directions:** With reference to the diagram, match the numbered structures with their corresponding lettered items.

(A)  arch of aorta
(B)  left recurrent laryngeal nerve
(C)  left vagus nerve
(D)  esophagus
(E)  trachea
(F)  brachiocephalic artery

(G)  pericardium
(H)  inferior vena cava
(I)  left common carotid artery
(J)  left subclavian artery
(K)  azygos vein
(L)  right vagus nerve

**Esophagus, Trachea, and Aorta: Thoracic Parts**
(Questions 333 through 349)

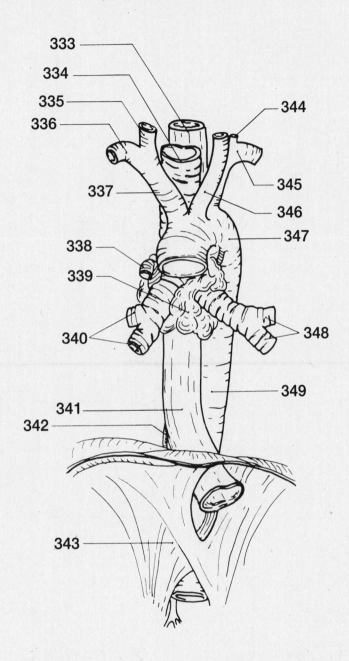

333
334
335
336
337
338
339
340
341
342
343
344
345
346
347
348
349

**Directions:** With reference to the diagram, match the numbered structures with their corresponding lettered items.

(A)  aorta
(B)  diaphragm
(C)  trachea
(D)  azygos arch
(E)  aortic arch
(F)  lymph glands
(G)  left bronchus
(H)  thoracic duct

(I)  esophagus
(J)  right bronchus
(K)  right common carotid artery
(L)  vertebral artery
(M)  left common carotid artery
(N)  left subclavian artery
(O)  right subclavian artery
(P)  innominate artery

**Vertebra, from Above and Side View**
(Questions 350 through 365)

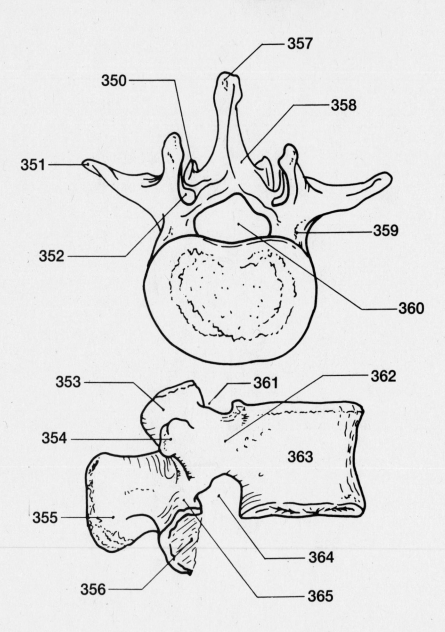

**Directions:** With reference to the diagram, match the numbered structures with their corresponding lettered items.

(A)  superior articular process
(B)  superior vertebral notch
(C)  inferior articular process
(D)  lamina
(E)  pedicle

(F)  body
(G)  spinous process
(H)  vertebral foramen
(I)  inferior vertebral notch
(J)  transverse process

## Thoracic Musculature
(Questions 366 through 370)

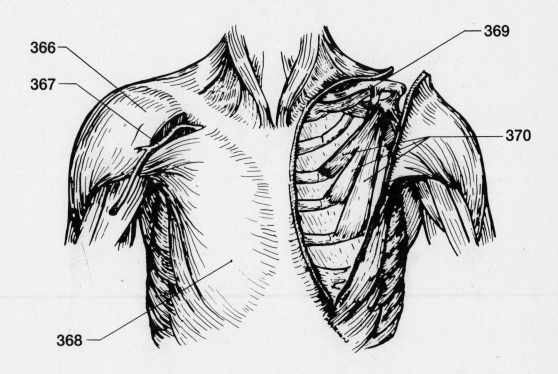

**Directions:** With reference to the diagram, match the numbered structures with their corresponding lettered items.

(A)  subclavius

(B)  deltoid

(C)  cephalic vein

(D)  pectoralis minor

(E)  pectoralis major

**Musculature of the Back**
(Questions 371 through 377)

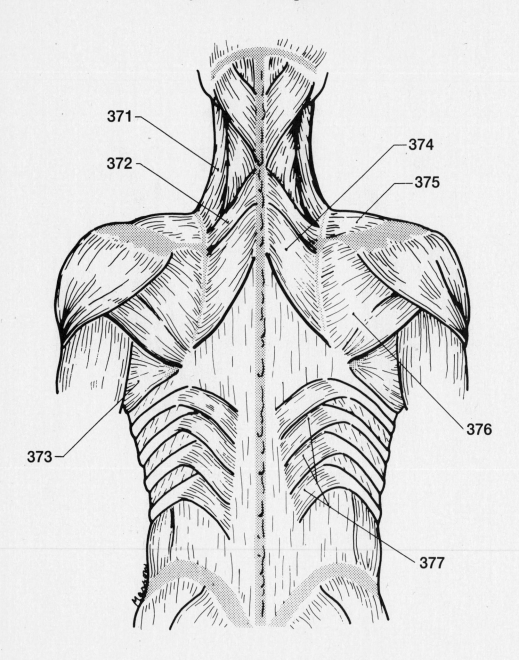

**Directions:** With reference to the diagram, match the numbered structures with their corresponding lettered items.

(A)  infraspinatus
(B)  levator scapulae
(C)  serratus anterior
(D)  rhomboideus minor

(E)  rhomboideus major
(F)  serratus posterior inferior
(G)  supraspinatus

**Musculature of the Back (Continued)**
(Questions 378 through 382)

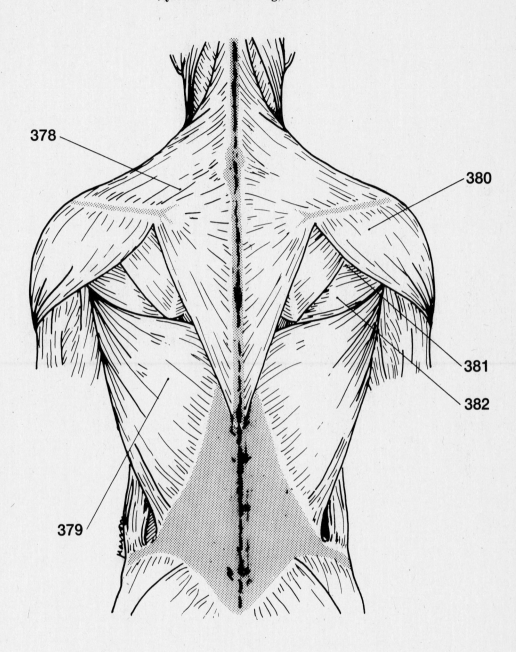

**Directions:** With reference to the diagram, match the numbered structures with their corresponding lettered items.

(A)  trapezius
(B)  teres minor
(C)  teres major

(D)  latissimus dorsi
(E)  deltoideus

# Answers

**Anterior Aspect of the Skull**

1. (C)
2. (I)
3. (F)
4. (H)
5. (D)
6. (J)
7. (K)
8. (L)
9. (A)
10. (G)
11. (B)
12. (E)

**Branches of the Maxillary Artery**

13. (L)
14. (G)
15. (F)
16. (C)
17. (B)
18. (A)
19. (J)
20. (I)
21. (K)
22. (H)
23. (E)
24. (D)

**Triangles of the Neck**

25. (A)
26. (F)
27. (B)
28. (G)
29. (H)
30. (C)
31. (I)
32. (E)
33. (J)
34. (D)

**Muscles of the Face, Deep Layer**

35. (H)
36. (K)
37. (G)
38. (A)
39. (B)
40. (I)
41. (J)
42. (D)
43. (C)
44. (F)
45. (E)

**Muscles of the Head and Deep Face**

46. (A)
47. (B)
48. (H)
49. (J)
50. (C)
51. (G)
52. (F)
53. (D)
54. (E)
55. (I)

**Lateral Aspect of the Skull**

56. (T)
57. (H)
58. (C)
59. (F)
60. (L)
61. (P)
62. (M)
63. (Q)
64. (O)
65. (E)
66. (N)
67. (A)

68. (K)
69. (D)
70. (I)
71. (J)
72. (S)
73. (R)
74. (G)
75. (B)

**The Mandible, External and Internal**

76. (J)
77. (C)
78. (A)
79. (G)
80. (I)
81. (M)
82. (H)
83. (N)
84. (O)
85. (G)
86. (L)
87. (F)
88. (P)
89. (E)
90. (Q)
91. (B)
92. (K)
93. (D)

**Ophthalmic Artery**

94. (H)
95. (K)
96. (I)
97. (B)
98. (M)
99. (F)
100. (G)
101. (L)
102. (D)
103. (E)
104. (A)
105. (J)
106. (C)

**Sagittal Section Through the Male Pelvis**

107. (A)
108. (G)
109. (B)
110. (C)
111. (D)
112. (I)
113. (J)
114. (H)
115. (F)
116. (E)

**Midsagittal Section Through the Female Pelvis**

117. (I)
118. (H)
119. (D)
120. (E)
121. (F)
122. (G)
123. (J)
124. (A)
125. (B)
126. (C)

**Boundaries and Subdivisions of the Perineum**

127. (C)
128. (J)
129. (B)
130. (H)
131. (E)
132. (K)
133. (I)
134. (F)

135. (L)          137. (A)          224. (D)          226. (G)
136. (D)          138. (E)          225. (E)

## Gluteal Region: Ligaments and Bony Parts
139. (J)          147. (G)
140. (F)          148. (C)
141. (K)          149. (L)
142. (O)          150. (M)
143. (N)          151. (H)
144. (D)          152. (A)
145. (I)          153. (E)
146. (B)

## A Relationship of the Pancreas to the Duodenum
227. (A)          234. (D)
228. (C)          235. (E)
229. (B)          236. (G)
230. (L)          237. (H)
231. (M)          238. (I)
232. (J)          239. (K)
233. (N)          240. (F)

## Right Hip Bone
154. (A)          161. (I)
155. (C)          162. (K)
156. (E)          163. (M)
157. (F)          164. (N)
158. (D)          165. (L)
159. (B)          166. (J)
160. (G)          167. (H)

## Position and Structure of the Large Intestine
241. (E)          247. (H)
242. (A)          248. (I)
243. (B)          249. (J)
244. (C)          250. (K)
245. (D)          251. (L)
246. (F)          252. (G)

## Anteromedial Aspect of the Arm
168. (B)          173. (D)
169. (J)          174. (F)
170. (I)          175. (E)
171. (C)          176. (H)
172. (A)          177. (G)

## Posterior Abdominal Wall
253. (E)          259. (K)
254. (F)          260. (A)
255. (G)          261. (B)
256. (H)          262. (C)
257. (I)          263. (D)
258. (J)

## Posterior Aspect of the Thigh
178. (B)          183. (J)
179. (D)          184. (C)
180. (E)          185. (G)
181. (F)          186. (H)
182. (A)          187. (I)

## Celiac Artery Distribution
264. (J)          270. (E)
265. (F)          271. (B)
266. (I)          272. (K)
267. (H)          273. (D)
268. (C)          274. (G)
269. (A)

## Posterior Aspect, Muscles of the Leg
188. (I)          197. (B)
189. (R)          198. (P)
190. (N)          199. (A)
191. (O)          200. (D)
192. (C)          201. (G)
193. (M)          202. (E)
194. (L)          203. (J)
195. (F)          204. (K)
196. (H)          205. (Q)

## Abdominal Musculature
275. (A)          279. (D)
276. (E)          280. (C)
277. (H)          281. (G)
278. (F)          282. (B)

## Anatomic Position of Esophagus and the Stomach
206. (A)          212. (C)
207. (B)          213. (H)
208. (D)          214. (I)
209. (G)          215. (J)
210. (L)          216. (K)
211. (F)          217. (E)

## Anterior View of the Lungs and Pericardium
283. (A)          288. (E)
284. (B)          289. (F)
285. (H)          290. (C)
286. (D)          291. (J)
287. (I)          292. (G)

## Diaphragm View From Below
293. (A)          297. (G)
294. (D)          298. (B)
295. (E)          299. (C)
296. (F)          300. (H)

## The Liver
218. (A)          221. (F)
219. (H)          222. (B)
220. (I)          223. (C)

## Ventral View of the Heart
301. (M)          304. (E)
302. (I)          305. (C)
303. (G)          306. (J)

307. (A)
308. (L)
309. (B)
310. (H)

311. (F)
312. (D)
313. (K)

### Dorsal View of the Heart
314. (A)
315. (D)
316. (G)
317. (E)

318. (B)
319. (C)
320. (F)

### Semidiagrammatic Representation of the More Anterior Structures of the Posterior Mediastinum
321. (E)
322. (F)
323. (L)
324. (K)
325. (D)
326. (H)

327. (J)
328. (C)
329. (I)
330. (B)
331. (A)
332. (G)

### Esophagus, Trachea, and Aorta: Thoracic Parts
333. (I)
334. (C)
335. (K)
336. (O)
337. (P)

338. (D)
339. (F)
340. (J)
341. (I)
342. (H)

343. (B)
344. (L)
345. (N)
346. (M)

347. (E)
348. (G)
349. (A)

### Vertebra, from Above and Side View
350. (C)
351. (J)
352. (A)
353. (A)
354. (I)
355. (G)
356. (C)
357. (G)

358. (D)
359. (E)
360. (H)
361. (B)
362. (E)
363. (F)
364. (I)
365. (D)

### Thoracic Musculature
366. (B)
367. (C)
368. (E)

369. (A)
370. (D)

### Musculature of the Back
371. (B)
372. (D)
373. (C)
374. (E)
375. (G)
376. (A)

377. (F)
378. (A)
379. (D)
380. (E)
381. (B)
382. (C)

# CHAPTER 10

# Practice Test

**Carefully read the following instructions before taking the Practice Test.**

1. This examination consists of 116 questions, covering the subject areas listed in the Table of Contents.
2. The Practice Test simulates an actual examination in question types and integration of subject areas.
3. You should set aside 1 hour and 40 minutes of *uninterrupted,* distraction-free time to take the Practice Test. This averages out to 50 seconds per question.
4. Be sure you have a clock (to time and pace yourself) and an adequate number of No. 2 pencils and erasers.
5. You should tear out and use the answer sheet that is provided on page 209.
6. Be sure to answer all of the questions, and be sure the number on the answer sheet corresponds to the question number in the Practice Test.
7. Use any remaining time to review your answers.
8. After completing the Practice Test, you can check all of your answers on pages 199 to 207. A score of 75% or higher should be considered as a passing score (87 correct answers).
9. After checking your answers and your score, you can analyze your strengths and weaknesses on the Practice Test Subspecialty List on page 208. To do this, you should check off your incorrect Practice Test answers on the Subspecialty List. You may find a pattern developing. For example, you may find you do well on the thorax but poorly on the back. In such an instance, you can go back and review the thorax section of this book and supplement your review with your texts and with the references cited in that section.

# Questions

1. Which of the following is a structural characteristic of the left atrium of the heart?

   (A) crista terminalis
   (B) thick walls
   (C) valvule of the foramen ovale
   (D) aortic valve
   (E) tricuspid valve

2. All of these statements describe the left ventricle of the heart EXCEPT

   (A) it lies in front of the left atrium
   (B) it shows fine trabeculae carneae
   (C) it holds the pulmonary orifice
   (D) the interventricular septum forms its anterior wall
   (E) it has chordae tendinae attaching to papillary muscles

3. Which of the following is a correct description of cardiac veins?

   (A) the great cardiac vein runs in the posterior interventricular sulcus
   (B) the great cardiac vein terminates in the coronary sinus
   (C) anterior cardiac veins empty directly into the left atrium
   (D) the coronary sinus empties into the left atrium
   (E) the coronary sinus occupies the anterior interventricular sulcus

4. All the following statements are correct about the cardiac nerves EXCEPT

   (A) they are responsible for initiating heart beat
   (B) they reach the heart along coronary arteries
   (C) they are found in the cardiac plexus
   (D) they include parasympathetic nerves
   (E) they modify rate and strength of heart beat

5. All of these statements are true about parasympathetic innervation of the heart EXCEPT

   (A) it is provided by the vagus nerve
   (B) efferent fibers entering the cardiac plexus are preganglionic axons
   (C) cardiac ganglia are found in the walls of the atria
   (D) it accelerates the heart rate
   (E) it is concerned with cardiac reflexes

6. Which of the following is true regarding sympathetic innervation of the heart?

   (A) efferents entering the cardiac plexus are preganglionic axons
   (B) it decreases the rapidity and strength of the heart beat
   (C) it causes constriction of coronary vessels
   (D) their afferents are the only source of cardiac reflexes
   (E) sympathetic cardiac afferents are the sole conductors of pain from the heart

7. Which of these structures is NOT a part of the conducting system of the heart?

   (A) sinoatrial node
   (B) atrioventricular node
   (C) internodal fasciculi
   (D) glomus aorticum
   (E) atrioventricular bundle

8. Great vessels directly entering or leaving the heart include all the following EXCEPT the

   (A) left subclavian artery
   (B) ascending aorta
   (C) pulmonary trunk
   (D) two venae cavae
   (E) pulmonary veins

9. Which of these statements is correct regarding the pulmonary trunk?

   (A) anatomically it is a vein
   (B) it begins at the aortic orifice
   (C) it conveys venous blood from the heart
   (D) it terminates in the right lung
   (E) it has no major divisions

10. In relation to the normal positions of the heart, all the following are true EXCEPT

   (A) it rests with its inferior surface on the central portion of the diaphragm
   (B) its position does not vary with body position
   (C) its position may vary according to body build
   (D) one third of its mass lies to the right of the median plane
   (E) position of the inferior border roughly corresponds to the level of the xiphisternal junction

11. The primary abnormality found in tetralogy of Fallot is

   (A) stenosis of the right ventricular outflow
   (B) ventricular septal defect
   (C) over-riding aorta
   (D) right ventricular hypertrophy
   (E) tricuspid atresia

12. Of the following structures, the only one confined to the mediastinum is the

   (A) esophagus
   (B) trachea
   (C) pericardial sac
   (D) aorta
   (E) brachiocephalic artery

13. All of these statements are correct about the thymus gland EXCEPT

   (A) it is the most superficial structure in the superior mediastinum
   (B) it is a primary lymphoid organ
   (C) it generates potentially immunocompetent cells
   (D) it functions until old age
   (E) it consists of the two lobes

14. All of the statements about the arch of the aorta are true EXCEPT

   (A) it gives rise to the left common carotid artery
   (B) it gives rise to the brachiocephalic artery
   (C) it connects with the ligamentum arteriosum
   (D) it gives rise to the left and right subclavian arteries
   (E) it incorporates the embryonic fourth aortic arch of the left side

15. The azygos vein empties into which of the following?

   (A) superior vena cava
   (B) inferior vena cava
   (C) right brachiocephalic vein
   (D) left brachiocephalic vein
   (E) left superior intercostal vein

16. Recurrent laryngeal nerves originate from the

   (A) phrenic nerves
   (B) vagus nerves
   (C) sympathetic trunks
   (D) cardiac plexus
   (E) pulmonary plexus

17. Which of the following is NOT a place along the esophagus where foreign bodies are prone to lodge?

   (A) at the beginning of the esophagus in the neck
   (B) at the region of esophageal contact with the aortic arch
   (C) where the esophagus is crossed by the left bronchus
   (D) at the esophageal hiatus of the diaphragm
   (E) at the lower esophageal sphincter

18. All of these branches are given off by the descending thoracic aorta EXCEPT the

   (A) bronchial arteries
   (B) esophageal arteries
   (C) posterior intercostal arteries
   (D) supreme intercostal arteries
   (E) subcostal arteries

19. Which area of the digestive system shows little fat in the mesentery and "windows" of translucency between the blood vessels of the mesentery?

   (A) duodenum
   (B) cecum
   (C) jejunum
   (D) ileum
   (E) transverse colon

20. Which of the following arteries should be avoided when a surgical passage is made through the transverse mesocolon?

   (A) right gastroepiploic
   (B) left gastroepiploic
   (C) gastroduodenal
   (D) right colic
   (E) middle colic

21. Which of the following intestinal segments is most likely to be involved in volvulus?

    (A) duodenum
    (B) ascending colon
    (C) descending colon
    (D) rectum
    (E) sigmoid

22. Diverticulosis is fairly common in which of the intestinal segments?

    (A) duodenum
    (B) jejunum
    (C) ileum
    (D) cecum
    (E) sigmoid

23. Which of the following statements correctly applies to the horseshoe kidney?

    (A) it is usually ectopic
    (B) it is usually fused at the superior poles
    (C) it is rotated
    (D) it has only one ureter
    (E) it has only one renal artery

24. The commonest congenital diaphragmatic defect is which of the following?

    (A) a persisting pleuroperitoneal canal
    (B) an esophageal hiatal hernia
    (C) an aortic hiatal hernia
    (D) a caval hiatal hernia
    (E) a muscular deficiency in the vertebrocostal trigone

25. Which of the following structures is retroperitoneal?

    (A) cecum
    (B) transverse colon
    (C) sigmoid colon
    (D) splenic flexure
    (E) descending colon

26. Which of the following statements correctly applies to the inferior mesenteric artery?

    (A) it arises from the aorta just below the third portion of the duodenum
    (B) it arises at the level of the fifth lumbar vertebra
    (C) it supplies the right one third of the transverse colon
    (D) it gives rise to the right colic artery
    (E) it provides branches to the ileocecal junction

27. The superior rectal artery is a branch of which of the following arteries?

    (A) right colic
    (B) ileocolic
    (C) left colic
    (D) inferior mesenteric
    (E) sigmoidal

28. The parasympathetic supply of the distal colon and of the pelvic viscera arises from which of the following?

    (A) lumbar nerves 2, 3, and 4
    (B) lumbar nerves 4 and 5
    (C) sacral nerves 2, 3, and 4
    (D) sacral nerves 4 and 5
    (E) lumbar nerves 4 and 5 and sacral nerves 1, 2, and 3

29. The portal vein is the large collecting vein into which empties all the venous blood from the gastrointestinal tract except for which of the following veins?

    (A) inferior mesenteric
    (B) right gastric
    (C) superior mesenteric
    (D) inferior rectal
    (E) left gastroepiploic

30. The superior extremities of the kidneys reach the upper border of which of the following structures?

    (A) third lumbar vertebra
    (B) ninth thoracic vertebra
    (C) 12th thoracic vertebra
    (D) fifth lumbar vertebra
    (E) pancreas

31. The renal papilla empties into which of the following structures?

    (A) glomerulus
    (B) proximal convoluted tubule
    (C) straight tubule
    (D) distal convoluted tubule
    (E) minor calyx

32. The afferent arteriole is a branch of which of the following arteries?

    (A) interlobar
    (B) arcuate
    (C) interlobular
    (D) efferent
    (E) glomerulus

33. The arteries of the abdominal portion of the ureter are branches of the renal artery and which of the following arteries?

    (A) inferior mesenteric
    (B) testicular or ovarian
    (C) aorta
    (D) lumbar
    (E) obturator

34. The renal arteries arise from the aorta at the level of which of the following vertebrae?

    (A) the 12th thoracic
    (B) the second lumbar
    (C) the fifth lumbar
    (D) the third sacral
    (E) the fifth sacral

35. The left renal artery lies behind which of the following structures?

    (A) the inferior vena cava
    (B) the testicular or ovarian artery
    (C) the pancreas
    (D) the left ureter
    (E) the inferior mesenteric artery

36. The superior suprarenal arteries are derived from which of the following arteries?

    (A) aorta
    (B) renal
    (C) superior mesenteric
    (D) inferior mesenteric
    (E) inferior phrenic

37. In moderate expiration the diaphragm reaches as high as which of the following levels?

    (A) the jugular notch
    (B) the third intercostal space
    (C) the fifth rib on the right side
    (D) the xiphisternal junction
    (E) the second rib on the left side

38. Which of the following statements correctly applies to the right crura of the diaphragm?

    (A) it is smaller than the left crura
    (B) it usually splits to enclose the esophagus
    (C) it takes origin from the lower lumbar vertebra
    (D) it is located in the central tendon
    (E) it is innervated by the vagus nerve

39. Which of the following statements correctly applies to the vena caval foramen?

    (A) it is an opening in the central tendon
    (B) it occurs at the lower border of the 12th thoracic vertebra
    (C) it is bounded at the sides by the crura of the diaphragm
    (D) the anterior and posterior vagal trunks occupy this opening
    (E) it transmits the superior vena cava

40. Which of the following statements correctly applies to the lesser thoracic splanchnic nerve?

    (A) it terminates in the aorticorenal ganglion
    (B) it is a postganglionic nerve

    (C) it is a parasympathetic nerve
    (D) it arises from TS-9 or 10
    (E) it carries no visceral afferent fibers

41. Which of the following statements correctly applies to the femoral nerve?

    (A) it is the principal preaxial nerve of the lumbar plexus
    (B) it arises from the anterior branches of lumbar nerves 2, 3, and 4
    (C) it passes under the inguinal ligament
    (D) it is a small nerve that is usually absent
    (E) it passes across the superior pubic ramus

42. The left testicular vein empties into which of the following veins?

    (A) inferior vena cava
    (B) left renal
    (C) left internal iliac
    (D) inferior mesenteric
    (E) portal

43. Which of the following muscles is innervated by the first cervical nerve?

    (A) stylohyoid
    (B) mylohyoid
    (C) hyoglossus
    (D) geniohyoid
    (E) anterior belly of the digastric

44. Which of the following structures is developed from the mesoderm of the second pharyngeal arch?

    (A) cricoid cartilage
    (B) styloid process
    (C) tongue
    (D) mandible
    (E) thyroid cartilage

45. The submandibular duct has an intimate relation to which of the following nerves?

    (A) hypoglossal
    (B) first cervical
    (C) marginal mandibular branch of the facial
    (D) mental
    (E) lingual

46. Which of the following nerves conveys the special sense of taste from the anterior two thirds of the tongue and parasympathetic fibers associated with both the pterygopalatine and submandibular ganglia?

    (A) facial
    (B) glossopharyngeal
    (C) vagus
    (D) oculomotor
    (E) great auricular

47. Which of the following nerves leaves the petrous portion of the temporal bone through the petrotympanic fissure?

(A) lingual
(B) buccal
(C) hypoglossal
(D) chorda tympani
(E) inferior alveolar

48. The submandibular ganglion contains which of the following types of cell bodies?

(A) sensory
(B) preganglionic sympathetic
(C) postganglionic parasympathetic
(D) preganglionic parasympathetic
(E) postganglionic sympathetic

49. The sympathetic innervation to the sublingual gland is associated with which of the following structures?

(A) vagus nerve
(B) facial nerve
(C) submandibular ganglion
(D) lingual nerve
(E) plexus along the facial artery

50. Which of the following muscles helps form the floor of the posterior cervical triangle?

(A) platysma
(B) sternocleidomastoid
(C) trapezius
(D) middle scalene
(E) cricothyroid

51. Which of the following muscles draws the mastoid process down toward the shoulder of the same side and thus turns the chin upward and to the opposite side?

(A) mylohyoid
(B) sternocleidomastoid
(C) trapezius
(D) omohyoid
(E) sternohyoid

52. The cranial portion of the accessory nerve unites with which of the following?

(A) vagus nerve
(B) hypoglossal nerve
(C) superior cervical ganglion
(D) glossopharyngeal nerve
(E) ansa cervicalis

53. The spinal portion of the accessory nerve enters the cranial cavity through which of the following foramina?

(A) jugular
(B) hypoglossal
(C) foramen magnum
(D) stylomastoid
(E) foramen spinosum

54. Which of the following nerves contributes to the formation of the subtrapezial plexus?

(A) brachial plexus
(B) accessory nerve
(C) facial
(D) glossopharyngeal
(E) trigeminal

55. Which of the following nerves provides the innervation of the infrahyoid muscles?

(A) lesser occipital
(B) subtrapezial plexus
(C) ansa cervicalis
(D) great auricular
(E) transverse cervical

56. The phrenic nerve arises from which of the following cervical nerves?

(A) second
(B) fourth
(C) sixth
(D) first
(E) seventh

57. The phrenic nerve makes its appearance at the lateral border of which of the following muscles?

(A) sternocleidomastoid
(B) rectus capitis anterior
(C) longus colli
(D) rectus capitis lateralis
(E) anterior scalene

58. The brachial plexus begins to form as the ventral rami pass between which of the following muscles?

(A) posterior and middle scalenes
(B) anterior and middle scalenes
(C) longus capitis and longus colli
(D) rectus capitis anterior and rectus capitis lateralis
(E) omohyoid and the sternohyoid

59. The vertebral artery arises from which of the following?

(A) brachiocephalic artery
(B) arch of the aorta
(C) axillary artery
(D) subclavian artery
(E) highest intercostal artery

60. Which of the following arteries usually arises from the second or third part of the subclavian artery?

    (A) internal thoracic
    (B) dorsal scapular
    (C) vertebral
    (D) thyrocervical
    (E) transverse cervical

**DIRECTIONS (Questions 61 through 72): Each group of items in this section consists of lettered headings followed by a set of numbered words or phrases. For each numbered word or phrase, select the ONE lettered heading that is most closely associated with it. Each lettered heading may be selected once, more than once, or not at all.**

**Questions 61 through 64**

    (A) structure giving rise to oocytes
    (B) structure attaching the ovary to the uterus
    (C) structure attaching the ovary to the broad ligament
    (D) structure attaching the uterine tube to the ovary
    (E) endocrine structures developed from ovarian follicles that expel their oocytes

61. corpora lutea

62. mesovarium

63. ligament of the ovary

64. ovarian fimbria

**Questions 65 through 68**

    (A) pelvic diaphragm
    (B) visceral pelvic fascia
    (C) deep perineal space
    (D) perineal membrane
    (E) central perineal tendon

65. levator ani

66. coccygeus

67. puborectalis

68. bulbourethral glands

**Questions 69 through 72**

    (A) coracoacromial ligament
    (B) annular ligament
    (C) coracohumeral ligament
    (D) coracoclavicular ligament
    (E) acromioclavicular ligament

69. prevents the scapula from being driven medially

70. prevents upward displacement of the humerus

71. prevents downward displacement of the adducted humerus

72. resists independent upward movement of the clavicle

**DIRECTIONS (Questions 73 through 80): Each group of items in this section consists of lettered headings followed by a set of numbered words or phrases. For each numbered word or phrase, select**

    A if the item is associated with (A) only,
    B if the item is associated with (B) only,
    C if the item is associated with both (A) and (B),
    D if the item is associated with neither (A) nor (B).

**Questions 73 through 76**

    (A) spina bifida occulta
    (B) spondylolisthesis
    (C) both
    (D) neither

73. low back pain

74. failure of halves of the vertebral arch to undergo osseous fusion

75. a defect in the vertebral arch between the superior and inferior facets

76. displacement of the anterior piece of L5 vertebra

**Questions 77 through 80**

    (A) anal region
    (B) urogenital diaphragm
    (C) both
    (D) neither

77. ischiorectal fossa

78. perineal body

79. pubic arch

80. iliolumbar artery

**DIRECTIONS (Questions 81 through 116):** For each of the items in this section, ONE or MORE of the numbered options is correct. Choose the answer

A if only 1, 2, and 3 are correct,
B if only 1 and 3 are correct,
C if only 2 and 4 are correct,
D if only 4 is correct,
E if all are correct.

81. Which of the following statements is true of the suprascapular nerve?

    (1) it crosses under the scapular (suprascapular) ligament
    (2) it passes from the supraspinous fossa to the infraspinous fossa through the spinoglenoid notch
    (3) it innervates the infraspinatus muscle
    (4) it has no cutaneous branches

82. An injury to the long thoracic nerve would usually result in

    (1) paralysis of the trapezius
    (2) projection of the inferior angle of the scapula on attempts to raise the arm forward
    (3) inability to adduct the scapula
    (4) paralysis of the serratus anterior

83. The axillary artery has which of the following characteristics?

    (1) it gives rise to the suprascapular artery
    (2) it is a direct continuation of the subclavian artery
    (3) it gives rise to the cephalic artery
    (4) it gives rise to the thoracoacromial artery

84. The brachial plexus is formed by

    (1) five ventral nerve rami uniting to form three trunks
    (2) three trunks bifurcating to form six divisions
    (3) divisions uniting to form three cords
    (4) cords dividing to end in nine terminal nerves

85. Which of the following statements provides correct information on the scheme of the brachial plexus?

    (1) the lateral cord becomes the musculocutaneous nerve
    (2) the median nerve is formed by nerve fibers from both lateral and medial cords
    (3) the axillary nerve is a branch of the posterior cord
    (4) the ulnar nerve innervates no muscles in the arm

86. Distribution of fibers from cords of the brachial plexus is correctly described by which of the following?

    (1) the posterior cord provides innervation to the subscapularis muscle through the subscapular nerve
    (2) the medial cord provides innervation to the deltoid muscle through the radial nerve
    (3) the medial cord provides innervation to the flexor digitorum profundus muscle through the ulnar nerve
    (4) the lateral cord provides innervation to the biceps muscle through the median nerve

87. The lesser tubercle of the upper end of the humerus receives muscle insertion of the

    (1) supraspinatus
    (2) infraspinatus
    (3) teres minor
    (4) subscapularis

88. The triceps brachii can be described correctly by which of the following statements?

    (1) it has a long head originating from the infraglenoid tubercle of the scapula
    (2) it inserts on the olecranon process of the ulna
    (3) it has a lateral head that is largely tendinous
    (4) it is supplied by the axillary nerve

89. Which of the following structures normally enters the gluteal region below the piriformis muscle?

    (1) sciatic nerve
    (2) inferior gluteal arteries and nerves
    (3) posterior cutaneous nerve of the thigh
    (4) superior gluteal vessels and nerve

90. Which of the following structures attach to the ischial tuberosity?

    (1) short head of the biceps femoris
    (2) obturator internus
    (3) gluteus maximus
    (4) sacrotuberous ligament

91. The great (long) saphenous vein can be described correctly by which of the following statements?

    (1) at the knee it passes just lateral to the patella
    (2) it begins on the medial end of the dorsal venous arch of the foot
    (3) it has no anastomoses with the short saphenous vein
    (4) it joins the femoral vein inferolateral to the pubic tubercle

92. Which of these statements is true about fractures of the upper end of the femur?

    (1) the usual injury is fracture of the head of the femur
    (2) they are common in persons over 60 years of age

(3) the blood supply usually is not interrupted
(4) such fractures are commoner in women than in men

93. Which of the following are true statements relating to veins of the lower limb?

(1) reverse flow is the result of incompetent valves
(2) saphenous vein grafts are used for bypass operations
(3) the great saphenous vein lies immediately anterior to the medial malleolus
(4) varicose veins are common in the posterior and medial part of the lower limb

94. Cutaneous nerves that supply skin of the thigh include the

(1) ilioinguinal
(2) lateral femoral cutaneous
(3) genitofemoral
(4) femoral

95. The iliotibial tract has attachment to the

(1) greater trochanter of the femur
(2) head of the fibula
(3) ischial spine
(4) tubercle of the iliac crest

96. Paralysis of the quadriceps muscle of one limb would result in

(1) complete loss of voluntary extension of the knee
(2) loss of the ability of the patient to stand unassisted
(3) some weakness of the ipsilateral hip flexion
(4) no loss of muscle bulk

97. The oculomotor nerve innervates which of the following muscles?

(1) the levator palpebrae superioris
(2) the lateral rectus
(3) the superior rectus
(4) the superior oblique

98. In addition to supplying most of the ocular muscles, the occulomotor nerve provides parasympathetic fibers to which of the following?

(1) the sphincter pupillae muscle
(2) the lacrimal gland
(3) the ciliary muscle of accommodation
(4) the nasal gland

99. Which of the following statements apply to the trochlear nerve?

(1) it is the smallest of the cranial nerves
(2) it contains somatic efferent fibers

(3) it is the only cranial nerve that emerges from the dorsal aspect of the brain stem
(4) it pierces the dura mater in the free border of the tentorium cerebelli just behind the posterior clinoid process

100. Which of the following statements apply to the ophthalmic division of the trigeminal nerve?

(1) it is a mixed nerve
(2) it is entirely sensory
(3) it arises from the geniculate ganglion
(4) it passes forward in the lateral wall of the cavernous sinus

101. Which of the following statements apply to the frontal nerve?

(1) it enters the orbit through the inferior orbital fissure
(2) it passes forward between the levator palpebrae superioris muscle and the periorbita
(3) it divides into the lacrimal and nasociliary nerves
(4) it is the largest branch of the ophthalmic nerve

102. Which of the following statements apply to the ophthalmic artery?

(1) it is a branch of the internal carotid artery
(2) it passes through the optic canal
(3) it terminates at the medial angle of the eye as the supratrochlear and dorsal nasal branches
(4) it is a terminal branch of the maxillary artery

103. Between the pubic tubercle and the anterior superior iliac spine is stretched

(1) the inguinal ligament
(2) the fundiform ligament
(3) the rolled-under inferior margin of the aponeurosis of the external abdominal oblique muscle
(4) the tunica dartos

104. The lineae transversae are usually located at the levels of which of the following structures?

(1) the jugular notch
(2) the umbilicus
(3) the symphysis pubis
(4) a little below the xiphoid process

105. Which of the following arteries arise from the femoral artery?

(1) the superficial epigastric
(2) the superficial circumflex
(3) the superficial external pudendal
(4) the inferior epigastric

| SUMMARY OF DIRECTIONS |
|---|

| A | B | C | D | E |
|---|---|---|---|---|
| 1, 2, 3 only | 1, 3 only | 2, 4 only | 4 only | All are correct |

106. The deep external pudendal artery may arise from which of the following arteries?

(1) the inferior epigastric
(2) the femoral
(3) the superior epigastric
(4) the medial circumflex femoral

107. Which of the following statements correctly apply to the superficial inguinal nodes?

(1) these nodes are arranged in the form of a T in the subcutaneous tissue of the groin
(2) these nodes form a chain parallel to the inguinal ligament
(3) these nodes are located about 1 cm below the inguinal ligament
(4) these nodes receive lymph from the lower abdominal wall

108. Which of the following statements describes the superior vena cava correctly?

(1) it enters the right atrium
(2) it is formed by union of the brachiocephalic veins
(3) it returns blood from all structures above the diaphragm except the lungs
(4) it ends at the level of the first right costal cartilage

109. Which of these items is true regarding the aortic arch?

(1) it begins posterior to the right half of the sternal angle
(2) it passes to the left of the trachea
(3) it casts a shadow called the aortic knob
(4) it has the ligamentum arteriosum on its inferior concave surface

110. Branches of the aortic arch include the

(1) brachiocephalic trunk
(2) right subclavian artery
(3) left common carotid
(4) right common carotid

111. Which of these items correctly describes the vagus nerves?

(1) they are part of the sympathetic system
(2) they control action of the diaphragm
(3) they cause dilation of coronary arteries
(4) they contribute to the pulmonary plexus of nerves

112. Which of the following correctly describes the thoracic duct?

(1) it is the main lymphatic duct
(2) it drains the cisterna chyli
(3) it empties into the venous system at the junction of the left internal jugular and subclavian veins
(4) it passes through the diaphragm via the esophageal hiatus

113. Which of the following is true regarding the azygos venous system?

(1) the azygos vein receives the thoracic duct
(2) these veins are the continuation of the ascending lumbar veins
(3) the hemiazygos vein empties into the inferior vena cava
(4) the azygos vein arches over the root of the right lung

114. The anterior mediastinum contains the

(1) sternopericardial ligaments
(2) esophagus
(3) branches of the internal thoracic artery
(4) heart

115. Correct description of costovertebral joints includes which of the following?

(1) during respiration, they permit movement of ribs and sternum
(2) typically the head of a rib articulates with the bodies of two vertebrae
(3) they are synovial joints allowing gliding movement
(4) the heads of the last three ribs articulate only with their own vertebral bodies

116. In a dissection of the left side of the mediastinum, which of these structures can be seen?

(1) common carotid
(2) brachiocephalic trunk
(3) subclavian artery
(4) azygos vein

# Answers and Explanations

1. **(C)** The interatrial septum forms part of the anterior wall of the left atrium. The thin area on the septum is the valvule of the foramen ovale, visible in the right atrium as the floor of the fossa ovalis. In the fetus this structure functioned as a valve for blood coming into the right atrium; the blood could push the valve aside and enter the left atrium. Blood attempting to pass in the opposite direction would force the valvule against the rigid rim of the foramen and prevent passage in that direction. *(Hollinshead, pp 528–531)*

2. **(C)** The pulmonary orifice is a structure of the right, not the left, ventricle. The conus arteriosus, or infundibulum, leads into the pulmonary trunk through the pulmonary orifice. The other statements are correct regarding the left ventricle. *(Hollinshead, pp 528–531)*

3. **(B)** The great cardiac vein terminates in the coronary sinus, a large vein that empties into the right atrium. The coronary sinus occupies the coronary sulcus located on the posterior heart, not in the anterior interventricular sulcus. The great cardiac vein becomes continuous with the coronary sinus at the point where the oblique vein of the left atrium enters it. *(Hollinshead, pp 536–537)*

4. **(A)** All of the statements about cardiac nerves are true except that they initiate the heart beat. The atria and ventricles contract in orderly sequence before nerves contact the embryonic heart, and the same is true when the transplanted heart is severed from its nervous connections. These nerves reach the heart along the coronary arteries and serve to modify the rate and strength of heart beat; they are not responsible for maintaining it. *(Hollinshead, pp 537–538)*

5. **(D)** Parasympathetic input to the cardiac plexus is provided by the two vagi. Vagal stimulation slows the heart beat and constricts the coronary arteries. The vagal afferents are concerned with cardiac reflexes. *(Hollinshead, p 538)*

6. **(E)** Sympathetic cardiac afferents are the sole conductors of pain from the heart. Vagal afferents are concerned chiefly with cardiac reflexes, but afferents in the sympathetic nerves also may be involved in reflexes. Sympathetic efferents increase the rapidity and strength of the heart beat and dilate the coronary arteries. The sympathetic efferents that enter the cardiac plexus are postganglionic, their axons having relayed in the ganglia of the sympathetic trunk from which the cardiac nerves issue. *(Hollinshead, pp 538–539)*

7. **(D)** All of the items about the conducting system of the heart are correct except that of the glomus aorticum. These structures, also called the aortic or para-aortic bodies, are chemoreceptors located on the outside of the great vessels, primarily between the ascending aorta and the pulmonary trunk; they initiate respiratory and cardiac reflexes in response to lowered oxygen tension in aortic blood. All of the components listed as parts of the conducting system of the heart are specially differentiated cardiac muscle fibers separated from ordinary myocardium by delicate envelopes of connective tissue. *(Hollinshead, pp 539–541)*

8. **(A)** The left subclavian artery leaves the arch of the aorta; it does not leave the heart directly. All of the other vessels listed are great vessels transporting blood into the heart from the body (superior and inferior venae cavae) or into the heart from the lungs (four pulmonary veins). The pulmonary trunk carries venous blood from the heart to the lungs, and the ascending aorta carries oxygenated blood from the heart out to the body. *(Hollinshead, pp 84,541–542)*

9. **(C)** Although the pulmonary trunk conveys blood from the heart to the lungs, anatomically it is an artery (as are its branches, the pulmonary arteries). The pulmonary trunk begins at the pulmonary orifice of the right ventricle. It terminates by dividing into right and left pulmonary arteries that pass to each lung. *(Hollinshead, pp 541–542)*

10. **(B)** All of the answers are correct except B. The heart position shifts significantly with change of body position. For example, when a recumbent person turns from side to side, the heart shifts toward the side the person is lying on. The position of the heart also changes on deep inspiration or expiration. *(Hollinshead, p 542)*

11. **(A)** The primary abnormality of tetralogy of Fallot is a narrowing (stenosis) of the right ventricular outflow tract caused by an unequal division of the bulbus cordis of the heart. The other features of this congenital abnormality are ventricular septal defects (VSD), over-riding aorta, and right ventricular hypertrophy. This hypertrophy results because the right ventricle has to pump blood directly into the high-pressure systemic circulation through the aorta as well as into its own abnormally narrow outflow tract. *(Hollinshead, p 550)*

12. **(C)** The only one of the structures listed that is confined to the mediastinum is the pericardial sac and its contents. The sac and its contents comprise the middle mediastinum. Other structures named extend beyond the mediastinum. *(Hollinshead, p 550)*

13. **(D)** The thymus is the most superficial structure in the superior mediastinum. Its endocrine function is much outweighed by its importance in generation of potentially immunocompetent cells. Primarily it is a lymphoid organ. The thymus continues to grow up to the age of 5 to 6 years and involutes progressively thereafter. In the adult it is largely replaced by fat and connective tissue. *(Hollinshead, pp 556–557)*

14. **(D)** All the items are correct except D. The arch of the aorta gives off the brachiocephalic, left common carotid, and left subclavian arteries, in that order. The brachiocephalic artery divides into the right subclavian and right common carotid arteries. *(Hollinshead, p 561)*

15. **(A)** The superior vena cava receives the azygos vein. This vein arches forward over the root of the right lung and enters the vena cava from the back; it is the component of the azygos system located on the right side. *(Hollinshead, pp 563,565)*

16. **(B)** A recurrent laryngeal branch arises from each of the vagus nerves. These recurrent nerves ascend to supply the larynx. The left recurrent laryngeal nerve arises from the left vagus nerve at the lower border of the aortic arch, skirts the convexity of the arch at its junction with the ligamentum arteriosum, and then ascends on the medial side of the aortic arch. The right recurrent laryngeal nerve arises from the right vagus in the base of the neck and hooks around the right subclavian artery. *(Hollinshead, p 566)*

17. **(E)** All of the items listed are described as sites of narrowing of the esophagus except E. The lower esophageal sphincter cannot be demonstrated anatomically or histologically; however, there is physiological evidence for the existence of a sphincter mechanism around the lower portion of the esophagus. The function of this "physiological sphincter" is to prevent regurgitation of gastric contents into the esophagus. *(Hollinshead, pp 567–568)*

18. **(D)** All of the arteries listed are given off by the descending thoracic aorta, with the exception of supreme intercostal arteries. The posterior arteries of the first two intercostal spaces originate, as a rule, by a common stem, the supreme intercostal artery, a branch of the costocervical branch of the subclavian artery, not from the aorta. *(Hollinshead, p 570)*

19. **(C)** Little fat exists in the mesentery of the upper jejunum, and "windows" of translucency between the blood vessels of the mesentery are numerous. Such windows become progressively less clear in the ileum. *(Woodburne, p 429)*

20. **(E)** The middle colic artery must be carefully avoided when a surgical passage is being made through the transverse mesocolon for short-circuiting operations such as gastroenterostomy. *(Woodburne, p 431)*

21. **(E)** A long, mobile sigmoid segment with a long mesentery creates a situation susceptible to a twisting of the sigmoid colon on its mesentery (volvulus).

22. **(E)** Diverticulosis is fairly common in the sigmoid colon. This is a herniation of the mucous membrane lining through the circular layer of muscle between teniae coli. Weakness of the wall where the blood vessels penetrate is thought to be a predisposing cause. *(Woodburne, p 437)*

23. **(A)** The horseshoe kidney is usually ectopic in position. It is characterized by a fusion across the midline of the lower poles of the two primordia. They are unrotated, with their hila directed forward. *(Woodburne, p 447)*

24. **(A)** The commonest congenital diaphragmatic defect is a persisting pleuroperitoneal canal located at the extremity of the left of the central tendon. Varying amounts of abdominal viscera may herniate through this route. *(Woodburne, p 450)*

25. **(E)** Both the ascending and descending colon are retroperitoneally located along the posterior abdominal wall. *(Woodburne, p 436)*

26. **(A)** The inferior mesenteric artery arises from the aorta about 3 to 4 cm above its bifurcation. The

origin of the inferior mesenteric artery is just below the third portion of the duodenum and at the level of the third lumbar vertebra. It supplies the left one third of the transverse colon. *(Woodburne, p 437)*

27. **(D)** The superior rectal artery is the continuation of the inferior mesenteric artery. Crossing the left common iliac vessels, it descends into the pelvis between the layers of the sigmoid mesocolon. *(Woodburne, p 438)*

28. **(C)** The parasympathetic supply of the distal colon and of the pelvic viscera arises in sacral segments of the spinal cord. From sacral nerves 2, 3, and 4, the pelvic splanchnic nerves pass forward on either side of the rectum to the inferior hypogastric plexus. *(Woodburne, p 439)*

29. **(D)** The portal vein is the large collecting vein into which empties all the venous blood from the gastrointestinal tract (except the lower part of the anal canal), from the spleen, and from the pancreas. *(Woodburne, p 440)*

30. **(C)** The superior extremities of the kidneys reach the upper border of the body of the 12th thoracic vertebra; their inferior extremities lie at the level of the third lumbar vertebra. *(Woodburne, p 441)*

31. **(E)** The renal pyramids taper to form, by their apices, the eight to 12 renal papillae on which open the major collecting ducts of the kidney. The renal papillae are received into usually eight cup-shaped minor, which represent the beginning of the duct system of the kidney. *(Woodburne, p 444)*

32. **(C)** The glomerulus is a network of capillaries that arises from an afferent arteriole, a branch of an interlobular renal artery. *(Woodburne, p 442)*

33. **(B)** The arteries of the abdominal portion of the ureter are branches of the renal artery and of the testicular or ovarian artery. *(Woodburne, p 444)*

34. **(B)** The renal arteries arise, one on each side of the aorta, at the level of the upper border of the second lumbar vertebra. Their origin is about 1 cm below that of the superior mesenteric artery. *(Woodburne, p 446)*

35. **(C)** The left renal artery lies behind the left renal vein, the pancreas, and the splenic vein. *(Woodburne, p 446)*

36. **(E)** An average of from six to eight superior suprarenal arteries are derived from the inferior phrenic artery. *(Woodburne, p 448)*

37. **(C)** In moderate expiration the diaphragm reaches as high as the fifth rib on the right side and the fifth intercostal space on the left. *(Woodburne, p 449)*

38. **(B)** The right crus is larger and longer than the left crus. It takes origin from the upper three lumbar vertebrae, whereas the left crus arises from only the upper two. The right crus usually splits to enclose the esophagus as it pierces the diaphragm. *(Woodburne, p 450)*

39. **(A)** The vena caval foramen is an opening in the central tendon. It is the highest of the three openings and lies at the level of the eighth thoracic vertebra. It transmits the inferior vena cava together with branches of the right phrenic nerve. *(Woodburne, p 451)*

40. **(A)** The lesser thoracic splanchnic nerve (T10,11) terminates in the aorticorenal ganglion. This nerve contains preganglionic and visceral afferent fibers. *(Woodburne, pp 453–454)*

41. **(C)** The femoral nerve is the largest branch of the lumbar plexus. It is formed by the posterior branches of the second, third, and fourth lumbar nerves. It passes under the inguinal ligament. *(Woodburne, p 468)*

42. **(B)** The left testicular vein is longer than the right. It ends in the left renal vein. *(Woodburne, p 464)*

43. **(D)** The geniohyoid is innervated by a branch of the first cervical nerve, conducted to it by the hypoglossal nerve. The muscles of the two sides are often blended together. *(Woodburne, p 171)*

44. **(B)** The stylohyoid ligament, the styloid process, and the stapes of the middle ear arise from the second arch mesoderm. The lesser horn and the upper median part of the body also come from the second arch. *(Woodburne, p 172)*

45. **(E)** The submandibular duct has an intimate relation to the lingual nerve which crosses it twice in this intermuscular interval. It passes forward and medialward, lying at first on the hyoglossus and then on the genioglossus muscles, and opens at the summit of the sublingual caruncle. *(Woodburne, p 173)*

46. **(A)** Peripherally, the facial nerve is motor to the muscles of facial expression. Intracranially, it has a portion which is sensory and parasympathetic and is designated as the nervus intermedius. Included in this nerve are fibers which convey the special sense of taste from the anterior two-thirds of the tongue and parasympathetic fibers associated with both the pterygopalatine and submandibular ganglia. *(Woodburne, p 173)*

47. **(D)** The fibers to the tongue and the submandibular ganglion are collected into the chorda tympani branch of the facial nerve. This nerve leaves the petrous portion of the temporal bone through the petrotympanic fissure and joins the posterior aspect of the lingual nerve. *(Woodburne, p 173)*

48. **(C)** The submandibular ganglion is a peripheral parasympathetic ganglion concerned with secretory and afferent nerves of the sublingual and submandibular glands. *(Woodburne, p 174)*

49. **(E)** The parasympathetic innervation is conveyed by the chorda tympani, a branch of the facial and the sympathetic, by way of the plexus along the facial artery. *(Woodburne, p 174)*

50. **(D)** The floor of the posterior cervical triangle is formed by the middle and the posterior scalene, the levator scapulae, and the splenius muscles, covered by the prevertebral layer of cervical fascia. *(Woodburne, p 175)*

51. **(B)** The sternocleidomastoid muscle draws the mastoid process down toward the shoulder of the same side and thus turns the chin upward and to the opposite side. Acting together, the muscles project the head forward and the chin upward. *(Woodburne, p 175)*

52. **(A)** The cranial portion of the accessory nerve unites with the vagus to form, principally, its inferior laryngeal branch. *(Woodburne, p 175)*

53. **(C)** The nerve enters the cranial cavity through the foramen magnum. Leaving the cranium by the jugular foramen, the accessory nerve at first lies between the internal carotid artery and the internal jugular vein. *(Woodburne, p 175)*

54. **(B)** In the posterior cervical triangle the accessory nerve is joined by branches of the third and fourth cervical nerves. The junction of these nerves and their distribution deep to the trapezius is designated as the subtrapezial plexus. *(Woodburne, p 177)*

55. **(C)** The ansa cervicalis complex provides the innervation of the infrahyoid muscles and the geniohyoid. *(Woodburne, p 177)*

56. **(B)** The phrenic nerve arises by a large root from the fourth cervical nerve, reinforced by smaller contributions from the third and fifth nerves. *(Woodburne, p 178)*

57. **(E)** The phrenic nerve makes its appearance at the lateral border of the anterior scalene muscle, descends vertically over the ventral surface of this diverging muscle, and enters the chest along its medial border. *(Woodburne, p 178)*

58. **(B)** The brachial plexus begins to form as the ventral rami pass between the anterior and middle scalene muscles. In this position, gray rami communicates are added to the ventral rami of the fifth and sixth nerves from the middle cervical ganglion and to the ventral rami of the seventh and eighth cervical and first thoracic nerves from the cervicothoracic. *(Woodburne, p 178)*

59. **(D)** The vertebral artery arises from the dorsosuperior aspect of the ascending portion of the subclavian artery. The artery passes anterior to the ventral rami of the seventh and eight cervical nerves and enters, from below, the costotransverse foramen of the sixth cervical vertebra. *(Woodburne, p 179)*

60. **(B)** The dorsal scapular artery occurs in 70% of specimens. It is usually a branch of the second or the third part of the subclavian artery, from which parts it arises with equal frequency. *(Woodburne, p 181)*

61. **(E)**

62. **(C)**

63. **(B)**

64. **(D)**

61–64. Each ovary is attached to the posterosuperior aspect of the broad ligament and is suspended from the posterior layer of this ligament by a fold of peritoneum, the mesovarium; the ovarian vessels pass to and from the ovary through this structure. The ovary also is attached to the uterus by a band of fibrous tissue, the ligament of the ovary, which runs in the broad ligament. The ovarian fimbria of the infundibulum attaches the uterine tube to the ovary; hence during ovulation the oocyte is picked up and is carried into the ampulla of the uterine tube by the action of the cilia on the mucosa of the fimbriae. Before puberty the surface of the ovary is smooth, whereas after puberty it becomes progressively scarred and distorted. This is the result of successive corpora lutea degeneration. The corpora lutea are endocrine structures that develop from follicles that expel their oocytes. If the oocyte shed each month is not fertilized, the corpus luteum degenerates and gradually is replaced by a scar, the corpus albicans. *(Moore, p 391)*

65. **(A)**

66. **(A)**

67. **(A)**

**68. (C)**

**65–68.** The two levatores ani muscles and the two coccygeus muscles, with their superior and inferior investing fasciae, form the funnel-shaped pelvic diaphragm. The pelvic diaphragm forms the fibromuscular floor of the abdominopelvic cavity and supports the contents of the pelvis. The levator ani forms the largest and most important part of the clinically significant pelvic diaphragm. For descriptive purposes it is convenient to describe three parts of the levator ani: the pubococcygeus, the puborectalis, and the iliococcygeus. Of these, the pubococcygeus and the puborectalis are the most important. The pubococcygeus encircles the urethra, the vagina, and the anus; it merges into the central perineal tendon. The puborectalis muscle is the part of the levator ani that lies medial to, but at a lower level, than the pubococcygeus. The puborectalis arises from the pubis and passes backward; the muscles of the two sides loop around the posterior surface of the anorectal junction, forming a U-shaped rectal sling. The deep perineal space is the fascial space enclosed by the superior and inferior fasciae of the urogenital diaphragm. In the male it contains the sphincter urethrae, the deep transverse perineal muscles, and the bulbourethral glands. *(Moore, pp 308,349–351,392–393)*

**69. (D)**

**70. (A)**

**71. (C)**

**72. (D)**

**69–72.** The coracoclavicular ligament is the chief bracing ligament at the acromial end of the clavicle; it consists of two parts, the trapezoid ligament and the conoid ligament. These ligaments resist independent, upward movement of the clavicle or downward movement of the scapula. The ligament also prevents the scapula from being driven medially. The coracoacromial ligament is a strong band arising from the posterior edge of the coracoid process and passing upward and laterally above the shoulder joint to attach to the free end of the acromion. It helps to prevent upward displacement of the humerus. The coracohumeral ligament is a band that arises from the lateral edge of the coracoid process and extends over the top of the shoulder to attach to the greater tubercle; it prevents downward displacement of the adducted humerus. The acromioclavicular ligament is rather thin and lax to permit a slipping movement between the scapula and clavicle. The annular ligament is a band that forms about four fifths of a circle around the

head and upper part of the neck of the radius so that distal displacement of the head of the radius cannot occur easily. *(Basmajian, pp 394–395,397, 400; Hollinshead, pp 168–169,171,205,272)*

**73. (C)**

**74. (A)**

**75. (B)**

**76. (B)**

**73–76.** Spina bifida occulta of either L5 or S1 vertebra is present in about 10% of people with no back problems. Also it may occur anywhere along the vertebral column. This defect results from failure of the halves of the vertebral arch to grow enough to meet each other and to undergo synostosis (osseous union). Although most persons with spina bifida have no specific complaints, some have low back pain.

A defect in the vertebral arch between the superior and inferior facets is called spondylolysis. The posterior fragment, consisting of the spinous process, the lamina, and the inferior articular processes, remains in normal relation to the arch of the sacrum. The anterior fragment of this divided vertebra and the superimposed vertebra occasionally may move forward. The anterior displacement of most of the vertebral column is called spondylolisthesis. If the anterior part of the bone does not move forward, the condition is called spondylolysis. *(Moore, pp 404–405, 623, 638,660–661)*

**77. (C)**

**78. (C)**

**79. (B)**

**80. (D)**

**77–80.** The ischiorectal fossa is related to both the anal region and the urogenital diaphragm. This fossa is a large wedge-shaped space located on each side of the anal canal and rectum. Anteriorly the ischiorectal fossa continues superior to the urogenital diaphragm as the anterior recess of the ischiorectal fossa. This fossa is filled with loose areolar tissue and fat.

The central perineal tendon, or perineal body, is a fibromuscular node located at the center of the perineum between the anal canal and the bulb of the penis. The central perineal tendon is the landmark of the perineum. It gives attachment to the transverse perineal muscles, the bulbospongiosus, some fibers of the external anal sphincter, and the levatores ani muscles of both sides.

The urogenital diaphragm is a thin sheet of striated muscle stretching between the two sides of the pubic arch, which is formed by the converging ischiopubic rami. The urogenital diaphragm covers the anterior part of the inferior pelvic aperture. The most anterior and most posterior fibers of the urogenital diaphragm run transversely, whereas the middle fibers encircle the urethra (sphincter urethrae muscle).

The iliolumbar artery is a posterior branch of the internal iliac artery, particularly related to neither item (A) nor (B). This vessel runs superolaterally to the iliac fossa, passing anterior to the sacroiliac joint and posterior to the psoas major muscle, where it separates the obturator nerve from the lumbosacral trunk. In the iliac fossa, it divides into an iliac branch, supplying the iliacus muscle, and a lumbar branch, supplying the psoas major and the quadratus lumborum. *(Moore, pp 294,297,301–303, 364–365)*

81. (E) On the dorsum of the scapula the suprascapular nerve and vessels lie in contact with the bone. The scapular notch is bridged by the suprascapular ligament: the vessels cross the ligament, and the nerve passes below it. They pass from the supraspinous fossa to the infraspinous fossa through the spinoglenoid notch. The suprascapular nerve supplies the supraspinatus and infraspinatus muscles and sends twigs to the shoulder joint. It has no cutaneous branches. *(Basmajian, p 321)*

82. (C) An injury to the long thoracic nerve, from C5, C6 and C7 roots of the brachial plexus, results in paralysis of the serratus anterior muscle. This is the chief muscle that protracts, or pulls, the scapula forward, as in a pushing movement. If it is paralyzed, the inferior angle of the scapula projects when attempts are made to raise the arm forward. It abducts, rather than adducts, the scapula. The trapezius would not be involved. *(Basmajian, pp 332,334–335)*

83. (C) The subclavian artery becomes the axillary artery at the lower border of the first rib, and it continues through the axilla to become the brachial artery at the lower border of the teres major. One of its major branches is the thoracoacromial artery that gives off several branches to the shoulder region. The suprascapular artery is a branch of the subclavian artery via the thyrocervical trunk. There is no cephalic artery but a cephalic vein. *(Basmajian, pp 333–334,429,467–469)*

84. (A) The brachial plexus is formed by ventral rami of five nerves uniting to form three trunks. The three trunks bifurcate to form three cords. The plexus begins as five nerves and ends in five major

nerves, not nine nerves, as is stated in item 4. *(Basmajian, p 334)*

85. (E) The lateral cord becomes the musculocutaneous nerve that innervates the biceps, brachialis, and coracobrachialis muscles. The median nerve is formed by nerve fibers from both lateral and medial cords and innervates muscles on the flexor aspect of the forearm (all except 1½ muscles innervated by the ulnar nerve). The axillary nerve is a branch of the posterior cord. The ulnar nerve passes through the arm region but does not innervate any muscles there. *(Basmajian, p 335)*

86. (B) The posterior cord through the subscapular nerve provides innervation to the subscapularis muscle. Nerve fibers from the medial cord provide innervation to half of the flexor digitorum profundus through the ulnar nerve; this muscle also receives fibers from the median nerve. The axillary nerve, from the posterior cord, supplies the deltoid muscle; and the musculocutaneous nerve, from the lateral cord, supplies the biceps muscle. *(Basmajian, p 335)*

87. (D) The subscapularis muscle has its insertion on the lesser tubercle of the humerus. The supraspinatus, infraspinatus, and teres minor insert on the three flat contiguous facets of the greater tubercle. *(Basmajian, pp 339–342)*

88. (A) The triceps brachii has its long or scapular head originating from the infraglenoid tubercle of the scapula. It has two humeral heads, a medial and a lateral head. The lateral head largely is tendinous. The common tendon of insertion is attached to the olecranon process of the ulna. It is supplied by the radial, not the axillary nerve. *(Basmajian, p 357)*

89. (A) The structures normally entering the gluteal region below the piriformis muscle consist of the following: sciatic nerve, inferior gluteal nerve and vessels, posterior cutaneous nerve of the thigh, nerve to the obturator internus, internal pudendal vessels, and pudendal nerve. The superior gluteal nerve and superior gluteal vessels pass through the greater sciatic foramen above the piriformis. They run as described in the answer to question 87. *(Basmajian, pp 258–259)*

90. (D) The sacrotuberous ligament is the only structure listed that attaches to the ischium. This ligament is a broad band running from the medial part of the tuberosity of the ischium to the sides of the sacrum and coccyx and adjacent dorsal surfaces. The short head of the biceps arises from the linea aspera and adjacent parts of the femur. The obturator internus arises from the pelvic surface of the obturator membrane and most of the periphery

of the bone. The gluteus maximus arises from bone and ligament along a strip between the posterior superior spine of the ilium and the tip of the coccyx; it is inserted into the gluteal tuberosity of the femur and the iliotibial tract. There is a bursa where the gluteus maximus plays across the greater trochanter. The lower border of the gluteus maximus extends from the tip of the coccyx across the tuberosity of the ischium and onward to the shaft of the femur; the ischial tuberosity, however, is uncovered when one sits down. *(Basmajian, pp 215,260–261,251)*

91. **(C)** The great saphenous vein ascends throughout the length of the limb. It begins at the medial end of the dorsal venous arch of the foot and passes in front of the medial malleolus. At the knee it passes behind the medial border of the patella. It joins the femoral vein just below and lateral to the pubic tubercle. It anastomoses freely with the short saphenous vein. *(Basmajian, pp 245–246)*

92. **(C)** Fractures of the neck of the femur or between the lesser and greater trochanters are common in persons over 60 years of age. They are commoner in older women than in older men because womens' bones may become weakened owing to postmenopausal osteoporosis. Blood vessels of the proximal part of the femur may be torn with femoral neck fractures. Generally the more proximal the fracture, the greater the chances of interruption of the vascular supply. A poor blood supply may result in nonunion and avascular necrosis of the femoral head. *(Moore, pp 428,596)*

93. **(E)** When the valves of the distal set of perforating veins become "incompetent" (dilated so that their cusps do not meet and close the vein), contractions of calf muscles that normally propel the blood upward cause a reverse flow through the perforating veins, causing the perforating and superficial veins to become tortuous and dilated (varicose veins). Vein grafts, using the great saphenous vein, have been employed for coronary and femoral operations. The great saphenous vein lies immediately anterior to the medial malleolus. Varicose veins are common in the posterior and medial parts of the lower limb, particularly in older persons, and cause considerable discomfort and pain. *(Moore, p 432)*

94. **(E)** Several cutaneous nerves supply skin on the anterior, medial, and lateral aspects of the thigh. The ilioinguinal nerve is distributed to the skin of the superior medial area of the thigh. Femoral branches of the genitofemoral nerve supply the skin just below the middle part of the inguinal ligament. The lateral femoral cutaneous nerve enters the region near the anterior superior iliac spine to supply skin on the anterior and lateral aspects of the thigh. The anterior cutaneous branches of the

femoral nerve supply the skin on the anterior and medial aspects of the thigh. The saphenous nerve, the longest branch of the femoral nerve, supplies skin on the medial side of the leg and foot. *(Moore, pp 432–433)*

95. **(D)** The iliotibial tract is an extremely strong lateral portion of the deep fascia of the thigh. It runs from the tubercle of the iliac crest to the tibia. The iliotibial tract receives tendinous reinforcements from the tensor fasciae latae and gluteus maximus muscles. The distal end of the tract is attached to the lateral condyle of the tibia, not the fibular head. The ischial spine separates the greater sciatic notch superiorly from the lesser sciatic notch inferiorly; it has no particular relationship to the iliotibial tract. *(Moore, pp 433,425)*

96. **(B)** If the quadriceps femoris muscle is paralyzed, the knee cannot be extended voluntarily; but the patient can stand erect because the body weight tends to overextend the knee. The rectus femoris portion of the quadriceps passes anterior to the hip joint, and thus it can help to flex at this joint; therefore hip flexion would be weakened somewhat by loss of this muscle. Atrophy is characteristic of the paralyzed quadriceps or as a result of its disuse (e.g., while the limb is in a cast). *Moore, pp 447,573)*

97. **(B)** Three cranial nerves supply the ocular muscles, but two of these nerves innervate only one muscle each. The abducens nerve innervates the abductor muscle, the lateral rectus; the trochlear nerve supplies the muscle, which operates through a pully (trochlea), the superior oblique. The levator palpebrae superioris, superior rectus, medial rectus, inferior rectus, and inferior oblique muscles are innervated by the oculomotor nerve. *(Woodburne, p 249)*

98. **(B)** In addition to supplying most of the ocular muscles, the oculomotor nerve provides the parasympathetic innervation for the sphincter pupillae muscle of the iris and for the ciliary muscle of accommodation. *(Woodburne, p 250)*

99. **(E)** The trochlear nerve is the smallest of the cranial nerves and supplies only one muscle, the superior oblique. It contains somatic efferent fibers for this muscle. The trochlear is the only cranial nerve emerging from the dorsal aspect of the brain stem. It pierces the dura mater in the free border of the tentorium cerebelli just behind the posterior clinoid process. *(Woodburne, p 250)*

100. **(C)** The ophthalmic nerve is entirely sensory. It passes forward in the lateral wall of the cavernous sinus inferior to the trochlear nerve. It arises from the upper portion of the semilunar ganglion. *(Woodburne, p 251)*

101. **(C)** The frontal nerve is the largest branch of the ophthalmic nerve. It enters the orbit through the superior orbital fissure and passes forward between the levator palpebrae superioris muscle and the periorbita. Midway through the orbit it divides into its supratrochlear and supraorbital branches. *(Woodburne, p 251)*

102. **(A)** The ophthalmic artery is a branch of the intercranial portion of the internal carotid artery, which arises from it just as the internal carotid emerges from the cavernous sinus. It passes through the optic canal to enter the orbit. Its terminal branches at the medial angle of the eye are the supratrochlear and dorsal nasal branches. *(Woodburne, p 252)*

103. **(B)** Between the pubic tubercle and the anterior superior iliac spine is stretched the inguinal ligament. This ligament is the rolled-under inferior margin of the aponeurosis of the external abdominal oblique muscle and marks, in the groin, the separation between the abdominal wall and the lower limb. *(Woodburne, p 367)*

104. **(C)** The lineae transversae are produced by tendinous bands that interrupt the continuity of the fibers of the rectus abdominis muscle. One is located a little below the xiphoid process, one is at the level of the umbilicus, and the third is halfway between the other two. *(Woodburne, p 368)*

105. **(A)** The femoral artery gives rise to the superficial epigastric, the superficial circumflex iliac, and the superficial external pudendal approximately 1 or 2 cm below the inguinal ligament. *(Woodburne, pp 369–370)*

106. **(C)** The deep external pudendal is a branch of the femoral or of the medial circumflex femoral. *(Woodburne, p 370)*

107. **(E)** The superficial inguinal nodes are from 12 to 20 lymph nodes arranged in the form of a T in the subcutaneous tissue of the groin. They form a chain parallel to and about 1 cm below the inguinal ligament. These nodes receive lymph from the lower abdominal wall, the buttocks, the external genitalia, and the perineum. *(Woodburne, p 371)*

108. **(A)** The superior vena cava ends at the level of the third costal cartilage by entering the right atrium. This large vein is formed by the union of the right and left brachiocephalic veins at the level of the first right costal cartilage. It returns blood from everything above the diaphragm (head, neck, upper limb, and thoracic wall) except the lungs. *(Moore, p 91)*

109. **(E)** All the listed items about the aortic arch are true. This curved continuation of the ascending aorta begins posterior to the right half of the sternal angle. It passes to the left of the trachea and esophagus. The terminal part of the aortic arch can be observed in anteroposterior radiographs of the chest. The shadow it casts is called the aortic knob. The ligamentum arteriosum passes from the root of the left pulmonary artery to the inferior concave surface of the aortic arch. *(Moore, pp 91–92)*

110. **(B)** The aortic arch has three branches: the brachiocephalic trunk, the left common carotid artery, and the left subclavian artery. The brachiocephalic trunk is the first and largest of the three branches; it divides into the right common carotid and the right subclavian arteries. On the left, these arteries come directly off the aortic arch. *(Moore, pp 92–94)*

111. **(D)** The vagus nerves are parasympathetic components of the pulmonary, esophageal, and cardiac plexuses. They do not control action of the diaphragm, as that is the function of the phrenic nerve. Their stimulation results in the constriction, not dilation of the coronary arteries. *(Moore, pp 81,95)*

112. **(A)** The thoracic duct is the main lymphatic duct that conveys most of the lymph of the body to the venous system. It drains the cisterna chyli, which lies in front of the 12th thoracic vertebra, posterior and to the right of the aorta. It empties into the venous system at the union of the left internal jugular vein with the subclavian. It passes through the aortic hiatus of the diaphragm, not the esophageal hiatus. *(Moore, pp 99,254–256)*

113. **(C)** The azygos veins are the superior continuation of the ascending lumbar veins. The azygos vein arches over the root of the right lung. The hemiazygous vein crosses the midline and enters the azygos vein, not the vena cava. The thoracic duct empties as described in answer to question 112; it does not enter the azygos vein. *(Moore, pp 100–101)*

114. **(B)** The anterior mediastinum lies between the body of the sternum anteriorly and the fibrous pericardium posteriorly. It is very narrow above the level of the fourth costal cartilage owing to the closeness of the right and left pleurae. The anterior mediastinum contains some loose areolar tissue, fat, lymph vessels, two or three lymph nodes, the sternopericardial ligaments, and a few branches of the internal thoracic artery. In infants and children the anterior mediastinum also may contain the lower part of the thymus gland. It does not contain the esophagus or the heart. *(Moore, p 101)*

**115.** **(E)** Typically the head of a rib articulates with the sides of the bodies of two vertebrae, and the tubercle of the rib articulates with the tip of a transverse process. Hence there are two articulations of the ribs with the vertebral column. They are the plane type of synovial joint allowing gliding movements. Exceptions to this general arrangement are the heads of the first rib and of the last three ribs that articulate only with bodies of their own vertebra. *(Moore, p 101)*

**116.** **(B)** On looking into a dissection of the left side of the mediastinum, the common carotid and the subclavian arteries are seen. Structures seen on the right side of the mediastinum and not on the left are the brachiocephalic trunk and the azygos vein. Diagrams showing these structures are found on pages 94 and 100 of Moore's text. *(Moore, pp 94,100)*

# Practice Test Subspecialty List

**HEAD AND NECK**

43, 44, 45, 46, 47, 48, 49, 50, 51, 52, 53, 54, 55, 56, 57, 58, 59, 60, 97, 98, 99, 100, 101, 102

**ABDOMINAL REGION**

25, 26, 27, 28, 29, 30, 31, 32, 33, 34, 35, 36, 37, 38, 39, 40, 41, 42, 103, 104, 105, 106, 107

**CLINICAL HEAD AND NECK AND ABDOMINAL**

19, 20, 21, 22, 23, 24

**THE THORAX**

1, 2, 3, 4, 5, 6, 7, 8, 9, 10, 11, 12, 13, 14, 15, 16, 17, 18, 108, 109, 110, 111, 112, 113, 114, 115, 116

**PELVIS AND PERINEUM**

61, 62, 63, 64, 65, 66, 67, 68, 73, 74, 75, 76, 77, 78, 79, 80

**THE UPPER LIMB**

69, 70, 71, 72, 81, 82, 83, 84, 85, 86, 87, 88

**THE LOWER LIMB**

89, 90, 91, 92, 93, 94, 95, 96

NAME _____
    Last                    First                    Middle

ADDRESS _____
        Street

_____
City                        State              Zip

**DIRECTIONS** Mark your social security number from top to bottom
in the appropriate boxes on the right. Refer to the
section "HOW TO TAKE THE PRACTICE TEST"
in the introduction to the book for more information.
PLEASE USE NO.2 PENCIL ONLY.

MAKE
ERASURES
COMPLETE

**PAGE 1**

SOC SEC NUMBER

0 1 2 3 4 5 6 7 8 9
0 1 2 3 4 5 6 7 8 9
0 1 2 3 4 5 6 7 8 9
0 1 2 3 4 5 6 7 8 9
0 1 2 3 4 5 6 7 8 9
0 1 2 3 4 5 6 7 8 9
0 1 2 3 4 5 6 7 8 9
0 1 2 3 4 5 6 7 8 9
0 1 2 3 4 5 6 7 8 9

PAGE 1 2
TYPE 1 2 3

1 A B C D E   2 A B C D E   3 A B C D E   4 A B C D E   5 A B C D E   6 A B C D E   7 A B C D E   8 A B C D E

9 A B C D E   10 A B C D E   11 A B C D E   12 A B C D E   13 A B C D E   14 A B C D E   15 A B C D E   16 A B C D E

17 A B C D E   18 A B C D E   19 A B C D E   20 A B C D E   21 A B C D E   22 A B C D E   23 A B C D E   24 A B C D E

25 A B C D E   26 A B C D E   27 A B C D E   28 A B C D E   29 A B C D E   30 A B C D E   31 A B C D E   32 A B C D E

33 A B C D E   34 A B C D E   35 A B C D E   36 A B C D E   37 A B C D E   38 A B C D E   39 A B C D E   40 A B C D E

41 A B C D E   42 A B C D E   43 A B C D E   44 A B C D E   45 A B C D E   46 A B C D E   47 A B C D E   48 A B C D E

49 A B C D E   50 A B C D E   51 A B C D E   52 A B C D E   53 A B C D E   54 A B C D E   55 A B C D E   56 A B C D E

57 A B C D E   58 A B C D E   59 A B C D E   60 A B C D E   61 A B C D E   62 A B C D E   63 A B C D E   64 A B C D E

65 A B C D E   66 A B C D E   67 A B C D E   68 A B C D E   69 A B C D E   70 A B C D E   71 A B C D E   72 A B C D E

73 A B C D E   74 A B C D E   75 A B C D E   76 A B C D E   77 A B C D E   78 A B C D E   79 A B C D E   80 A B C D E

81 A B C D E   82 A B C D E   83 A B C D E   84 A B C D E   85 A B C D E   86 A B C D E   87 A B C D E   88 A B C D E

89 A B C D E   90 A B C D E   91 A B C D E   92 A B C D E   93 A B C D E   94 A B C D E   95 A B C D E   96 A B C D E

97 A B C D E   98 A B C D E   99 A B C D E   100 A B C D E   101 A B C D E   102 A B C D E   103 A B C D E   104 A B C D E

105 A B C D E   106 A B C D E   107 A B C D E   108 A B C D E   109 A B C D E   110 A B C D E   111 A B C D E   112 A B C D E

113 A B C D E   114 A B C D E   115 A B C D E   116 A B C D E   117 A B C D E   118 A B C D E   119 A B C D E   120 A B C D E

121 A B C D E   122 A B C D E   123 A B C D E   124 A B C D E   125 A B C D E   126 A B C D E   127 A B C D E   128 A B C D E

129 A B C D E   130 A B C D E   131 A B C D E   132 A B C D E   133 A B C D E   134 A B C D E   135 A B C D E   136 A B C D E

137 A B C D E   138 A B C D E   139 A B C D E   140 A B C D E   141 A B C D E   142 A B C D E   143 A B C D E   144 A B C D E

145 A B C D E   146 A B C D E   147 A B C D E   148 A B C D E   149 A B C D E   150 A B C D E   151 A B C D E   152 A B C D E

153 A B C D E   154 A B C D E   155 A B C D E   156 A B C D E   157 A B C D E   158 A B C D E   159 A B C D E   160 A B C D E

| | | 0 | 1 | 2 | 3 | 4 | 5 | 6 | 7 | 8 | 9 |
|---|---|---|---|---|---|---|---|---|---|---|---|
| S | N | 0 | 1 | 2 | 3 | 4 | 5 | 6 | 7 | 8 | 9 |
| O | U | 0 | 1 | 2 | 3 | 4 | 5 | 6 | 7 | 8 | 9 |
| C | M | 0 | 1 | 2 | 3 | 4 | 5 | 6 | 7 | 8 | 9 |
| | B | 0 | 1 | 2 | 3 | 4 | 5 | 6 | 7 | 8 | 9 |
| S | E | 0 | 1 | 2 | 3 | 4 | 5 | 6 | 7 | 8 | 9 |
| E | R | 0 | 1 | 2 | 3 | 4 | 5 | 6 | 7 | 8 | 9 |
| C | | 0 | 1 | 2 | 3 | 4 | 5 | 6 | 7 | 8 | 9 |
| | | 0 | 1 | 2 | 3 | 4 | 5 | 6 | 7 | 8 | 9 |

PAGE 1 2
TYPE 1 2 3

161 A B C D E 162 A B C D E 163 A B C D E 164 A B C D E 165 A B C D E 166 A B C D E 167 A B C D E 168 A B C D E

169 A B C D E 170 A B C D E 171 A B C D E 172 A B C D E 173 A B C D E 174 A B C D E 175 A B C D E 176 A B C D E

177 A B C D E 178 A B C D E 179 A B C D E 180 A B C D E 181 A B C D E 182 A B C D E 183 A B C D E 184 A B C D E

185 A B C D E 186 A B C D E 187 A B C D E 188 A B C D E 189 A B C D E 190 A B C D E 191 A B C D E 192 A B C D E

193 A B C D E 194 A B C D E 195 A B C D E 196 A B C D E 197 A B C D E 198 A B C D E 199 A B C D E 200 A B C D E

201 A B C D E 202 A B C D E 203 A B C D E 204 A B C D E 205 A B C D E 206 A B C D E 207 A B C D E 208 A B C D E

209 A B C D E 210 A B C D E 211 A B C D E 212 A B C D E 213 A B C D E 214 A B C D E 215 A B C D E 216 A B C D E

217 A B C D E 218 A B C D E 219 A B C D E 220 A B C D E 221 A B C D E 222 A B C D E 223 A B C D E 224 A B C D E

225 A B C D E 226 A B C D E 227 A B C D E 228 A B C D E 229 A B C D E 230 A B C D E 231 A B C D E 232 A B C D E

233 A B C D E 234 A B C D E 235 A B C D E 236 A B C D E 237 A B C D E 238 A B C D E 239 A B C D E 240 A B C D E

241 A B C D E 242 A B C D E 243 A B C D E 244 A B C D E 245 A B C D E 246 A B C D E 247 A B C D E 248 A B C D E

249 A B C D E 250 A B C D E 251 A B C D E 252 A B C D E 253 A B C D E 254 A B C D E 255 A B C D E 256 A B C D E

257 A B C D E 258 A B C D E 259 A B C D E 260 A B C D E 261 A B C D E 262 A B C D E 263 A B C D E 264 A B C D E

265 A B C D E 266 A B C D E 267 A B C D E 268 A B C D E 269 A B C D E 270 A B C D E 271 A B C D E 272 A B C D E

273 A B C D E 274 A B C D E 275 A B C D E 276 A B C D E 277 A B C D E 278 A B C D E 279 A B C D E 280 A B C D E

281 A B C D E 282 A B C D E 283 A B C D E 284 A B C D E 285 A B C D E 286 A B C D E 287 A B C D E 288 A B C D E

289 A B C D E 290 A B C D E 291 A B C D E 292 A B C D E 293 A B C D E 294 A B C D E 295 A B C D E 296 A B C D E

297 A B C D E 298 A B C D E 299 A B C D E 300 A B C D E 301 A B C D E 302 A B C D E 303 A B C D E 304 A B C D E

305 A B C D E 306 A B C D E 307 A B C D E 308 A B C D E 309 A B C D E 310 A B C D E 311 A B C D E 312 A B C D E

313 A B C D E 314 A B C D E 315 A B C D E 316 A B C D E 317 A B C D E 318 A B C D E 319 A B C D E 320 A B C D E